Nanomedicine and the Nervous System

Nanomedicine and the Nervous System

Editors
Colin R. Martin PhD
Chair in Mental Health, School of Health
Nursing and Midwifery
University of West of Scotland
UK

Victor R. Preedy PhD DSc
Professor of Nutritional Biochemistry
School of Medicine
King's College London
and
Professor of Clinical Biochemistry
King's College Hospital
UK

Ross J. Hunter MBBS MRCP PhD
Cardiology Research Fellow
St Bartholomew's Hospital
London
UK

CRC Press
Taylor & Francis Group
an informa business
www.taylorandfrancisgroup.com

6000 Broken Sound Parkway, NW
Suite 300, Boca Raton, FL 33487
711 Third Avenue
New York, NY 10017
2 Park Square, Milton Park
Abingdon, Oxon OX14 4RN, UK

Science Publishers
Jersey, British Isles
Enfield, New Hampshire

Published by Science Publishers, an imprint of Edenbridge Ltd.
• St. Helier, Jersey, British Channel Islands
• P.O. Box 699, Enfield, NH 03748, USA
E-mail: *info@scipub.net* Website: *www.scipub.net*

Marketed and distributed by:

	6000 Broken Sound Parkway, NW Suite 300, Boca Raton, FL 33487
CRC Press Taylor & Francis Group an **informa** business www.taylorandfrancisgroup.com	711 Third Avenue New York, NY 10017
	2 Park Square, Milton Park Abingdon, Oxon OX14 4RN, UK

Copyright reserved © 2012

ISBN 978-1-57808-728-0

Cover Illustrations: Reproduced by kind courtesy of the undermentioned authors:
• Figure No. 4 from Chapter 3 by Edward de Asis, Russell J. Andrews and Jun Li
• Figure Nos. 5 and 6 from Chapter 11 by Šárka Kubinová and Eva Syková
• Figure Nos. 3 and 6 from Chapter 15 by Pauline Resnier, Anne Clavreul and Catherine Passirani

```
             Library of Congress Cataloging-in-Publication Data
Nanomedicine and the nervous system / editors, Colin R. Martin,
Victor R. Preedy, Ross J. Hunter.
       p. cm.
   Includes bibliographical references and index.
   ISBN 978-1-57808-728-0 (hardback)
  1.  Nanomedicine.  2.  Neurosciences--Materials.  3.  Biomedical
materials.  4.  Nanostructured materials.   I. Martin, Colin R., 1964-
II. Preedy, Victor R. III. Hunter, Ross J. (Ross Jacob), 1977-
   R857.N34N354 2012
   610.28'4--dc23
                                                          2011043774
```

The views expressed in this book are those of the author(s) and the publisher does not assume responsibility for the authenticity of the findings/conclusions drawn by the author(s). No responsibility is assumed by the publisher for any injury and/or damage to persons or property as a matter of products liability, negligence or otherwise, or from any use or operation of any methods, products, instructions or ideas contained in the material herein. Because of rapid advances in the medical sciences, in particular, independent verification of diagnoses and drug dosages should be made.

All rights reserved. No part of this publication may be reproduced, stored in a retrieval system, or transmitted in any form or by any means, electronic, mechanical, photocopying or otherwise, without the prior permission of the publisher, in writing. The exception to this is when a reasonable part of the text is quoted for purpose of book review, abstracting etc.

This book is sold subject to the condition that it shall not, by way of trade or otherwise be lent, re-sold, hired out, or otherwise circulated without the publisher's prior consent in any form of binding or cover other than that in which it is published and without a similar condition including this condition being imposed on the subsequent purchaser.

Printed in the United States of America

Dedicated to Miss Caragh Brien

Foreword

Diseases of the central nervous system represent complex clinical entities usually accompanied by general decline in function ultimately requiring long term care. Consequently, such presentations impact significantly on the quality of life of the individual concerned and their significant family members and carers. Injury to the nervous system, while no less devastating in terms of impact to the individual than overt disease, is often characterised by similar issues, including chronicity and impaired function. With an ageing population, the risk for disease of the central nervous system such as Alzheimer's disease and stroke is estimated to increase substantially. We now know more about risk and protective factors associated with these disorders, but current treatment is unsatisfactory. Innovations in treatment approaches to both diseases and injury to the central nervous system offers the patient hope within the context of a situation often defined by despair. The field of nanomedicine has the potential to facilitate novel and contemporary interventions that may yield significant improvements in the health outcomes of patients experiencing pathology related to the nervous system. A fundamental problem for those working in this area is the availability of contemporary, relevant and applied information. Colin Martin, Victor Preedy and Ross Hunter have done an excellent job bringing together experts in the nanomedicine field to produce an accessible, evidence-based and clinically relevant account of the issues. Readers with either a clinical or research interest in this area will undoubtedly find this volume a valuable and welcome resource. In summary, I would highly recommend this text as an informative and authoritative account.

PROFESSOR EEF HOGERVORST

Professor Hogervorst worked as a research scientist at the Universities of Oxford and Cambridge, as well as in the USA and Indonesia to investigate risk factors, treatment and early diagnoses of age related disease of the nervous system, with a focus on dementia. She currently holds Chairs at Loughborough University and the University of Indonesia where she teaches Biological Psychology, Epidemiology and Research Methods. She was on the Expert Board for the Medical Research Council (MRC) Neurosciences and Mental Health, is Associate Editor for the

Journal of Alzheimer's disease, is regularly invited as a keynote speaker and has written over 100 peer reviewed articles and book chapters in her field.

Preface

The nanosciences are a rapidly expanding field of research with a wide applicability to all areas of health. They encompass a variety of technologies ranging from particles to networks and nanostructures. For example, nanoparticles have been proposed to be suitable carriers of therapeutic agents whilst nanostructures provide suitable platforms for sub-micro bioengineering. However, understanding the importance of nanoscience and technology is somewhat problematical as a great deal of text can be rather technical in nature with little consideration to the novice. In this collection of books on *Nanoscience Applied to Health and Medicine* we aim to disseminate the information in a readable way by having unique sections for the novice and expert alike. This enables the reader to transfer their knowledge base from one discipline to another or from one academic level to another. Each chapter has an abstract, key facts, applications to other areas of health and disease and a "mini-dictionary" of key terms and phrases within each chapter. Finally, each chapter has a series of summary points. In this book Nanomedicine and the nervous system we focus on nanomedicine and naontechnology as applied to the nervous system which include the brain and peripheral nervous system. We cover an introduction to the field, general aspects, safety of nanostructures, nanoparticle-based immunoassays, carbon nanofiber microbrush arrays, nanoelectrode arrays, protein nanoassemblies, nanoparticle-assisted laser desorption/ionization (nano-paldi)-based imaging, nanomaterials and ion channels, nanobiomolecular engineering, alpha-synuclein nano-structures, nanomaterials for stem cell imaging, neuronal performance, treatment of stroke and spinal cord injury, sirna and nanoparticles, transferrin anchored nanoparticles, lipid nanostructures, nanocarriers and brain tumours, electrospinning of nanofibres, nanofiber scaffolds, nanoparticle-iron chelator conjugates and toxicological profiles of carbon nanotubes.

Contributors to **Nanomedicine and the nervous system** are all either international or national experts, leading authorities or are carrying out ground breaking and innovative work on their subject. The book is essential reading for neurologists, psychologists, mental health workers, research scientists, medical doctors, pathologists, biologists, biochemists,

chemists and physicists, general practitioners as well as those interested in disease and nano sciences in general. **Nanomedicine and the Nervous System** is part of a collection of books on *Nanoscience Applied to Health and Medicine*.

The Editors

Nanomedicine and Brain Diseases: The Way Forward

The impact of neurological disease process on all aspects of social, psychological and physical functioning can be both devastating and far-reaching, culminating in a huge burden on the individual's quality of life and a significant impact on family life. The triad of Parkinson's disease, Multiple Sclerosis and Motor Neurones Disease present a public portrayal of clinical entities often epitomised by despair and hopelessness. However, these high profile presentations represent merely the 'tip of the iceberg' in terms of the broad range of devastating and debilitating neurological diseases with unique and often poorly understood etiological pathways.

Against a background of over 100 major definable neurological disease diagnoses, there exists many interactions between other domains, for example, neuromuscular disorders, cardiovascular disease, renal disease etc, and those where neurological involvement is suspected, but not confirmed, for example, chronic fatigue syndrome and Capgras syndrome. Neurological disorders with an established genetic etiology, for example, Huntington's disease, in many respects present to the patient a terrifying future of increasing disability and incapacity culminating in an unpleasant, often undignified and invariably premature death. However, even before the advent of definitive genetic testing for Huntington's disease, the established 50% familial risk rate led many public health professionals, researchers and policy makers to speculate that with appropriate genetic counselling, Huntington's disease would decrease in prevalence and ultimately, become a disease of historical relevance only. This has not proved to be the case and indeed the incidence of Huntington's disease in many geographical locations is unchanged, thus revealing that even within the context of major neurological disorders that can be modified significantly (in terms of prevalence) with simple evidenced-based counselling interventions, the individual and societal deleterious impact remains undiminished.

The breadth, diversity and complexity of clinical presentations does not diminish in any way however an overarching theme of contemporary intervention, that being the focus on symptom control and management pharmacotherapy ethos rather than a focus on cure. Given the chronicity, misery and indeed, clinical and economic burden associated with many diseases of the brain, the limitations of traditional approaches within medicine to this group of patients is thrown into sharp relief. Augmenting, developing and innovating in the current context within which the therapeutic armamentarium is traditionally circumscribed is both a challenging and exciting task, yet represents an unmissable opportunity to transform the interventions and consequential outcomes of patients with a neurological disease diagnosis. The advent of the nanosciences furnishes the scientific community working within the brain disease and disorders arenas such scope for radical innovation in treatment development and intervention.

It is therefore clear that the clinical and experimental application of nanosciences may offer many unique insights into the understanding and treatment of brain diseases and the promise to reflect and consider these devastating clinical presentations in a reflective and novel way. Nanoscience, as an approach and sub-discipline that forms one aspect of the comprehensive clinical research battery within the sphere of neurological disease impact, strengthens the veracity, integrity and scope of the evidence base for treatment and fosters a forum for interdisciplinary research collaboration with common goals of disease impact amelioration and consequential improvement in symptom profile and quality of life, and ultimately, a contributory dimension to cure. It remains to be seen how and when the substantive contribution of applied nanoscience research will impact on clinical outcome, as this modern speciality develops and innovates, however, it is hoped that this volume will contribute, in at least a modest way, toward a path of basic clinical science enhancement and positive clinical outcomes for patients.

Professor Colin R Martin
Chair in Mental Health
Faculty of Education, Health and Social Sciences
University of the West of Scotland
UK

Professor Victor R Preedy
Professor of Nutritional Biochemistry
School of Medicine
King's College London
        and
Professor of Clinical Biochemistry
King's College Hospital
UK

Ross J. Hunter MRCP
Cardiology Research Fellow
St Bartholomew's Hospital
London
UK

Contents

Dedication	v
Foreword	vii
Preface	ix
Nanomedicine and Brain Diseases: The Way Forward	xi

Section 1: General Aspects

1. **Safety of Carbon Nanotubes for Neuronal Tissue** — 3
   Giuseppe Bardi

2. **Nanoparticle-Based Immunoassays and their Applications in Nervous System Biomarker Detection** — 17
   Chunping Jia, Qinghui Jin and Jianlong Zhao

3. **Electrical Stimulation of Brain Tissue with Carbon Nanofiber Microbrush Arrays** — 38
   Edward de Asis, Russell J. Andrews and Jun Li

4. **Nanoelectrode Arrays for Monitoring and Modulating Nervous System Electrical and Chemical Activity** — 60
   Russell J. Andrews and Jun Li

5. **Probing Protein Nanoassemblies in Neurodegenerative Disease** — 75
   Min S. Wang and Michael R. Sierks

6. **Nanoparticle-assisted Laser Desorption/ionization (nano-PALDI)-based Imaging Mass Spectrometry (IMS) and its Application to Brain Sciences** — 97
   Saira Hameed, Yuki Sugiura, Yoshishige Kimura, Kalmesh Kumar Shrivas and Mitsutoshi Setou

7. **Nanomaterials and Ion Channels: Observed Effects and Possible Mechanisms** — 119
   Lorin M. Jakubek and Robert H. Hurt

8. Biomolecular Engineering for the Regulation of Alpha-synuclein Nanostructure—toward the Development of Alpha-synuclein Targeting Nanomedicine ... 139
Natsuki Kobayashi, Jihoon Kim and Koji Sode

9. Nanomaterials for Stem Cell Imaging in the Central Nervous System ... 162
Aniruddh Solanki, Shreyas Shah, Michael H. Koucky and Ki-Bum Lee

10. Carbon Nanotubes and Neuronal Performance ... 183
Alessandra Fabbro, Francesca Maria Toma, Giada Cellot, Maurizio Prato and Laura Ballerini

Section 2: Therapeutics

11. Nanotechnologies for Treatment of Stroke and Spinal Cord Injury ... 209
Šárka Kubinová and Eva Syková

12. Delivering siRNA Using Nanoparticles to the Central Nervous System ... 228
S. Neslihan Alpay, Bulent Ozpolat, Hee-Dong Han, Gabriel Lopez-Berestein, Anil K. Sood and Hui-Lin Pan

13. Targeting Drug Delivery to the Brain via Transferrin Anchored Nanoparticles ... 250
Jiang Chang and Didier Betbeder

14. Therapeutic-loaded Lipid Nanostructures and Brain Diseases ... 264
Maria Luisa Bondì and Emanuela Fabiola Craparo

15. Nanocarriers for the Treatment of Brain Tumours ... 286
Pauline Resnier, Anne Clavreul and Catherine Passirani

16. Electrospinning of Nanofibres for Repair of the Injured Peripheral Nervous System ... 310
Dorothee Hodde, José Luis Gerardo-Nava, Ronald Deumens, Jörg Mey and Gary Anthony Brook

17. Bridging the Gap for Regeneration after CNS Injury by Nanofiber Scaffolds: use of SAPNS for Traumatic Brain and Spinal Cord Injury ... 330
Jiasong Guo, Gilberto K.K. Leung and Wutian Wu

18. Nanoparticle-iron Chelator Conjugates as Multifunctional   347
Disease-modifying Drugs for Prevention and Treatment of
Alzheimer's Disease
Gang Liu, Ping Men, George Perry and *Mark A. Smith*

19. The Pharmacological and Toxicological Profiles of Carbon   367
Nanotubes as Drug Carriers Toward Central Nervous System
Yingge Zhang and *Changxiao Liu*

Index   389
About the Editors   395
Color Plate Section   397

Section 1: General Aspects

1

Safety of Carbon Nanotubes for Neuronal Tissue

Giuseppe Bardi

ABSTRACT

Carbon nanotubes (CNTs) are allotropic forms of carbon with a cylindrical structure. CNTs have high length-to-diameter ratio exhibiting extraordinary strength, unique electrical properties and efficient thermal conduction. The internal and external surfaces of CNTs can be independently modified with chemical moieties in order to employ them for several applications in electronics, optics, different fields of material science, as well as life sciences and medicine. To appreciate whether CNTs may be useful to develop new technologies for medical application, scientists must understand deeply their potential toxicity and the mechanism of interaction with tissues and specific cells. The research for possible CNT applications to the interaction, restoration and improvement of the biology of the nervous system is still in its infancy but seminal studies have already shed light on the path. However, the toxicity of CNTs in the neural tissue is not defined so clearly. Whether the presence of CNTs impairs the physiological cell functions depends on the purity of the administered nanotubes,

---

Center for Nanotechnology Innovation, IIT@NEST Italian Institute of Technology, Piazza San Silvestro 12, 56126 Pisa, Italy; E-mail: giuseppe.bardi@iit.it

List of abbreviations after the text.

their diverse chemical modifications and last but not least the type of neural cell the toxicity tests have been performed on. Moreover, the difference between *in vitro* cell cultures and *in vivo* models create further discrepancies in the interpretation of the data. To date, more data on *in vivo* models are required to recognise the safety of CNTs and fix standard procedures to measure precisely their potential harmfulness.

INTRODUCTION

Material scientists have produced several organic and inorganic nanostructures with potential application to medicine such as specific drug delivery into tumours or tissue reconstruction following degenerative disorders of the neuronal tissue. Among these nanostructures, carbon nanotubes (CNTs) have been studied since the beginning of the last decade as possible substrate for neuronal growth (Mattson et al. 2000), or to improve neuronal electrical signalling while supporting dendrite elongation and cell adhesion (Lovat et al. 2005; Ni et al. 2005; Cellot et al. 2009). One of the next chapters will fully describe the employment of carbon nanotubes to improve neuronal performance (**Carbon nanotubes and neuronal performance by** Prof. L. Ballerini).

CNTs (Fig. 1) are composed of a single graphene sheet rolled into a cylinder (Single-Walled Carbon Nanotubes, SWCNTs) or multiple concentric cylinders (Multi-Walled Carbon Nanotubes, MWCNTs) with diameters ranging from 0.4 to 2 nm for SWCNTs and 1.4 to 100 nm for MWCNTs (Lacerda et al. 2006). Conversely, their length can reach several μm and form long fibres rising up some toxicological concerns (Donaldson and Poland 2009; Donaldson et al. 2010). Despite administration of high doses of CNT through inhalation or systemic injection can destabilise

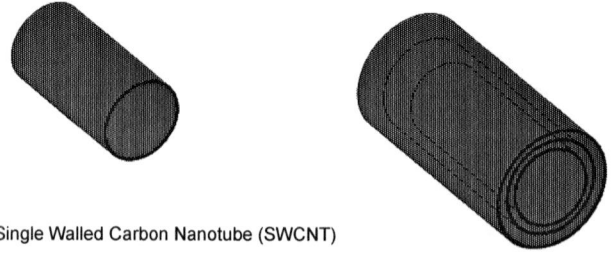

Single Walled Carbon Nanotube (SWCNT)

Multi Walled Carbon Nanotube (MWCNT)

Fig. 1. Scheme of SWCNT and MWCNT structure. Single Walled Carbon Nanotube (SWCNT) and Multi Walled Carbon Nanotube (MWCNT) structure is schematically represented.

immune response (Mitchell et al. 2007; Mitchell et al. 2009), the opportunities to employ CNTs in the nervous system are very promising (Malarkey and Parpura 2007; Cellot et al. 2009).

Unfortunately, data on CNT toxicity for neurons and glial cells, mainly from *in vitro* observations, have been controversial and confusing (McKenzie et al. 2004; Liopo et al. 2006; Kostarelos 2008; Belyanskaya et al. 2009). The reasons for this disagreement can be found in the different types of CNTs, their purity and surface oxidation (Vittorio et al. 2009), doses and variety of surface functionalisation used in the performed studies. Moreover, we should distinguish between results obtained from the interaction of CNTs with growing neuronal cell lines and results of CNTs in cultures of post-mitotic, differentiated primary neurons in presence or absence of glial cells. The species and the nervous system areas from which the neurons have been dissociated must also be taken in consideration.

CARBON NANOTUBES AS PERMISSIVE SUBSTRATE FOR NEURONS

In the first published work on CNTs as substrate for neurons (Mattson et al. 2000), Mattson and colleagues clearly demonstrated that primary hippocampal neurons could grow on non functionalised MWCNTs previously placed on polyethyleneimine (PEI) coated coverslips but the carbon nanotubes did not stimulate neuron branch formation. On the other hand, the same authors showed that length and number of branches were increased when neurons were placed on a substrate of MWCNTs coupled to aldehyde 4-hydroxynonenal (4-HNE), known to increase intracellular Ca^{2+} and modify cytoskeletal proteins (Mark et al. 1997). Subsequent investigations on the use of CNTs as substrate, functionalised either with branched PEI, or other moieties resulting in surface charge variations (Hu et al. 2005; Malarkey and Parpura 2010), described CNTs as a permissive substrate for neuronal *in vitro* maturation. An interesting outcome of these results was the direct relationship between amount of positive charge present on the substrate and length and branching increasing of the neurons. Indeed, our group also found a certain degree of PEI-CNTs biocompatibility with neuronal tissue *in vitro* and *in vivo* that will be described in the next paragraphs of this chapter.

Another confirmation for the feasibility to use CNTs as substrate for neurons comes from the experiments of Matsumoto and collaborators (Matsumoto et al. 2007), who covalently functionalised MWCNTs with neurotrophins, namely nerve growth factor (NGF) and brain-derived growth factor (BDNF). They added functionalised-CNTs to culture medium and showed neurite outgrowth, similar to the effects of these neurotrophins administrated as soluble factors.

Carbon nanotubes also have the advantage to be organised, if needed, in several cm-long fibres (CNFs) with diameters up to 100 µm. CNFs appear as porous scaffolds of aggregated carbon nanotubes which can be functionalised on purpose to allow the growth of neural projections (Dubin et al. 2008). This possibility seems very promising for regenerative medicine, allowing restoration of inter neuronal circuitry, neuronal-muscle connections or neuron-to other organs after injuries.

So carbon nanotubes are very attractive tools to study or restore the organisation of neural networks, however more basic research is required in order to validate these nanostructured materials for safe therapeutically oriented applications (Sucapane et al. 2009).

CARBON NANOTUBES FOR DRUG OR GENE DELIVERY: IMPORTANCE OF THE UPTAKE PATHWAY

One of the main goals for the delivery of therapeutics is the possibility to specifically release most of the cargo only in the target tissue or, more precisely, into the target cell. In this way, a reduction of drug administered dose would be required, decreasing the potential side effects on the surrounding healthy tissues.

Diverse types of possible chemical modifications of CNT surfaces allow them to be applied as drug or gene carriers for cell or tissue specific delivery (Bianco et al. 2005; Lacerda et al. 2006). Nevertheless, this enormous versatility of surface chemical modifications and the physicochemical properties of CNTs needs a thorough study of the toxicity induced by the functionalisation itself (Firme and Bandaru 2010; Raffa et al. 2010). Since release of a certain active molecule within a target cell would, most of the times, require cellular internalisation of the nano-carrier, potential toxicity provoked by the intracellular faith of it could stop the advantages offered by the nano-technological tools exploited for drug delivery.

Dispersion of CNTs is a necessary first step in order to achieve cellular uptake, as the cell membrane avoids diffusion of external molecules larger than ~1 kDa (Kostarelos et al. 2007). The formation of CNT complexes, nanotube length and diameters, and the above mentioned functionalisation are crucial factors to understand the mechanism of cellular internalisation and the subsequent path of trafficking through various possible subcellular compartments. The ways used by the cell to facilitate the internalisation of a carrier can be specific such as those mediated by several diverse receptor proteins or non-specific mediated by different mechanisms. Large particles >1µm can be phagocytosed by competent cells or by non phagocytic cells via macropinocytosis. So, these uptake mechanisms appear to be the pathway for internalisation of single dispersed nanotubes longer than 1µm or small CNT aggregates. Probably, shorter (< 120 nm) single dispersed CNTs could also penetrate the cells through clathrin-mediated endocytosis or even through a caveolin-mediated process, if the nanotube would be less than 60 nm in length. On the other hand, the presence of soluble CNTs could disrupt the regulated-not constitutive depolarisation-dependent plasma membrane/vesicular recycling (endocytotic mechanism) of neurons, which has been suggested to underlie the alteration of morphology and neurite growth of these cells in presence of nanotubes (see above) (Malarkey et al. 2008).

Intracellular pathway together with stability of the nanoparticle-drug/gene should be carefully evaluated in order to perform the delivery. For example, if functionalised CNT-drug particles were destined to a lysosomal compartment, CNTs engaged by this endocytotic mechanisms would go towards fusion with a lysosome with the high risk of degradation for the cargo (Bareford and Swaan 2007). Nevertheless, if the goal were, for example, to address lysosomal storage diseases, CNT engagement in the lysosome would be an advantage.

OXIDATIVE STRESS AND APOPTOSIS

When CNTs are dispersed in solutions for therapeutic purposes CNT purity becomes a major issue to be addressed. Industrial preparation of CNTs often involves metals like Fe, Cr or Ni (MacKenzie et al. 2008), potent catalysts of redox oxidations. Even if these metals were present in elemental non-ionic form, once uptaken by the cell they could move into the lysosomes and converted in pro-oxidant ions by the acidic environment.

Direct induction of oxidative stress by CNTs is still debated (Fenoglio et al. 2006). Metal impurities and structural defects seem to be playing a major role in induction of oxygen reactive species (ROS) (Pulskamp et al. 2007; Fenoglio et al. 2008), whereas the carbon structure of nanotube could behave as scavenger for free radicals, so protecting the cell. The generation of ROS and induction of apoptosis have been correlated to engineered nanoparticles, but whether NPs are the cause of cell death is still not well understrood (Shvedova et al. 2010).

Cell oxidative state regulates and controls cell death and survival and it is correlated to apoptosis. Signal transduction pathways leading to apoptosis are clearly defined during brain injury and neurodegenerative conditions, particularly in ischemic stroke and Alzheimer's disease (Loh et al. 2006). To study the effects that CNT interaction with neural tissue may cause, it is essential to investigate whether the presence of such nanoparticles could lead to programmed cell death in the long term, besides the direct physical damage potentially provoked by the possible insult due to the administration that usually elicits a necrotic type of cell death.

However, literature on CNT induced cell death in the nervous system is still poor. Most of the observations on CNT induced toxicity and inflammatory reactions emphasise the responses of immune system towards inhaled or intravenously injected SWCNTs or MWCNTs from different sources. This ambiguity and systemic administration do not properly indicate potential toxicity of CNTs in the nervous system. Few papers until now have shown mechanisms underneath interaction and possible toxic damages that CNT administration can cause in the nervous system and even less is known in the brain *in vivo*.

Table 1. Key facts of Apoptosis.

1. Apoptosis is the process of programmed cell death (PCD) occurring in multicellular organisms.
2. A series of biochemical events lead to characteristic cell morphology changes and subsequent death.
3. Typical changes are loss of cell membrane asymmetry, blebbing, cell shrinkage, nuclear fragmentation, chromatin condensation, and DNA fragmentation.
4. Apoptosis is different from necrosis, which is a form of traumatic cell death that results from acute cellular injury.
5. Apoptosis is an important biological process during the development of an organism and its entire life.
6. Differentiation of fingers in a developing human embryo is driven by the apoptosis of cells between the fingers resulting in the separation of the digits.
7. Apoptosis plays a major role during the development of nervous system outlining the structure of the nervous circuitry.
8. The unregulated triggering of PCD in the adult brain can be correlated to neurodegenerative diseases and aging.

This table lists the key facts of apoptosis including its role in nervous system.

IN VIVO: WHAT DO WE KNOW?

To employ CNTs as therapeutical devices for drug or gene delivery in the nervous system we should find a way to vehicle them across the Blood Brain Barrier (BBB) which separates this tissue from the blood stream, filtering and regulating the entrance of nutrients, hormones and molecules of several types.

Unfortunately, no published papers have shown intravenously injected CNTs able to cross the BBB and subsequently found in the brain. So, the very few observations of *in vivo* administration come from intracranial microinjection of CNT suspensions (Bardi et al. 2009; VanHandel et al. 2009; Ciofani et al. 2010; Zhang et al. 2010).

Our group has shown that intracranial administration of non-covalently functionalised MWCNTs in living mouse cerebral cortex did not induce degeneration of the neurons surrounding the site of injection. To avoid bundles and clusters formation 97%-pure MWCNTs were coated with the non-ionic surfactant PF-127 to promote CNT dispersion in water solution. A careful investigation in parallel *in vivo-in vitro* of the biocompatibility features of our MWCNTs and their surfactant revealed that a concentration of PF-127 as low as 0.01% in the buffer/medium suspension can, by itself, induce apoptosis within 24h of mouse primary cortical neurons *in vitro*. As mentioned above in this chapter, the functionalisation dependent toxicity must be carefully evaluated to choose an appropriate chemical modification of CNTs. However, when PF-127 coated MWCNTs (3.5 µg/ml *in vitro*) were released in the medium toxicity was significantly reduced, suggesting that the mechanism triggering the apoptotic programme was not dependent on MWCNTs and their presence was tolerated by the primary cells. The PF-127/MWCNT suspension (0.1% PF-127/35 µg/ml MWCNTs/ PBS; 1µl volume) was injected *in vivo* showing no damages to the overall organisation of mouse brain cortical layers in the area surrounding the lesion after 3 days. The volume of injury induced by the MWCNT suspension was not different from the one induced by the injection of CNT-free controls. Moreover, a long term observation 18 days after the injection: MWCNT treated mice presented the formation of a glial scar in the cortex localised at the injection site without major alterations of the cortical structure. Albeit these observations were not performed with sophisticated microscopy, they demonstrated for the first time that brain cortex could tolerate MWCNTs injected *in vivo*.

Following the intriguing path of the possible induction of apoptotic cell death by CNTs in brain cortex, our group performed other *in vivo* explorations using covalently functionalised MWCNTs with branched PEI directly injected in living animals (Fig. 2). A certain degree of compatibility (Bardi et al. unpublished observation) was confirmed. We measured the apoptotic cell death by TUNEL neither detecting differences with control injection nor observing tissue degeneration of the surrounding area. Indeed, branched PEI CNTs were considered as permissive substrate for neurons *in vitro* by other groups (Hu et al. 2005; Malarkey and Parpura 2010), indicating that PEI-functionalised CNTs could have future applications in nanomedicine. Nevertheless, it is important to stress that investigators should always take into account that chemical moieties used to modify the CNT surfaces could be potentially dangerous for the cells. As shown in Fig. 3, *in vitro* experiments using PEI/PBS solutions administered to adherent primary neurons and to a lymphocytic cell line growing in suspension (Fig. 4) can be extremely toxic for cells. On the other hand, FITC-PEI functionalised CNTs administered to mixed

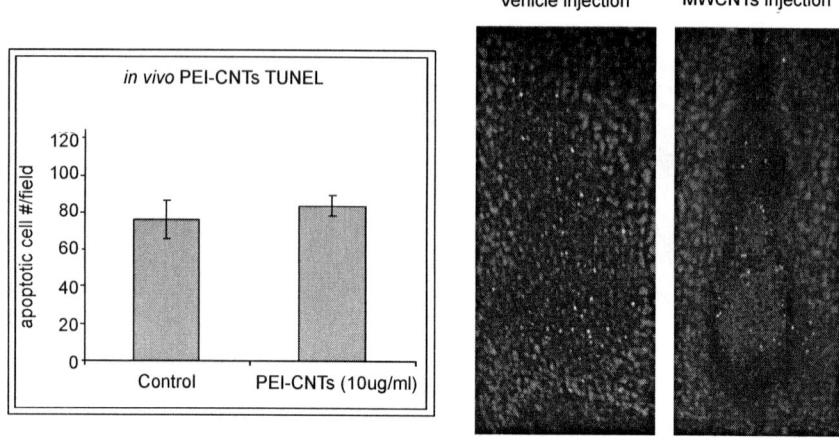

Fig. 2. PEI coated MWCNTs do not induce apoptotic cell death in cortical neurons *in vivo*. *In vivo* PEI-CNTs toxicity. TUNEL has been performed on fixed tissue 3 days after PEI-CNTs administration. The small bright spots in the center of the injection represent the apoptotic cells. The results graphs shows no statistically significant differences between control and PEI-CNTs treated cells *in vitro* and *in vivo*. Author unpublished observations (NINIVE, Non Invasive Nanotransducer for *In Vivo* gene thErapy-FP6-STRP 033378).

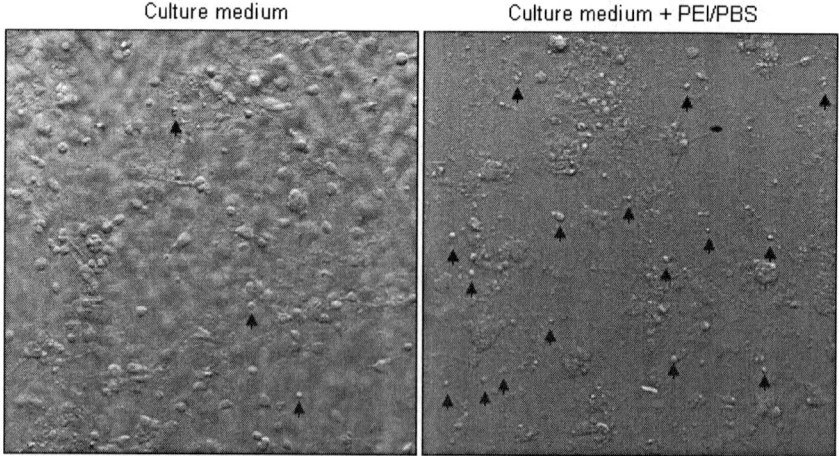

Fig. 3. PEI induces apoptotic cell death in cortical neurons *in vitro*. *In vitro* PEI solution toxicity. Primary cortical neurons 24h after administration of PEI-PBS solution. Arrows indicate apoptotic cells. Author's unpublished observations (NINIVE, Non Invasive Nanotransducer for *In Vivo* gene thErapy-FP6-STRP 033378).

primary neurons do not show the same toxicity of the coating alone and can be easily spotted by imaging techniques in localised areas of the cells (Fig. 5). These observations imply that special attention should be paid to the functionalisation-carrier system stability which must cope with the dynamic biological environment, avoiding the adverse effects that the release of chemicals from CNT surface could create.

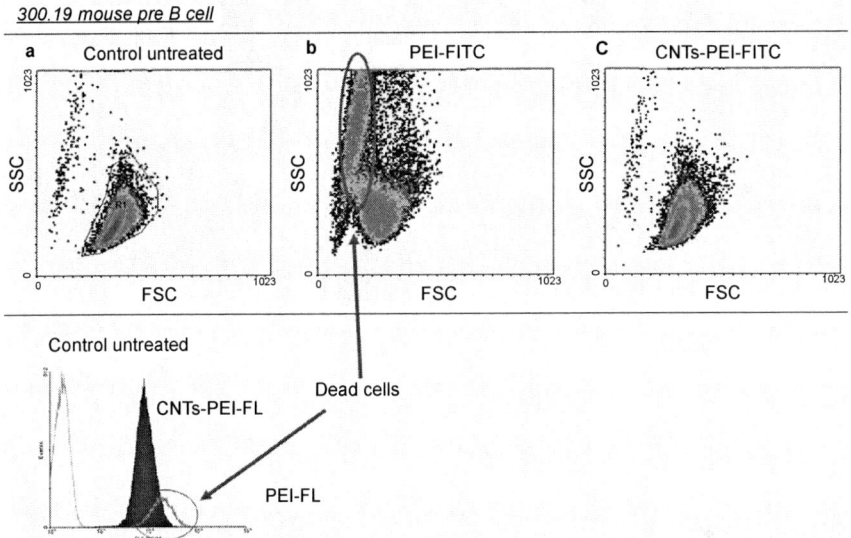

Fig. 4. PEI induces cell death but not when coupled to CNTs. PEI-FITC solution toxicity. FACS results of *in vitro* 300-19 pre B lymphocytes treated for 20min at 37°C 5% CO_2 with a) Culture medium, b) PEI-FITC/PBS solution, c) 2μg/ml PEI-FITC-CNTs. Author's unpublished observations (NINIVE, Non Invasive Nanotransducer for *In Vivo* gene thErapy-FP6-STRP 033378).

Another important issue, which has not been deeply investigated, is the possibility that CNTs might induce initiation of the inflammatory response by neural tissue resident immune system. Until now, the only *in vivo* observation regarding interaction of CNTs with microglial cell and their subsequent activation was described in a mouse glioma model (VanHandel et al. 2009; Zhang et al. 2010). The authors demonstrated that tumour-infiltrating macrophages (CD45high/CD11b/chigh) selectively phagocytosed CNTs with high efficiency. Moreover, the same investigators found that after 72 h no significant changes in the expression of inflammatory cytokines such as TNF-α or IL1-β was present if CNTs were injected in a wild type mouse suggesting minimal toxicity when analysed by quantitative RT-PCR.

CNTs-PEI-FL on primary mouse neurons

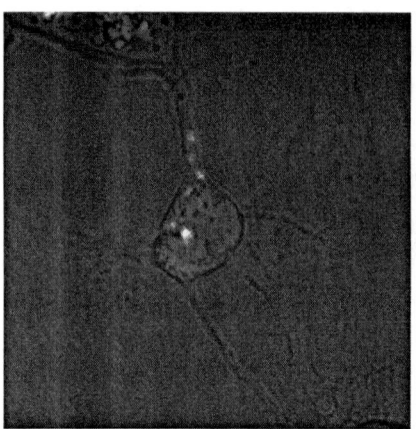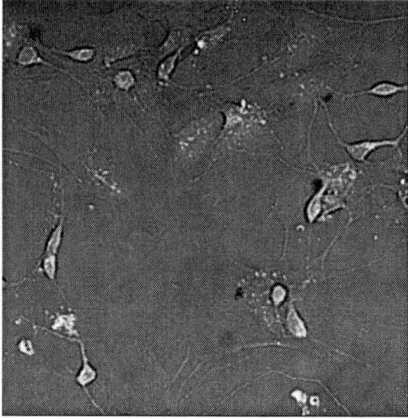

Fig. 5. FITC-PEI-CNTs on primary neurons. Confocal microscopy images of mixed primary neurons in presence of FITC-PEI-CNTs (bright spots). Author's unpublished observations (NINIVE, Non Invasive Nanotransducer for *In Vivo* gene thErapy - FP6-STRP 033378).

The missing piece of the puzzle is the potential long term CNT induced inflammation and a systematic study of the responses to many types of surface modifications. The nervous system is a "special" tissue where neurons and glial cells are well enclosed in the net of extracellular matrix. Resident immune system of neural tissue is represented by microglial cells with phagocytic behaviour. Inflammatory response in the brain should be studied carefully in future in order define a better picture of the long term CNT effects. Today we have no clue regarding the effect of a long term exposure of the nervous system to CNTs or other nanoparticles. It will be necessary to investigate which of the proposed nano-tools might hold such a level of biocompatibility allowing their use as part of a routine medical technology.

CONCLUSIONS

The brain is the core of the Central Nervous System (CNS) and the shelter of our thoughts, representing the symbol of evolution of the human species. Any kind of technological application designed to interact with the brain or the whole CNS must be of major concern for scientists. To set safety rules for medical or pharmaceutical purposes a series of standards will be required. Indeed, the mass of information in our database is sometimes perplexing since many data are controversial. Many institutions worldwide are trying to canalise this information flow. The huge differences among the CNTs present in the market are considered a few times when these technologies are applied to living cells or organisms. Then, we support the idea that a precise characterisation of the synthesised nanotechnologies followed by precise toxicological studies should be requested before medical applications.

ACKNOWLEDGEMENTS

The author thanks Dr Lisa Gherardini for the a careful reading of this Chapter and the wise writing advices.

Summary
- Carbon nanotubes are an allotropic form of the carbon with cylindrical structure and can be divided in Single Walled Nanotubes or Multi Walled Nanotubes.
- The functionalisation of the CNT surfaces is a key point of their exploitation.
- Most of the data on CNT related toxicity comes from different *in vitro* models, often leading to controversial results.
- Special attention should be paid to the toxicity of the chemical moieties used to functionalise the CNTs and the stability of their coupling.
- Missing information from *in vivo* models will be required to precisely understand the interaction of CNTs with the nervous system.

Dictionary

Blood-Brain Barrier (BBB): physical barrier between the central nervous system and the circulating blood, consisting of tight junctions in the endothelial cells of the capillaries that restrict the diffusion of large molecules.

Dendrite: branched projections of a neuron where the electrochemical stimulation from other neural cells is mainly received.

Drug delivery: the process of administering a pharmaceutical compound aimed at achieving a therapeutic effect.

Glia: non-neuronal cells that maintain homeostasis, providing support and protection to the neurons and forming myelin.

Graphene: one-atom-thick planar sheet of carbon atoms organised in a honeycomb crystal structure.

Neuron: the key cells of the nervous system that processes and transmits information by electrical and chemical signalling.

Phagocytosis: cellular process of engulfing solid particles.

Post-mitotic: a mature cell that is no longer capable of undergoing mitosis.

Surfactant (surface active agent): compound that lower the surface tension of a liquid, allowing dispersion of unsoluble compound in solution.

Toxicity: the degree of damage that a substance can create to a cell, tissue or the whole organism.

Abbreviations

BBB	:	Blood Brain Barrier
BDNF	:	Brain-Derived Growth Factor
CNF	:	carbon nanofibre
CNS	:	Central Nervous System
CNT	:	carbon nanotube
NGF	:	Nerve Growth Factor
NP	:	Nanoparticle
PBS	:	Phosphate Buffers Saline
PCD	:	Programmed Cell Death
PEI	:	polyethyleneimine
PF-127	:	Pluronic F-127

References

Bardi, G., P. Tognini, G. Ciofani, V. Raffa, M. Costa and T. Pizzorusso. 2009. Pluronic-coated carbon nanotubes do not induce degeneration of cortical neurons *in vivo* and *in vitro*. Nanomedicine 5: 96–104.

Bareford, L.M. and P.W. Swaan. 2007. Endocytic mechanisms for targeted drug delivery. Adv. Drug Deliv. Rev. 59: 748–758.

Belyanskaya, L., S. Weigel, C. Hirsch, U. Tobler, H.F. Krug and P. Wick. 2009. Effects of carbon nanotubes on primary neurons and glial cells. Neurotoxicology 30: 702–711.

Bianco, A., K. Kostarelos and M. Prato. 2005. Applications of carbon nanotubes in drug delivery. Curr. Opin. Chem. Biol. 9: 674–679.

Cellot, G., E. Cilia, S. Cipollone, V. Rancic, A. Sucapane, S. Giordani, L. Gambazzi, H. Markram, M. Grandolfo, D. Scaini, F. Gelain, L. Casalis, M. Prato, M. Giugliano and L. Ballerini. 2009. Carbon nanotubes might improve neuronal performance by favouring electrical shortcuts. Nat. Nanotechnol. 4: 126–133.

Ciofani, G., V. Raffa, O. Vittorio, A. Cuschieri, T. Pizzorusso, M. Costa and G. Bardi. 2010. *In vitro* and *in vivo* biocompatibility testing of functionalized carbon nanotubes. Methods Mol. Biol. 625: 67–83.

Donaldson, K. and C.A. Poland. 2009. Nanotoxicology: new insights into nanotubes. Nat. Nanotechnol. 4: 708–710.

Donaldson, K., F.A. Murphy, R. Duffin and C.A. Poland. 2010. Asbestos, carbon nanotubes and the pleural mesothelium: a review of the hypothesis regarding the role of long fibre retention in the parietal pleura, inflammation and mesothelioma. Part Fibre Toxicol. 7: 5.

Dubin, R.A., G. Callegari, J. Kohn and A. Neimark. 2008. Carbon nanotube fibers are compatible with Mammalian cells and neurons. IEEE Trans Nanobioscience 7: 11–14.

Fenoglio, I., M. Tomatis, D. Lison, J. Muller, A. Fonseca, J.B. Nagy and B. Fubini. 2006. Reactivity of carbon nanotubes: free radical generation or scavenging activity? Free Radic. Biol. Med. 40: 1227–1233.

Fenoglio, I., G. Greco, M. Tomatis, J. Muller, E. Raymundo-Pinero, F. Beguin, A. Fonseca, J.B. Nagy, D. Lison and B. Fubini. 2008. Structural defects play a major role in the acute lung toxicity of multiwall carbon nanotubes: physicochemical aspects. Chem. Res. Toxicol. 21: 1690–1697.

Firme, C.P., 3rd and P.R. Bandaru. 2010. Toxicity issues in the application of carbon nanotubes to biological systems. Nanomedicine 6: 245–256.

Hu, H., Y. Ni, S.K. Mandal, V. Montana, B. Zhao, R.C. Haddon, V. Parpura. 2005. Polyethyleneimine functionalized single-walled carbon nanotubes as a substrate for neuronal growth. J. Phys. Chem. B. 109: 4285–4289.

Kostarelos, K. 2008. The long and short of carbon nanotube toxicity. Nat. Biotechnol. 26: 774–776.

Kostarelos, K., L. Lacerda, G. Pastorin, W. Wu, S. Wieckowski, J. Luangsivilay, S. Godefroy, D. Pantarotto, J.P. Briand, S. Muller, M. Prato and A. Bianco. 2007. Cellular uptake of functionalized carbon nanotubes is independent of functional group and cell type. Nat. Nanotechnol. 2: 108–113.

Lacerda, L., A. Bianco, M. Prato and K. Kostarelos. 2006. Carbon nanotubes as nanomedicines: from toxicology to pharmacology. Adv. Drug Deliv. Rev. 58: 1460–1470.

Liopo, A.V., M.P. Stewart, J. Hudson, J.M. Tour and T.C. Pappas. 2006. Biocompatibility of native and functionalized single-walled carbon nanotubes for neuronal interface. J. Nanosci. Nanotechnol. 6: 1365–1374.

Loh, K.P., S.H. Huang, R. De Silva, B.K. Tan and Y.Z. Zhu. 2006. Oxidative stress: apoptosis in neuronal injury. Curr. Alzheimer Res. 3: 327–337.

Lovat, V., D. Pantarotto, L. Lagostena, B. Cacciari, M. Grandolfo, M. Righi, G. Spalluto, M. Prato and L. Ballerini. 2005. Carbon nanotube substrates boost neuronal electrical signaling. Nano Lett. 5: 1107–1110.

MacKenzie, K.J., O.M. Dunens, C.H. See and A.T. Harris. 2008. Large-scale carbon nanotube synthesis. Recent Pat. Nanotechnol. 2: 25–40.

Malarkey, E.B. and V. Parpura. 2007. Applications of carbon nanotubes in neurobiology. Neurodegener Dis. 4: 292–299.

Malarkey, E.B. and V. Parpura. 2010. Carbon nanotubes in neuroscience. Acta. Neurochir. Suppl. 106: 337–341.

Malarkey, E.B., R.C. Reyes, B. Zhao, R.C. Haddon and V. Parpura. 2008. Water soluble single-walled carbon nanotubes inhibit stimulated endocytosis in neurons. Nano Lett. 8: 3538–3542.

Mark, R.J., M.A. Lovell, W.R. Markesbery, K. Uchida and M.P. Mattson. 1997. A role for 4-hydroxynonenal, an aldehydic product of lipid peroxidation, in disruption of ion homeostasis and neuronal death induced by amyloid beta-peptide. J. Neurochem. 68: 255–264.

Matsumoto, K., C. Sato, Y. Naka, A. Kitazawa, R.L. Whitby and N. Shimizu. 2007. Neurite outgrowths of neurons with neurotrophin-coated carbon nanotubes. J. Biosci. Bioeng. 103: 216–220.

Mattson, M.P., R.C. Haddon and A.M. Rao. 2000. Molecular functionalization of carbon nanotubes and use as substrates for neuronal growth. J. Mol. Neurosci. 14: 175–182.

McKenzie, J.L., M.C. Waid, R. Shi and T.J. Webster. 2004. Decreased functions of astrocytes on carbon nanofiber materials. Biomaterials 25: 1309–1317.

Mitchell, L.A., F.T. Lauer, S.W. Burchiel and J.D. McDonald. 2009. Mechanisms for how inhaled multiwalled carbon nanotubes suppress systemic immune function in mice. Nat. Nanotechnol. 4: 451–456.

Mitchell, L.A., J. Gao, R.V. Wal, A. Gigliotti, S.W. Burchiel and J.D. McDonald. 2007. Pulmonary and systemic immune response to inhaled multiwalled carbon nanotubes. Toxicol. Sci. 100: 203–214.

Ni, Y., H. Hu, E.B. Malarkey, B. Zhao, V. Montana, R.C. Haddon and V. Parpura. 2005. Chemically functionalized water soluble single-walled carbon nanotubes modulate neurite outgrowth. J. Nanosci. Nanotechnol. 5: 1707–1712.

Pulskamp, K., S. Diabate and H.F. Krug. 2007. Carbon nanotubes show no sign of acute toxicity but induce intracellular reactive oxygen species in dependence on contaminants. Toxicol. Lett. 168: 58–74.

Raffa, V., G. Ciofani, O. Vittorio, C. Riggio and A. Cuschieri. 2010. Physicochemical properties affecting cellular uptake of carbon nanotubes. Nanomedicine (Lond) 5: 89–97.

Shvedova, A.A., V.E. Kagan and B. Fadeel. 2010. Close encounters of the small kind: adverse effects of man-made materials interfacing with the nano-cosmos of biological systems. Annu. Rev. Pharmacol. Toxicol. 50: 63–88.

Sucapane, A., G. Cellot, M. Prato, M. Giugliano, V. Parpura and L. Ballerini. 2009. Interactions Between Cultured Neurons and Carbon Nanotubes: A Nanoneuroscience Vignette. J. Nanoneurosci. 1: 10–16.

VanHandel, M., D. Alizadeh, L. Zhang, B. Kateb, M. Bronikowski, H. Manohara and B. Badie. 2009. Selective uptake of multi-walled carbon nanotubes by tumor macrophages in a murine glioma model. J. Neuroimmunol. 208: 3–9.

Vittorio, O., V. Raffa and A. Cuschieri. 2009. Influence of purity and surface oxidation on cytotoxicity of multiwalled carbon nanotubes with human neuroblastoma cells. Nanomedicine 5: 424–431.

Zhang, L., D. Alizadeh and B. Badie. 2010. Carbon nanotube uptake and toxicity in the brain. Methods Mol. Biol. 625: 55–65.

2

Nanoparticle-Based Immunoassays and their Applications in Nervous System Biomarker Detection

Chunping Jia,[1,a] *Qinghui Jin*[1,b,*] *and Jianlong Zhao*[1,c]

ABSTRACT

Diseases of the nervous system affect many people worldwide seriously. Biomarker detection for the nervous system in body fluids is important and is becoming a promising diagnostic method. Over the past two decades, the emergence of nanotechnology has opened up a new era for high sensitivity detection of biomolecules. Nanoparticles, such as gold nanoparticles and quantum dots, with high surface areas, with excellent biocompatibility and unique physicochemical properties are widely used for developing highly sensitive biomarker detection platforms. Gold nanoparticles, in particular, play an important role in immunoassay systems as

---

[1]State Key Lab of Transducer Technology, Shanghai Institute of Microsystem and Information Technology, Chinese Academy of Science, 865 Changning Road, Shanghai. 200050, PR China.
[a]E-mail: Jiachp@mail.sim.ac.cn.
[b]E-mail: Jinqh@mail.sim.ac.cn.
[c]E-mail: jlzhao@mail.sim.ac.cn.
*Corresponding author

List of abbreviations after the text.

labels or scaffolds. The lower detection limit for some nanoparticle probe-based immunoassays is much lower (about 6 orders of magnitude) than conventional enzyme-linked immunosorbent assays (ELISAs). Although substantial progress has been made on the development of highly sensitive protein marker detection methods based on nanomaterials, many of these innovations have not been used in clinics because of poor specificity and reproducibility. A study of the reasons for nonspecific absorption, the interface of nanoparticles with biomolecules and the effect of proteins within body fluids may aid the wider clinical adoption of these novel approaches. This chapter summarizes the progress on immunoassay systems for nervous biomarker detection based on nanoparticles probes, the hindrances to application of these advanced methods in clinical samples and related solutions. Much more effort should be put into helping these new tests reach the clinic earlier.

INTRODUCTION

Diseases of the nervous system, such as Alzheimer's disease (AD) and brain cancer, seriously affect many people worldwide. In particular, Alzheimer's disease is a progressive mental disorder and the leading cause of dementia characterized by memory loss and cognitive impairment in people over age 65 (Andreasen and Blennow 2002). Alzheimer's disease is currently the third most highly occurring and expensive disease following cancer and cardiocerebral vascular diseases. There have been no non-invasive early diagnostic methods for most of these diseases until now. Related biomolecules will be present in cerebrospinal fluid, blood or other body fluids even before symptoms appear and these biomolecules, or biomarkers, are the most important early indicators of these diseases. Thus, the detection of biomarkers in biological fluids for persons with high-risk is a promising diagnostic method. For example, amyloid beta derived diffusible ligands, ADDL, is a promising biomarker for AD and the detection of ADDL in biological fluids has been used widely in the diagnosis of AD (Andreasen and Blennow 2002).

Many detection methods exist for such biomarkers. ELISA is one of these most important biochemical techniques, used mainly to detect the presence of antibodies or antigens in a sample, based on antibody-antigen immunoreactions. Due to its simplicity, low-cost, easy readability, acceptability and safety (Lequin 2005), ELISA is widely used and is the gold standard for detection of protein markers, pathogens, and other molecules related to various diseases, with a detection limit from 0.1 ng to 1 µg mL^{-1}

(Koppelman et al. 2004). However, the relative levels of protein markers are very low in the early stages of diseases, especially nervous diseases, and are beyond the detection limit of ELISA. For example, normally the ADDL concentration in the brain or cerebrospinal fluid is below detection levels (<1 pM) by conventional diagnostic methods. There is a need to improve the sensitivity of the current ELISA method for highly sensitive detection protein biomarkers, which is important for early diagnosis of neurodegenerative, cancer and other diseases.

Developing a highly sensitive protein detection method remains a challenge for scientists, due to lack of direct amplification of the targets. With direct amplification of a target, such as the polymerase chain reaction (PCR) for nucleic acids, a recognition event (nucleic acid hybridization) can trigger a catalytic process and the amount of target molecules can grow exponentially. Therefore, several copies of DNA or RNA can be detected with PCR methods (Gijohann and Mirkin 2009). Using the ELISA method, hundreds or even thousands of protein molecules could not be detected directly until now. Many methods have been developed to amplify the ELISA signal, such as through the biotin-streptavidin reaction, which can increase the amount of signal molecules (horseradish peroxidase-HRP, fluorescence) per antigen-antibody reaction event.

Recently, the emergence of nanotechnology has opened a new era for highly sensitive detection of biomolecules, especially proteins. Nanoparticles of various shapes, sizes, and compositions have found broad applications in biomolecular detection. Gold nanoparticles, in particular, with their high surface areas and unique physicochemical properties, are used widely in the development of biomarker platforms (Alivisatos 2004; Rosi and Mirkin 2005; Castaneda et al. 2007). Gold nanoparticles have excellent biocompatibility and can be conjugated with DNAs, antibodies, enzymes and other biomolecules, which can enable promising applications in signal enhancement for biochemical detection (Ambrosi et al. 2007; Cui et al. 2008). Readily available gold nanoparticles have unique optical properties and can easily be detected with the naked eye or instruments. Furthermore, gold nanoparticles can strongly quench fluorescence and be detected by a fluorescence spectrometer. Great progress has been achieved on highly sensitive protein detection methods based on nanomaterials. However, many of these innovations have not been used in clinics because of poor specificity and reproducibility, amongst other reasons. Therefore, much more effort should be expended on enabling early clinical adoption. In this chapter, we aim to summarize the progress of biomarker detection methods for the nervous system based on nanoprobes, the factors hindering application of these advanced methods in clinical samples and related solutions.

Applications to Areas of Health and Disease

The highly sensitive biomarker detection method based on nanoparticle probes will be widely used in clinics to detect diseases such as cancer, nervous system diseases and some types of genetic diseases. The abnormal expression of several types of biomarker indicates the occurrence of disease. Therefore, detection of these biomarkers in biological fluids (serum, cerebrospinal fluid, etc.) in persons at high risk is a promising diagnostic method. Furthermore, the expression levels of some biomarkers can also indicate the treatment effect, which is the basis of molecular targeted therapy.

Key Facts on Nanoparticle-Based Immunoassays and Their Applications in Nanomedicine

- Biomarker detection in biological fluids is becoming a promising diagnostic method for cancer, nervous system diseases and genetic diseases.
- Nanoparticles have high ratio of surface area-to-volume, biocompatibility and unique physicochemical properties.
- Great progress has been made on highly or ultra-sensitive protein detection methods based on nanomaterials as labels or scaffolds. Low to 3 attomolar PSA in samples can be detected using a Bio Bar-code assay.
- Many novel, highly sensitive detection methods have not been used in clinics because of poor specificity and reproducibility.
- Study into the reasons for nonspecific absorption, the interface of nanoparticles to biomolecules, the effects of background proteins within body fluids, may help wide clinical uptake of novel nanoparticle-based approaches.

THE IMMUNOASSAY SYSTEM BASED ON NANOPARTICLE PROBES

Gold nanoparticles function mainly as labels or nanoscaffolds in immunoassays (Gómez-Hens et al. 2008). For example, gold immunochromatographic assays use gold nanoparticles (AuNPs) as optical labels and the bright red color of AuNPs can be detected directly and easily by the naked eye. Furthermore, silver deposition can cause the auto-catalytic reduction of silver on AuNPs, increasing the size of the structure, which results in optical signal enhancement. Mirkin's group showed that gold deposition from solution leads to the continuous nucleation of new AuNPs by the probe AuNPs, in addition to autocatalytic growth. Also, these newly nucleated particles aggregate on the probe

AuNPs, resulting in signal enhancement and gold microstructures that are larger than those developed by silver deposition (Kim et al. 2009). Due to their redox properties, AuNPs can also act widely as electrochemical labels in electrochemical immunoassays.

Multiple Labeled Nanoparticles Based Immunoassays

Most importantly, AuNPs act as nanoscaffolds to increase the sensitivity of many bioassays because of the properties of AuNPs, such as high ratio of surface area-to-volume, surface-energy and surface-activation. Mirkin's group introduced an ultra-sensitive nanoparticle-based assay system, the bio-bar-code assay (BCA), for protein detection (Nam et al. 2003; Georganopoulou et al. 2005; Bao et al. 2006). In this system, multiple biomolecules, detector antibody and DNA strands are immobilized on AuNPs. The detector antibody can sandwich the target protein that is captured by antibody on magnetic micro-particles (MMPs). DNA strands function as the barcode or surrogate amplification units and are unique to each target protein. For each target protein binding event, the sandwich complex of MMP-target-AuNP probes carries a large number of DNA strands so the signal is amplified (Fig. 1). The sensitivity of this assay relies on amplification resulting from thousands of DNA strands for each target protein. There is substantial amplification and so protein content low to 30 attomolar levels can be detected. If these DNA barcodes are further amplified with the polymerase chin reaction (PCR), the sensitivity of a BCA can attain 3 attomolar sensitivity, which is the highest sensitivity method for protein detection to date.

The sensitivity and principle of this BCA method is similar to the immuno-PCR method reported earlier. In immuno-PCR, the detector antibody or the second antibody is covalently coupled to DNA strands, analogous to the barcode DNA in BCA, while in immuno-PCR, this DNA is amplified with PCR and detected with gel electrophoresis or by real time PCR using Taqman or other probes. There are several advantages of BCA over immuno-PCR. First, the detector antibody and barcode DNA are both labeled on gold nanoparticles and this avoids the need for conjugation of the antibodies to DNA strands, which is difficult and boring. Second, the DNA strands can be detected with a gene chip and gold nanoparticle-based silver enhancement, which can enable further amplification of signal and improvement of the sensitivity. However, this detection scheme requires sophisticated instrumentation and procedures (e.g., microarrayer and light-scattering measurement), which complicates the assay implementation and limits its wide application in clinical diagnosis.

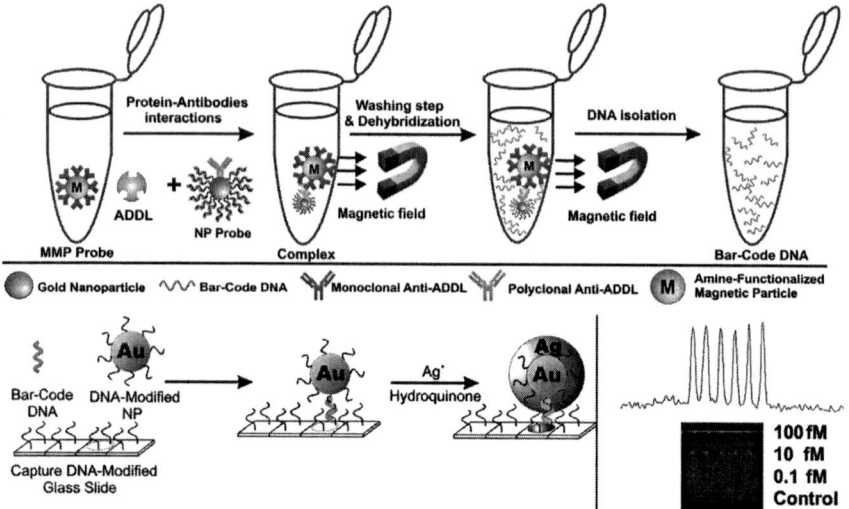

Fig. 1. Schematic of the bio-bar-code assay (BCA) procedure for detecting ADDL in cerebral spinal fluids. Magnetic Micro-particles (MMPs) were modified with Amyloid beta derived diffusible ligands (ADDL) capture antibody and gold nanoparticles (AuNP) probes were modified with ADDL detector antibody and Bar-code DNA strands. The sandwich structures (magnetic beads-target protein-AuNP probes) were formed when the target proteins were present. After several wash steps, the bar-code DNA was eluted into the solution and was then detected with silver enhancement and microarray or PCR. Down to 0.1 fM ADDL protein can be detected with this assay. Reprinted with permission from reference (Georganopoulou et al. 2005) © 2005, The National Academy of Sciences.

Color image of this figure appears in the color plate section at the end of the book.

A double-codified gold nanolabel for enhanced immunoanalysis was developed recently by Ambrosi et al. (2007). In their method, AuNPs were modified with anti-human IgG HRP-conjugated antibody. HRP was used here for signal amplification and could efficiently catalyze 10000 substrate turnovers, which significantly improved the detection limit. HRP can oxidize various chromogenic substrates and produce visible product, which can easily be detected. The detection limit using these nanolabels was 52 ng/L of human IgG with spectrophotometric detection, which was 50 times lower than that obtained using the HRP-labeled anti-human IgG of 2.4 mg/L. One major drawback of this method is that the detector antibody used in this method must be conjugated with HRP and the process of creating HRP-linked antibodies is very complex and expensive, which limits its wide application.

A highly sensitive protein detection method based on the advantages of gold nanoparticles and ELISA ("Nano-ELISA") using a novel enzyme-labeled AuNP has been developed by our group (Jia et al. 2009; Liu et al. 2010). In this method, AuNPs were immobilized with both detector antibody and thiolated-oligonucleotide strands (barcode DNA) and the end of the DNA strands opposite to AuNP was modified with biotin. HRP was introduced onto gold nanoparticles through the biotin-streptavidin reaction. The target protein was sandwiched by the enzyme-labeled AuNP probe and the capture probe through immunoreaction. The target immunoreaction event could then be sensitively transduced via the enzymatically amplified optical signal (Fig. 2). By this strategy, carcinoembryonic antigen (CEA), a model protein, was detected with high sensitivity and good specificity, which was approximately 130-fold more sensitive than conventional ELISA. Furthermore, Nano-ELISA is more effective and time-saving. The basic operation and instrumentation for Nano-ELISA is similar to those of ELISA, which can enable this method to be used more widely in clinics and labs than other methods, such as electrochemistry.

The sensitivity of any immunoassay is determined by the label, the antibody affinity, the assay background and the variance of the background measurement (Rissin et al. 2010). Therefore, for a given immunoassay system, the label method and quality of nanoparticle probes affect its sensitivity to a great extent. For the immunoassays mentioned above, AuNPs were immobilized with both detector antibody and barcode molecules (DNA, HRP). AuNPs act as the carrier or a "bridge" for the detector antibody and detection signals. When the target protein is sandwiched by the capture and detector antibody, the signal of one target protein is transformed to a signal by hundreds of barcode molecules (DNA, HRP). Thus, the more barcode molecules on AuNP probes, the more sensitive the method is. Because the diameter of DNA strands is about 2 nm and the size of a protein is about 10–100 nm, more DNA strands can be immobilized per AuNP. There are about 200–300 DNA strands on each 13 nm AuNP but only ten IgG molecules (10 nm) (Ambrosi et al. 2007; Jia et al. 2009; Liu et al. 2010). Therefore, use of AuNP probes immobilized with barcode DNA is a nice strategy for detecting low abundance molecules.

Nanoparticles Based Single-molecule ELISA

Another unique advantage of using nanoparticles is the ability to capture and enrich target molecules by magnetic collection of the nanoparticles. In fact, in BCA or Nano-ELISA system the nanoparticles also act to enrich the protein or nucleic acid. Furthermore, the isolation and detection of single protein molecules can provide a promising approach for measuring extremely low concentrations of proteins, similar to the Fluorescence Activating Cell Sorter (FACS).

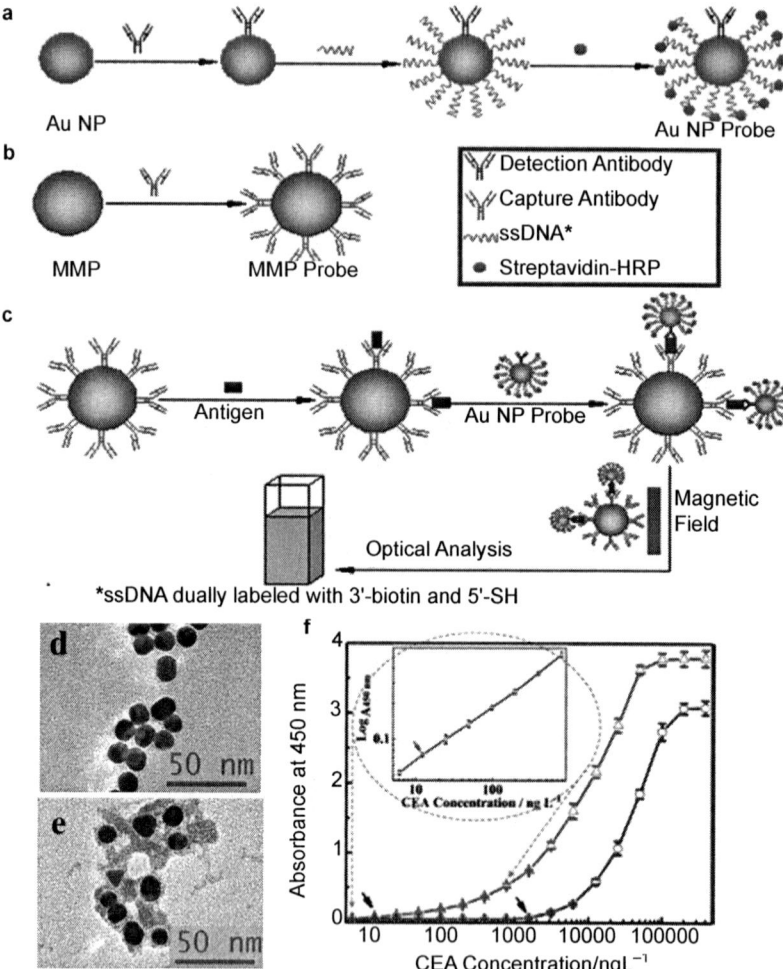

Fig. 2. Schematic of preparation of multi-component Au NP probes (a) and MMP probes (b), Nano-ELISA procedure (c), Transmission electron micrograph (TEM) of gold nanoparticles (d) and AuNP probes(e), and calibration curves for spectrophotometric detection of CEA protein using the multi-component Au NP probe based immunoassay (red light) and the conventional ELISA (blue light) (f). MMPs were modified with captured antibody and AuNP probes were modified with detector antibody and horseradish peroxidase (HRP), which was immobilized onto AuNPs through DNA strands and the streptavidin-biotin reaction. Sandwich structures (magnetic beads-target protein-AuNP probes) were formed and incubated with 3,3′,5,5′-tetramethylbenzidine (TMB) solution. Inset: log-log calibration curve of the multi-component AuNP probe-based immunoassay at a low CEA concentration; CEA concentration was varied from 6 to 781 ng/L. Reprinted with permission from reference (Liu et al. 2010) © 2010, The Royal Society of Chemistry.

Color image of this figure appears in the color plate section at the end of the book.

Based on these principals, Rissin et al. reported an ultrasensitive single-molecule ELISA using nanoparticles for capture and enrichment (Rissin et al. 2010). Their approach makes use of arrays of femtoliter-sized reaction chambers, called single-molecule arrays, which can isolate and detect single enzyme molecules (Fig. 3). The low-abundance proteins in blood were captured on microscopic beads decorated with specific capture antibodies and then the proteins were sandwiched with detector antibody labeled with biotin, through which the enzymatic reporter beta-galactosidase labeled with streptavidin was introduced into the detection system. Beads with or without sandwich structures were then loaded onto micro-wells and the fluorescence of beta-galactosidase catalyzed substrates was scanned with fluorescence imaging. There is only one bead in the majority of wells. Fluorescent signals in one well means that a bead has captured the single protein molecules. The concentration of protein is correlated to the percentage of beads that carry a protein molecule. This digital ELISA approach detected as few as ~10–20 enzyme-labeled complexes in 100 µl of sample (~10^{-19} M) and allowed detection of clinically

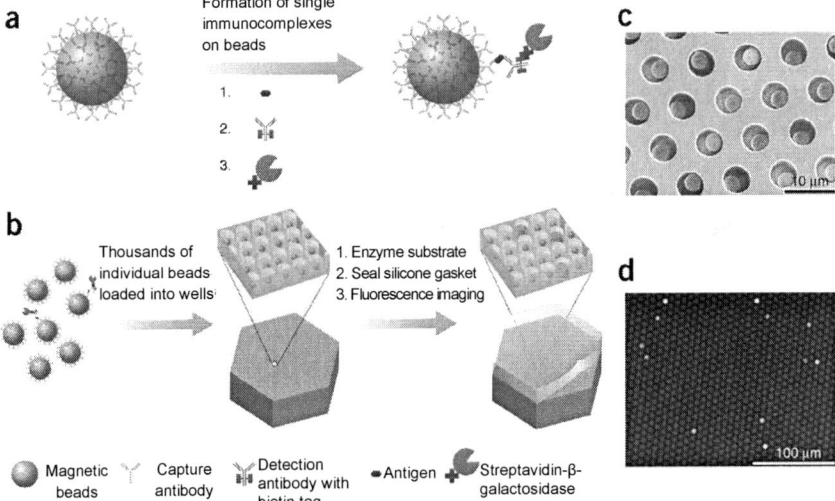

Fig. 3. Schematic of single-molecule ELISA, based on well arrays. (a) The captured antibody was immobilized onto magnetic beads and single proteins were captured through the formation of sandwich structures (magnetic beads-single protein molecules-second antibody with beta-galactosidase). (b) The beads, with or without sandwich structures, were loaded onto micro-wells and the fluorescence of beta-galactosidase catalyzed substrates was scanned with fluorescence imaging. (c) Result of scanning electron micrograph (SEM). There is one bead in the majority of wells. (d) Result of fluorescence imaging. Fluorescent signals in one well means that the bead has captured the single protein molecules. Reprinted with permission from reference (Rissin et al. 2010) ©2010, Nature Publishing Group.

Color image of this figure appears in the color plate section at the end of the book.

relevant proteins in serum at concentrations (<10^{-15} M) much lower than conventional ELISA (Rissin et al. 2010). Digital ELISA detected PSA in serum sample from patients who had undergone radical prostatectomy at concentrations low to 14 fg/ml (Rissin et al. 2010).

The sensitivity of this digital ELISA is lower than BCA, which can detect 330 fg/mL PSA (Thaxton et al. 2009). Furthermore, it seems that the sensitivity of protein detection could be increased at least 100-fold if nonspecific interactions that cause background signals could be minimized and the logistics of the assay can further be simplified. Thus, digital ELISA has the potential to facilitate earlier diagnosis and treatment of disease. However, for this digital ELISA, the process of creating wells with suitable average diameters and depths is exacting, as is the control of diffusing each bead into a well, which affects the sensitivity of this immunoassay.

Nanoparticles Based Multiplexed Determinations of Protein

For most diseases, such as of the nervous system and cancer, there have been no specific and sensitive single biomarkers so multiple biomarkers must be detected. Therefore, multiple detection methods, especially for protein, should be developed. For multiple immunoassay systems, fluorescence molecules are usually used as labels. Most fluorescent dyes in liquid are easily photobleached, which presents great challenges for versatile use. Nanoparticles with fluorescent dyes, such as quantum dots, chelated lanthanide incorporated fluorescent nanospheres, dyed polystyrene etc., can overcome these drawbacks and have stable emission profiles. Furthermore, the internal volume of each nanoparticle may contain numerous fluorescent dye molecules, which can lead to the enhancement of fluorescence signals. QDs with different diameter have different emission wavelengths so can usually be used as fluorescence labels for multiple sandwich immunoassays and thus there have many reports on protein and nucleic acid detection methods based on QDs. To detect fluorescence signals, very expensive instruments, such as lasers, must be developed, which limits their application severely.

High density microarrays can provide a large amount of information per unit time and microarray based immunoassay systems are a perfect multiple methodology. Combining the advantage of microarray and nanoparticles, many approaches have been reported. For example, Kim et al. described the use of gold deposition as a light scattering signal enhancer in a multiplexed, microarray-based scanometric immunoassay using gold nanoparticle probes (Kim et al. 2009). The use of gold deposition, especially multiple rounds of gold deposition, can result in greater signal enhancement than a typical silver development. Using this system, the assay was capable of detecting 300 aM (~ 9000 copies) of PSA in buffer and 3 fM in 10% serum. The use of gold deposition may have significant utility in scanometric detection schemes and broader clinical and research applications (Kim et al. 2009).

New Nanomaterials used in Immunoassays

Most immunoassay detection systems use nanoparticles as labels, amplifiers or scaffolds, such as gold or silver nanoparticles, QDs and magnetic metal nanoparticles. With the development of nanotechnology, attempts have been made to use many new nanomaterials in biomolecule detection biosensors, for example nanowires, carbon nanotubes (CNTs), and graphene. Graphene especially, a flat monolayer of carbon atoms tightly packed into a two-dimensional honeycomb lattice, for which its discoverers, Andrei Geim and Konstantin Novoselov, were awarded the 2010 Nobel Prize in Physics, has stimulated a vast amount of research, due to its fascinating properties. Graphene may enable a revolution in biomolecule detection. Graphene has a large specific surface area, good conductivity and fracture strength. An ultrasensitive electrochemical immunosensor based on a graphene sheet (GS) has been developed by Yang et al. (2010). In this assay, the capture antibody is immobilized onto a GS, which functions as micro-magnetic beads or microplates to capture and enrich the targets. Detector antibody and HRP were both labeled onto the GS by mediator thionine (TH) and the resulting nanostructure (GS-TH-HRP-Ab2) was used as the label for the immunosensor. The lower detection limit of this immunosensor is 1 pg/mL of PSA, with good reproducibility, selectivity and stability. The good performance of the immunosensor is attributed to the graphene sheet's high surface area-to-volume ratio, which allows the immobilization of high-levels of Ab1, Ab2, TH and HRP, and its good electrical conductivity, which can improve electron transfer amongst TH, HRP, H_2O_2 and the electrode (Yang et al. 2010).

The most significant report on the application of graphene concerns single DNA molecule detection (DNA sequencing) using graphene nanopores (Fig. 4.) (Garaj et al. 2010). Tiny holes, ranging in diameter from 5 to 23 nm, have been drilled through individual layers of graphene using an electron beam. The current carried by the salt ions when a voltage is applied across electrodes immersed in the solutions is measured and the change in electrical signals can be measured when different single DNA molecules pass through the nanopore. The report also shows that graphene may be an ideal substrate for very high resolution, high-throughput, nanopore-based single-molecule detectors, such as DNA, RNA and protein.

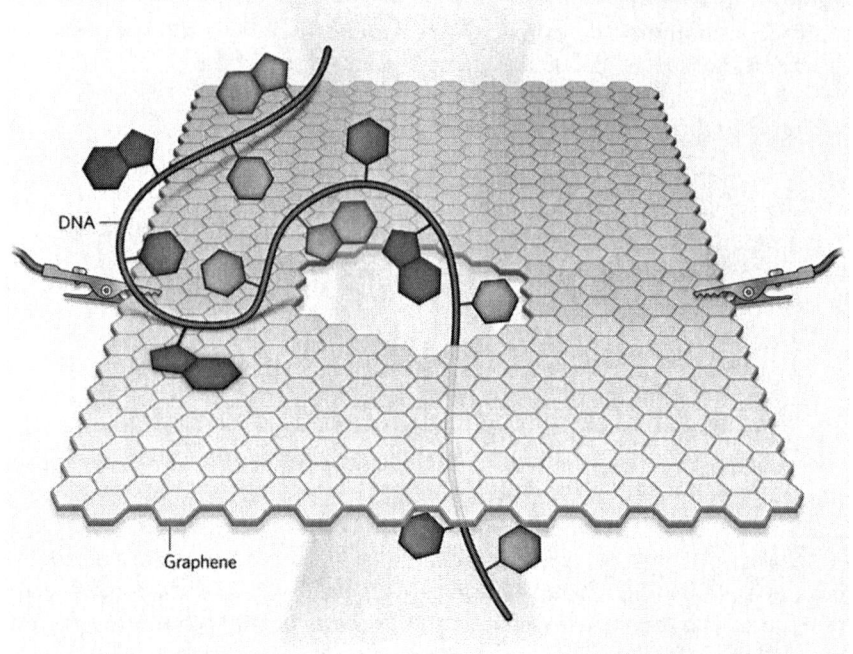

Fig. 4. Illustration of a graphene-based nanopore for single DNA sequencing. DNA fragments pass through the nanopore on graphene. Different colors represent different nucleotides. Reprinted with permission from references (Garaj et al. 2010 and Bayley 2010) © 2010, Nature Publishing Group.

Color image of this figure appears in the color plate section at the end of the book.

FACTORS HINDERING THE APPLICATION OF NANOPARTICLE-BASED IMMUNOASSAYS AND SOLUTIONS

Although there has been great progress on nanoparticle probe-based immunoassays, most reports describe only the principle of these methods. The detected samples are simple or synthetic single proteins, which cannot reflect a real clinical sample, especially serum. In fact, most of these methods have not been applied for detection in real samples. Real samples, especially serum, comprise thousands of proteins and have unique properties, which require detection methods to be stable and specific. However, most methods based on nanoparticles possess some disadvantages in repeatability and stability because of nonspecific adsorption, size variation, aggregation and other reasons (Gómez-Hens et al. 2008). In particular, AuNPs have excellent biocompatibility and can easily bind a great deal of biomolecules, which leads to nonspecific adsorption of other biomolecules in serum. Furthermore, antibody molecules can cross-react with some non-associated protein in serum samples and heterophile antibodies in serum can also cause serious false (negative or positive) results. Therefore, most measurements should be done with low nonspecific signals.

Nonspecific Absorption

Like classic ELISA, many blocking reagents should be added onto the surface of micro-plates, micro-magnetic beads, or other solid-phase substrates and AuNP probes to alleviate nonspecific binding. Unlike antigen-antibody specific interactions, the adsorption process is nonspecific. It is possible that any substance may adsorb to substrates and nanoparticles at any stage during the assay. Thus, a blocking agent should also be added into assay and dilution buffers. The commonly used classic ELISA blocking agents are bovine serum albumin, fetal calf serum, normal rabbit or horse serum, casein, and nonfat dried milk. Furthermore, detergents should be added to ease nonspecific binding, including sodium dodecyl sulfate, Tween-20, Triton-100, etc. For nanoparticles, especially AuNP probes, other agents should be added to avoid aggregation and size variation of nanoparticles, which can cause nonspecific binding or high background color. Polyethylene glycol (PEG) has been found to minimize nonspecific adsorption of proteins onto NPs and to reduce their uptake by cells. In preparation of AuNP probes, these types of agent, such as PEG or sucrose, are added and can be used to stabilize and passivate the gold nanoparticles.

Although the combination of nanotechnology and biotechnology has resulted in the rapid development of bio-diagnostics, these methods, especially for serum samples are still limited by non-specific adsorption and variability. In addition to using the blocking agents mentioned above, it is very important for us to understand the cause of non-specific adsorption, the interface between biomolecules and nanoparticles, the activity of biomolecules on nanoparticles, because the blocking reagent can only decrease non-specific adsorption to a certain extent and the adsorption cannot be eliminated completely. Reports that characterize the cause of non-specificity of nanoparticle-based immunoassays, especially for the clinical samples, are scarce. Tuomas Näreoja et al. described a study on non-specificity of an immunoassay using Eu-doped polystyrene nanoparticle labels to detect clinically important analytes thyroid stimulating hormone and PSA (Näreoja et al. 2009). Different particle sizes and high affinity monoclonal antibodies and their Fab and single-chain Fv recombinant antibody fragments were investigated. Their results showed that nonspecific binding in a nanoparticle immunoassay is not highly affected by the antibody constant domain, glycosylation or size of the nanoparticle label (Näreoja et al. 2009). These results were obtained only from simple serum-free immunoassays as model systems rather than clinical samples. Clinical samples may strongly alter a non-specific binding profile in an assay through, e.g., several antibodies or other unknown reasons (Näreoja et al. 2009).

Interface of Nanoparticles to Biomolecules

The interface of nanoparticles to biomolecules should be studied to enable the wide application of nanoparticle-based immunoassays. However, reports into the interface between nanoparticles and biomolecules are surprisingly scarce. The surface properties or ligands on nanoparticles and the type of biomolecules are the most critical aspects for the interface. For oligonucleotides, the DNA strands, usually linked with thio groups (HS-DNA) can be immobilized onto gold nanoparticles through the Au-S covalent bond. However, the non-specific adsorption of DNA in nanoparticle-DNA conjugates, adsorb nanoparticles via the nucleotides (Sunho and Kimberly 2010). DNA self-adsorption on nanoparticles can inhibit its ability to bind to the targets. Self-adsorption depends on coverage, where lower coverage increases self-adsorption because of larger exposed nanoparticle surface area. Increasing coverage decreases NSA but can also reduce hybridization ability due to steric hindrance. Self-adsorption of DNA also depends on the DNA sequence, because each nucleotide has a different affinity for gold surfaces. This self-adsorption can be decreased by thiol-modified gold nanoparticles but not thiol-modified DNA (Sunho and Kimberly 2010).

For proteins, the absorption to the gold NPs is mainly via non-covalent hydrophobic interaction. This absorption is mainly relevant to the bond on the surface of nanoparticles and the isoelectric point (pI), which differs according to the amino acid composition of proteins. When NPs are introduced to a biological fluid, protein molecules are adsorbed to their surface and compete for the nanoparticle surface, resulting in a dynamic protein "corona" that largely defines the biological identity of the particle (Cedervall et al. 2007). It is clear that identifying the proteins in the corona and understanding how adsorption occurs and evolves is critical for useful application of NPs (Cedervall et al. 2007). In fact, other biomolecules or molecules (e.g., oligonucleotides, enzymes and PEGs), immobilized specifically or adsorbed non-specifically, may also be the important components of the corona surrounding the nanoparticles (Fig. 5). Dobrovolskaia et al. developed an experiment to study the interaction of colloidal gold nanoparticles with human blood, the effect of particle size and analysis of plasma protein binding profiles (Dobrovolskaia et al.

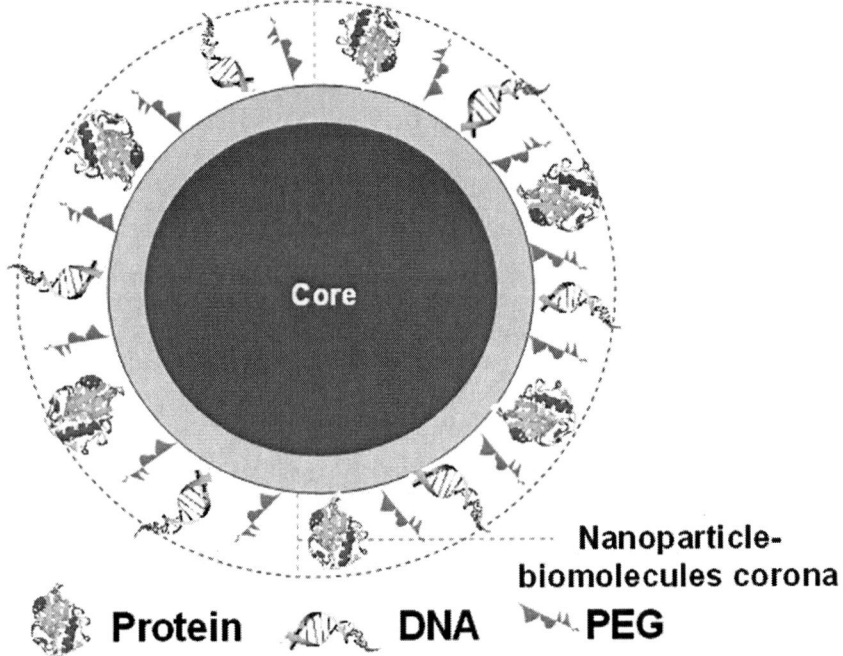

Fig. 5. Schematic of nanoprobes. Protein, oligonucleotide, PEG and other molecules, immobilized specifically or adsorbed non-specifically to the surface of nanomaterials, compete for the nanoparticle surface and result in a corona around the nanoparticles (unpublished material of author).

Color image of this figure appears in the color plate section at the end of the book.

2009). Their results revealed 69 different proteins bound to the surface of gold nanoparticles and 30 nm gold colloids bound almost twofold more protein mass than 50 nm colloids. This difference reflects the greater total surface area of the 30 nm colloids. Their data also suggests that interaction may be multilayered in that a cationic protein binds to the anionic gold surface at one site and brings another anionic protein to the other site (Dobrovolskaia et al. 2009).

The Effect of Heterophilic Antibody

As well as nonspecific binding or absorption in immunoassay systems, heterophilic antibodies in serum samples can also cause serious false (negative or positive) results. Heterophilic antibodies are antibodies induced by external antigens (heterophilic antigens, such as mouse, horse, etc.) that cross-react with self-antigens. In sandwich ELISA, heterophilic antibodies can sandwich capture and detector antibodies, which causes a false positive result. This kind of false positive is present for about 15% of serum samples. Therefore, blocking agents for heterophilic antibodies should be added into the immunoassay buffer to detect serum samples. Furthermore, serum samples can also be diluted to 1/4 or lower to alleviate their effects.

The Cross-reaction of Antibodies

For multiple protein detection methodology, there are several species of antibodies for different target proteins in a reaction unit. Therefore, some nonspecific signal can be present due to the significant cross-reactivity of antibodies with other target protein. To avoid this, antibodies with high affinity and selectivity should be used. Because of the disadvantages of poor selectivity, difficult production processes and high costs of antibodies, many chemical or other protein detection systems will replace antibody-based systems, such as aptamers and molecular imprinting techniques. In particular, application of aptamers for protein, drug molecule or cell detection has made some progress. Aptamers are short oligonucleotides with high selectivity and affinity to targets. Through many rounds of selection *in vitro*, the aptamer best recognizing the target can be identified and isolated. Thus, aptamers have high sensitivity and affinity to the targets. Once the nucleic acid sequence of the aptamer is identified for a particular target, it is produced synthetically—a distinct advantage over antibodies, which require biological processes (e.g., hybridoma). Furthermore, oligonucleotides can be easily immobilized onto nanoparticles and these types of label can be further amplified by PCR or other nucleic acid detection techniques. The main disadvantage of aptamers is that the selection of aptamers that bind to targets with high specificity and affinity from thousands of DNA pool is very difficult and labor-intensive. If the aptamer discovery system can be made rapid, highly efficient, automated, and applicable to a wide range of targets, the application of aptamers and nanotechnology will essentially drive the development of diagnostics.

DETERMINATION OF BIOMARKERS FOR THE NERVOUS SYSTEM

Normally the ADDL concentration in brain or cerebrospinal fluid is too low to be detected (<1 pM) by conventional diagnostic methods. In 2004, a nanobiosensor based on monitoring the localized surface plasmon resonance (LSPR) spectrum of Ag nanoparticles was reported for detecting anti-ADDLs, also implicated in AD, by monitoring the interaction between ADDLs and anti-ADDLs (Haes et al. 2004). The lower detection limit of anti-ADDLs is 100 pM. If the strong nonspecific binding could be alleviated, a lower concentration of proteins could be detected. With further improvements in both selectivity and sensitivity, the LSPR biosensor has the potential to become an accurate and economical alternative to traditional clinical assays.

In conjunction with the BCA of the Mirkin group, fluorescent and scanometric detection methodologies were used to detect ADDL with high sensitivity (Georganopoulou et al. 2005). Down to a 10 attomolar concentration of ADDL can be detected and the sensitivity of the scanometric method proved useful in showing a difference in plasma ADDL levels between Alzheimer patients and control samples, providing a promising beginning for the early detection and diagnosis of this debilitating neurodegenerative disease (Georganopoulou et al. 2005).

Neely et al. developed an ultrasensitive and highly selective detection for the Alzheimer's disease biomarker, tau protein, using the two-photon Rayleigh scattering (TPRS) properties of gold nanoparticle (Neely et al. 2009). Tau proteins play a very important role in the structure of the neuron and are a type of progressive biomarker or AD. Monoclonal anti-tau protein antibody was immobilized onto gold nanoparticles. When the tau protein is present in fluids, several gold nanoparticles probes can bind to each protein, thereby producing nanoparticle aggregates. As a result, the color of the gold NP probes solution changes from red to a bluish color and a new broadband appears around 150 nm, far from their plasmon absorption band. This bioassay takes less than 35 min and can provide a quantitative measurement of tau protein concentration in the pg/mL region, which is 3 orders of magnitude more sensitive than the usual colorimetric technique. There have been many other immunoassay systems for detecting biomarkers for the nervous system (Table 1). Because biomarkers for the nervous system are scarce, many more specific and sensitive biomarkers must be discovered, which can help nanotechnology to be used widely in diagnostics of the nervous system.

Summary Points

- This chapter summarizes the progress in immunoassay systems for nervous system biomarker detection based on nanoparticle probes, such as BCA, Nano-ELISA, nanochip and single molecule ELISA etc. These novel immunoassay systems have many advantages in terms of sensitivity, cost, flexibility, operation, time-saving and so on. The lower detection limit of some nanoparticle probe-based immunoassays, e.g., BCA, is down to zeptomolar level. Although great improvements have been made in nano-biotechnology, most of these systems have not been widely used in clinics or other fields. The chapter also summarizes the factors hindering application of

Table 1. List of detection methods for nervous system biomarkers.

Protein	Nanoparticles	Methodology	limit of detection	references
ADDL	Ag nanoparticles	LSPR	100pM	Haes et al. 2004
ADDL	Gold nanoparticles	BCA	10attomolar	Georganopoulou et al. 2005
ADDL	Polystyrene nanospheres	LSPR	7pM	Haes et al. 2005
beta-Amyloid	Gold nanoparticles and nanofluid	SERS	1nM	Chou et al. 2008
Tau protein	Gold nanoparticles immunochip	LSPR	10pg/mL	Vestergaard et al. 2008
Tau protein	Gold nanoparticles	TPRS	1pg/mL	Neely et al. 2009

This table lists immunoassay systems based on nanoparticles for nervous system biomarkers. ADDL, Amyloid beta derived diffusible ligands; LSPR, Localized Surface Plasmon Resonance; BCA, Bio Bar-code Assay; SERS, Surface-Enhanced Raman Spectroscopy; TPRS, Two-Photon Rayleigh Scattering (unpublished material of author).

these novel immunoassays, including the causes of nonspecific absorption, the interface between nanoparticles and biomolecules, the effect of background proteins within body fluids and the cross-reactivities of antibody. Much effort should be put into understanding these principles, which can contribute to the adoption of these novel nanoparticle-based immunoassay systems in clinical and other applications.

- Not only the sensitivity but also the selectivity and specificity of the immunoassay system should be emphasized. To some extent, the more sensitive, the less specific the immunoassay is because the procedures involved in system with ultra-high sensitivity usually include many tedious steps (including several washing steps), which require strict adherence to operational protocols.
- The production process for nanomaterials, preparation of nanoprobes, and the procedures of immunoassays should be standardized and automated as far as possible to alleviate variability.
- The nonspecific adsorption of nanoprobe-based immunoassays should be alleviated especially for analyzing complex clinical samples (e.g., serum). Understanding the interface between nanomaterials and biomolecules is an increasingly hot topic in enabling nanotechnology to be used widely in medical diagnostics.
- Due to the complexity of protein structure, poor selectivity, difficult product processing, cross-reactivation and strong nonspecific adsorption of protein molecules, many chemical, oligonucleotide-based or other protein detection systems should be developed to replace antibody-based systems, such as aptamers and molecular imprinting techniques.

- Nano-biotechnology is a type of interdisciplinary research, which requires many scientists from different fields to work closely together to enable nanotechnology to be used widely in medical diagnostics or other fields in a better way.

Definitions

Biomarker: The biomolecules (protein, DNA, RNA, etc.) present or expressed abnormally in cerebrospinal fluid, blood or other body fluids when or before symptoms appear. Such molecules can also be used to measure the progress of disease or the effects of treatment.

Nanoparticles: Materials with nanosizes within the range 1–100 nm. Nanomaterials are characterized by a high ratio of surface area to volume, high surface-energy and surface-activation.

Immunoassay: A biomolecule detection method that makes use of the binding reaction between an antigen and its homologous antibody.

ELISA: One of the most important biochemical techniques, used mainly to detect the presence of antibodies or antigens in a sample based on antibody-antigen immunoreactions. Traditional ELISA typically involves chromogenic reporters and substrates.

Microarray: A multiplex chip on a solid substrate (glass slide, nylon membrane or other substrates) that assays large amounts of biomolecules per unit of time with high-throughput screening methods.

Sensitivity: The lower detection limit of target biomolecules in samples.

Isoelectric point: The pH at which a protein takes on a net negative charge.

Abbreviations

AD	:	Alzheimer's Disease
ADDL	:	Amyloid beta derived diffusible ligands
AuNPs	:	gold nanoparticles
BCA	:	Bio Bar-code Assay
CNT	:	Carbon Nanotube
CEA	:	Carcinoembryonic Antigen
ELISA	:	Enzyme-Linked Immunosorbent Assays
FACS	:	Fluorescence Activating Cell Sorter
GS	:	Graphene Sheet
HRP	:	Horseradish Peroxidase
pI	:	Isoelectric Point
LSPR	:	Localized Surface Plasmon Resonance
MMP	:	Micro-Magnetic Particles
PCR	:	Polymerase Chain Reaction
PEG	:	Polyethylene glycol

QD   :   Quantum Dot
SERS   :   Surface-Enhanced Raman Spectroscopy
TPRS   :   Two-Photon Rayleigh Scattering

References

Alivisators, P. 2004. The use of nanocrystals in biological detection. Nat. Biotechnol. 22: 47–52.
Ambrosi, A., M.T. Castaneda, A.J. Killard, M.R. Smyth, S. Alegret and A. Merkoci. 2007. Double-Codified Gold Nanolabels for Enhanced Immunoanalysis. Anal. Chem. 79(14): 5232–5240.
Andreasen, N. and K. Blennow. 2002. β-Amyloid (Aβ) protein in cerebrospinal fluid as a biomarker for Alzheimer's disease. Peptides. 23: 1205–1214.
Bao,Y.P., T.F. Wei, P.A. Lefebvre, A. Hao, L.X. He, G.T. Kunkel and U.R. Muller. 2006. Detection of Protein Analytes via Nanoparticle-Based Bio Bar Code Technology. Anal. Chem. 78: 2055–2059.
Bayley, H. 2010. Holes with an edge. Nature 467: 164–165.
Castaneda, M.T., S. Alegret and I.A. Merkoc. 2007. Electrochemical Sensing of DNA Using Gold Nanoparticles. Electroanalysis 19: 743–753.
Cedervall, T., I. Lynch, S. Lindman, T. Berggård, E. Thulin, H. Nilsson, K.A. Dawson and S. Linse. 2007. Understanding the nanoparticle-protein corona using methods to quantify exchange rates and affinities of proteins for nanoparticles. Proc. Natl. Acad. Sci. USA 104: 2050–2055.
Chou, I.-H., M. Benford, H.T. Beier and G.L. Cot. 2008. Nanofluidic Biosensing for beta-Amyloid Detection Using Surface Enhanced Raman Spectroscopy. Nano Lett. 8: 1729–1735.
Cui, R., H. Huang, Z. Yin, D. Gao and J. Zhu. 2008. Horseradish peroxidase-functionalized gold nanoparticle label for amplified immunoanalysis based on gold nanoparticles/carbon nanotubes hybrids modified biosensor. Biosens. Bioelectron. 23: 1666–1673.
Dobrovolskaia, M.A., A.K. Patri, J. Zheng, J.D. Clogston, N. Ayub, P. Aggarwal, B.W. Neun, J.B. Hall and S.E. McNeil. 2009. Interaction of colloidal gold nanoparticles with human blood: effects on particle size and analysis of plasma protein binding profiles. Nanomedicine: Nanotechnology, Biology, and Medicine 5: 106–117
Garaj, S., W. Hubbard, A. Reina, J. Kong, D. Branton and J.A. Golovchenko. 2010. Graphene as a subnanometre trans-electrode membrane.Nature 467: 190–193.
Georganopoulou, D.G., L. Chang, J.M. Nam, C.S. Thaxton, E.J. Mufson, W.L. Klein, A. Chad and C.A. Mirkin. 2005. Nanoparticle-based detection in cerebral spinal fluid of a soluble pathogenic biomarker for Alzheimer's disease. Proc. Natl. Acad. Sci. USA 102: 2273–2276.
Gijohann, D.A. and C.A. Mirkin. 2009. Drivers of biodiagnostic development. Nature 462: 461–464.
Gómez-Hens, A., J.M. Fernández-Romero and M.P. Aguilar-Caballos. 2008. Nanostructures as analytical tools in bioassays. Trend. Anal. Chem. 27: 394–406.

Haes, A.J., P. Hall, L. Chang, W.L. Klein, P. Richard and V. Duyne. 2004. A Localized Surface Plasmon Resonance Biosensor: First Steps toward an Assay for Alzheimer's Disease. NANO. LET. 4: 1029–1034.
Haes, A.J., L. Chang, W.L. Klein and R.P. Van, Duyne. 2005. Detection of a Biomarker for Alzheimer's Disease from Synthetic and Clinical Samples Using a Nanoscale Optical Biosensor. J. Am. Chem. Soc. 127: 2264–2271.
Jia, C.H., X.Q. Zhong, H. Bao, M.Y. Liu, F.X. Jing, X.H. Lou, S.H. Yao, J.Q. Xiang, Q.H. Jin and J.L. Zhao. 2009. Nano-ELISA for highly sensitive protein detection. Biosens. Bioelectron. 24: 2836–2841.
Kim, D., W.L. Daniel and C.A. Mirkin. 2009. Microarray-Based Multiplexed scanometric Immunoassay for Protein Cancer Markers Using Gold Nanoparticle Probes. Anal. Chem. 81: 9183–9187.
Koppelman, S.J., C.M. Lakemond, R. Vlooswijk and S.L. Hefle. 2004. Detection of soy proteins in processed foods: literature overview and new experimental work. J. Aoac. Int. 87(6): 1398–1407.
Lequin, R. 2005. Enzyme Immunoassay (EIA)/Enzyme-Linked Immunosorbent Assay (ELISA). Clin. Chem. 51: 2415–2418.
Liu, M.Y., C.P. Jia, Y.Y. Huang, X.H. Lou, S.H. Yao, Q.H. Jin, J.L. Zhao and J.Q. Xiang. 2010. Highly sensitive protein detection using enzyme-labeled gold nanoparticle probes. Analyst. 135: 327–331.
Nam, J.M., C.S. Thaxton and C.A. Mirkin. 2003. Nanoparticle-based bio-barcodes for the ultrasensitive detection of proteins. Science 301: 1884–1886.
Näreoja, T., M. Vehniäinen, U. Lamminmäki, P.E. Hänninen and A. Härmä. 2009. Study on nonspecificity of an immuoassay using Eu-doped polystyrene nanoparticle labels. J. Immunol. Methods 345: 80–89.
Neely, A., C. Perry, B. Varisli, A.K. Singh, T. Arbneshi, D. Senapati, J.R. Kalluri and P.C. Ray. 2009. Ultrasensitive and Highly Selective Detection of Alzheimer's Disease Biomarker Using Two-Photon Rayleigh Scattering Properties of Gold Nanoparticle. ACS. NANO 3: 2834–2840.
Rissin, D.M., C.W. Kan, T.G. Campbell, S.C. Howes, D.R. Fournier, L. Song, T. Piech, P.P. Patel, L. Chang, A.J. Rivnak, E.P. Ferrell, J.D. Randall, G.K. Provuncher, D.R. Walt and D.C. Duffy. 2010. Single-molecule enzyme-linked immunosorbent assay detects serum proteins at subfemtomolar concentrations. Nat. Biotechnol. 28: 595–510.
Rosi, N.L. and R.C. Mirkin. 2005. Nanostructures in Biodiagnostics. Chem. Rev. 105: 1547–1550.
Sunho, P. and H.S. Kimberly. 2010. Nanoscale interfaces to biology. Curr. Opin. Chem. Biol. 14: 616–622.
Thaxton, C.S., R. Elghanian, A.D. Thomas, S.I. Stoeva, J.S. Lee, N.D. Smith, A.J. Schaeffer, H. Klocker, W. Horninger, G. Bartsch and C.A. Mirkin. 2009. Nanoparticle-based bio-barcode assay redefines "undetectable" PSA and biochemical recurrence after radical prostatectomy. Proc. Natl. Acad. Sci. USA 106: 18437–18442.
Vestergaard, M., K. Kerman, D.-K. Kim, H.M. Hiep and E. Tamiya. 2008. Detection of Alzheimer's Tau Protein Using Localised Surface Plasmon Resonance-Based Immunochip. Talanta 74: 1038–1042.
Yang, M.H., A. Javadi, H. Li and S. Gong. 2010.Ultrasensitive immunosensor for the detection of cancer biomarker based on graphene sheet. Biosens. Bioelectro. 26: 560–565.

Electrical Stimulation of Brain Tissue with Carbon Nanofiber Microbrush Arrays

Edward de Asis,[1] Russell J. Andrews[2] and Jun Li[3]

ABSTRACT

The development of reliable Neural Electrical Interfaces (NEI) is an important focus of contemporary neuroscience research for the treatment of neurological disorders utilizing implantable neuroprosthetic devices. State-of-the-art neural implants utilize metal microelectrodes for *in vivo* extracellular stimulation and recording of neural cells. However, metal electrodes are typically over 100 µm in size, precluding them from probing single neurons (which are less than 20 µm in diameter). In addition, the high impedance of the metal electrodes requires large driving voltages, thereby increasing the damage to tissues and triggering electrolysis of water—a process which is toxic to cells. Reducing electrode size is limited by the increase in impedance. Consequently, the

---

[1]School of Engineering, San Francisco State University, 1600 Holloway Ave., San Francisco, CA 94132; E-mail: edeasis@sfsu.edu
[2]NASA Ames Research Center, Moffett Field, CA 94035; E-mail: rja@russelljandrews.org
[3]Department of Chemistry, Kansas State University, Manhattan, KS 66506; E-mail: junli@ksu.edu

List of abbreviations after the text.

spatial resolution of metal electrodes is far from desired values. Vertically Aligned Carbon Nanofibers (VACNFs) fabricated into Microbrush Arrays (MBAs) are a promising material for next-generation neuroprosthetic devices due to: 1) a three-dimensional (3D) topography that is compatible with the natural tissue environment; 2) reduced electrode impedance by the high surface area of nanostructure, and 3) soft mechanical contact with neural cells by the elasticity of freestanding CNFs. In this chapter, we show the following progresses in utilizing these materials for neural stimulation and recording: 1) as a cell culture substrate—VACNFs coated with polypyrrole (PPy) reduce the mechanical stress on neuron-like cells and effect a mechanotransduction probe with high spatial resolution; and 2) as an neural stimulation electrode—a MBA of PPy-coated VACNFs achieves the highest stimulation efficacy compared to tungsten wire electrodes, a platinum microelectrode array (MEA), and an as-grown VACNF MBA. Significantly, the PPy-coated VACNF MBA achieves reduced impedance and enables the stimulation of neural tissues without electrolyzing water. The PPy-coated VACNF MBA can be miniaturized to microns in dimension that enables spatial resolution potentially down to single neurons.

INTRODUCTION

A current challenge in neuroscience research is the development of therapies for nonrecoverable injuries to the nervous system such as spinal cord injury and stroke, neurodegenerative diseases such as Parkinson's Disease (PD) and Huntington's Disease (HD) and mood disorders such as Major Depression and Obsessive Compulsive Disorder (OCD). An approach to treating these diseases is to stimulate neural tissues electrically with implantable devices. Current neuroprosthetic implants are typically constructed from metal electrodes that are at least tens of microns in diameter. Because the neurons in mammalian central nervous systems are on the order of a few µms, these metal electrodes are incapable of selectively probing single neurons. Furthermore, the impedance of the NEI between the electrode and the surrounding tissue is high, thus requiring large voltages for stimulation and causing large noises in recording. The large stimulating voltage may damage the neural tissue or trigger electrolysis of water (which is toxic to cells). The rigid mechanical property and flat surface of metal electrodes are also incompatible with soft 3D tissues and thus limit the stability of the NEI.

VACNFs are a new type of 3D nanostructured electrode. They consist of separated carbon nanofibers (CNFs) of high aspect ratio (radius of 50–100 nm and length of 5–10 µm in this application) which are highly conductive and mechanically robust. The internal CNF structure consists of a stack of conical graphitic cups (Melechko et al. 2005). Each CNF in the VACNF array behaves as a soft lever that can be bent elastically by a small lateral force. This chapter presents the use of VACNFs coated with electrical conductive polymers (ECPs) such as PPy as a novel electrode material for future neuroprosthetic devices. The merits of 3D topography, soft mechanical properties, biocompatibility by ECP modification, high electrical conductivity, low impedance, and miniaturization capability will be demonstrated.

Applications to Areas of Health and Disease

Functional Electrical Stimulation (FES) and Deep Brain Stimulation (DBS) are clinical therapies for injuries and disorders of the nervous system. In FES, implanted stimulating electrodes deliver small amplitude electric currents to motor nerves to restore control and movement of paralyzed extremities in spinal cord injury and stroke patients (Popovic et al. 2001). In DBS, an electrode is surgically implanted deep in the brain and programmed to deliver a stimulating pulse train to the surrounding neural tissue to treat neurological disorders such as PD, OCD, Major Depression, and HD (Lee et al. 2009). Cochlear implants incorporating stimulation electrodes that excite the auditory nerve have partially restored lost hearing, and retinal implants consisting of microelectrode arrays interfaced with retinal ganglion cells are under development to treat lost vision (Squire et al. 2003; Wang et al. 2006). The NEI between the electrode and the surrounding neural tissue must be engineered to achieve the following: 1) biocompatibility to preclude immune responses to the implanted device and to promote long-term stability of electrode performance, 2) miniaturization to a size comparable to neural cells to facilitate electrode placement close to target cells, and 3) reduced impedance to improve the stimulation efficiency. PPy-coated VACNFs in MBA configuration can be used as a flexible supporting substrate for neuronal growth with minimized mechanical stress, improved biocompatibility, reduced impedance, and enhanced stimulation efficiency compared to contemporary metal microelectrodes. The combination of these properties can make a much more reliable NEI for FES and DBS potentially.

PPy-COATED VACNF MBA FOR ELECTROPHYSIOLOGY

The PPy-coated VACNF MBA provides a viable alternative to traditional, commercially available metal MEAs by adding a 3D brush-like nanostructure which is more compatible to the natural extracellular matrix in tissues and provides many desired new properties. Herein, we summarize our research efforts in exploring these novel electrode materials for NEI applications over the past six years. It covers device fabrication, surface modification, materials and electrochemical characterization, biocompatibility study with PC12 cell culture and electrical stimulation with rat hippocampal brain slices.

Fabrication of PPy-coated VACNF MBA

The fabrication of the PPy-coated VACNF MBA involves selective growth of brush-like VACNFs on patterned metal electrodes which is achieved by coating a Ni catalyst layer at desired locations using UV lithography. The electrode array including the microelectrodes, interconnect lines, and peripheral bonding pads (as shown in Fig. 1) was fabricated with a lift-off process by depositing a 200 nm thick Cr or Ti layer on a lithographically patterned Si wafer which was covered with 400 nm thick silicon nitride (Si_3N_4). The resistance as measured between the microelectrode and bonding pad was approximately 800 Ω. After the liftoff process, a 400 nm Si_3N_4 layer was deposited to cover the electrode array via PECVD using silane (SiH_4, 2000 sccm) and ammonia (NH_3, 33.5 sccm) at a pressure of 250 mTorr at 350°C. Windows in the silicon nitride were defined right on top of the microelectrodes and the bonding pads via photolithography using a 1.6 µm thick Shipley 3612 resist and opened by removing Si_3N_4 via an 80-second reactive ion etch (RIE) using sulphur hexafluoride (SF_6, 25 sccm) at 125 W and 200 mTorr. A 30 nm thick Ni catalyst layer was then

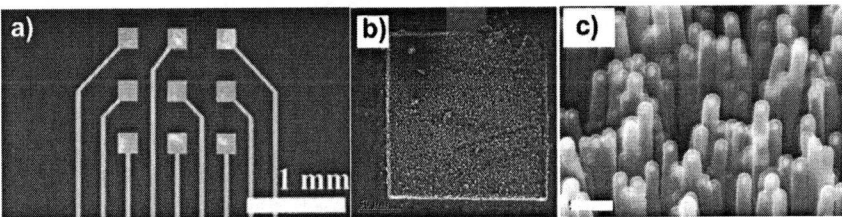

Fig. 1. PPy-coated VACNF MBA. SEM images of a) 3x3 PPy-coated VACNF MBA, b) enlarged view of one of the microbrush electrodes (scale bar = 50 µm), c) further enlarged view of PPy-coated VACNFs (scale bar = 500 nm).

deposited with an ion beam sputter on the microelectrode area. CNFs were grown via PECVD using a C_2H_2 feedstock (22.5 sccm) and NH_3 diluent (80 sccm) at a processing pressure of 4 Torr and processing temperature of 725°C for 10 minutes. The growth process yielded VACNFs of 2 to 4 µm in length and 100 to 250 nm in diameter. The as-grown VACNFs present Ni catalyst particles at the tip of each CNF, which were removed via a 5 minute chemical etching in 1 M nitric acid. PPy was electrochemically deposited from 50 mM pyrrole in 1.0 M KCl solutions by holding the electrode potential at 1.50 V (vs. Ag/AgCl (3M KCl)) for 120 seconds using a potentiostat (Nguyen-Vu et al. 2006).

Materials Benefits of Polypyrrole Coating

A limitation of bare as-grown VACNFs is their susceptibility to irreversible collapsing into a tepee structure upon drying (Fig. 2a), attributed to the capillary forces on the high-aspect-ratio CNFs. In cell culture, the stiff collapsed microbundles of CNFs tend to alter the morphology of PC12 cells (Nguyen-Vu 2007). We found that even a thin PPy coating of 10 nm in thickness can significantly increase the mechanical strength of CNFs and preserve their initial vertical alignment during repeated wet chemical treatment and drying processes (Fig. 2b).

Fig. 2. SEM Images of as-grown VACNFs vs. PPy-coated VACNFs after Wetting and Drying.

In addition, the active electrochemical properties of PPy provide a large pseudocapacitance which can be utilized to reduce the impedance of VACNF electrodes. The electrochemically deposited PPy possesses a positively charged backbone which is compensated by doping with anions. For example, electrochemical deposition of PPy on VACNF using KCl as the supporting electrolyte dopes PPy with Cl^-. As the electrode potential is changed, the PPy backbone is further oxidized or reduced, which effectively convert the electronic current into flux of Cl^- ions in and out of the PPy film. Notably, Cl^- anions are compatible with the physiological environment surrounding living cells which minimizes the toxic effects encountered with metal or metal oxide electrodes. As a result, PPy and other ECP coatings have been used to improve the metal stimulating electrodes in many implantable neuroprosthetic devices (Cui and Martin 2003).

PPy-coated VACNF as a Biocompatible Cell Culture Substrate

The biocompatibility of bare VACNFs and PPy-coated VACNFs with neurons including their effects on cell function and viability can be inferred from the growth rate and cell morphology in cell culture study. Chemical modifications to the PPy coating may optimize VACNF MBAs for cell culture. Since extensive studies have been carried out to investigate stimulation induced release of catecholamines (such as dopamine) from PC12 cells, these cells are chosen as an *in vitro* model system for cell culture study and developing PPy-coated VACNF sensors for treatment of PD.

PC 12 cells generally do not adhere to plastic cell culture dishes and thus grow in floating colonies instead. To promote PC12 adhesion, the plastic substrate must be coated with extracellular matrix (ECM) proteins such as Type IV collagen. This is also required for VACNF substrates. To study the effects of PPy and ECM coating on PC12 adhesion to VACNFs, PC12 cell culture was studied on four different substrates: 1) Uncoated as-grown VACNFs, 2) PPy-coated VACNFs, 3) ECM-coated VACNFs, and 4) ECM on PPy-coated VACNFs (Nguyen-Vu et al. 2007). The corresponding results are shown in Fig. 3a-d. Without ECM coating, PC12 cells do not adhere to neither as-grown VACNFs nor PPy-coated VACNFs. Consequently, only a few scattered single PC12 cells were observed on the substrate (Fig. 3a and b). With ECM coating, PC12 cells form monolayer patches on both bare VACNFs and PPy-coated VACNFs, indicating improved cell adhesion by collagen (Fig. 3c and d). However, the actin stain shows nonuniform micronetwork on VACNFs without PPy coating (Fig. 3a and c) likely due to mechanical stress of the bunched CNF tips to the cell body, while all cells on PPy-coated VACNFs (Fig. 3b and d) have more uniform actin distribution. The growth rate on ECM coated samples, however, strongly depended on the type of anion dopants in the PPy film. Doping with Cl^- was found to give much lower PC12 growth rate, but can be dramatically increased using polystyrenesulfonate (PSS^-) as dopants. These results indicate that the PPy film on VACNF may serve as a medium for chemical and biochemical modification by changing the anionic dopants, a process which is accomplished by substituting the supporting electrolyte in the electrochemical PPy deposition process. When cultured in the media containing nerve growth factor (NGF), PC12 cells grown on PSS^--doped PPy-coated VACNFs exhibit longer neurites (spreading over 100 μm) and

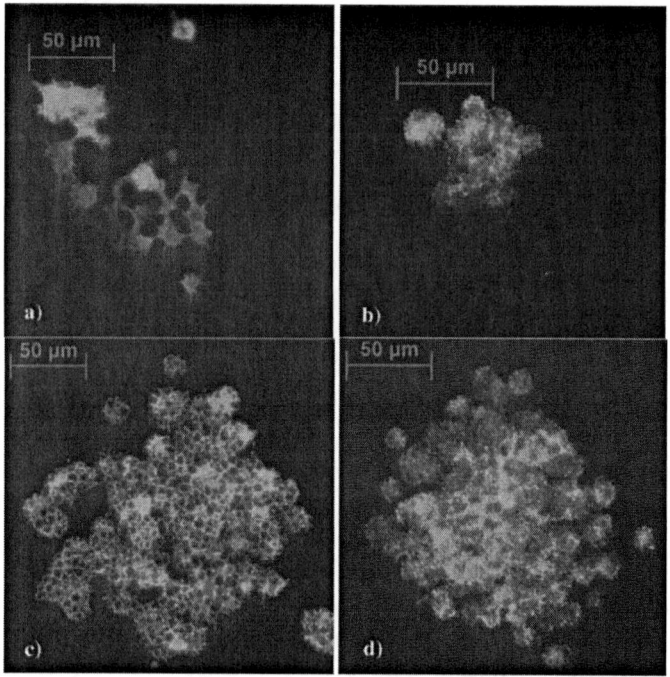

Fig. 3. **Actin Stained PC12 Cell Cultures on VACNF at Day 7.** Fluorescence images of VACNF arrays with (+) and without (-) PPy and collagen ECM coatings: a) PPy- and ECM-, b) PPy+ and ECM-, c) PPy- and ECM+, and d) PPy+ and ECM+.
Color image of this figure appears in the color plate section at the end of the book.

higher confluency than those on Cl^--doped counterparts. The viability tests using a fluorescence kit containing calcein and ethidium dyes indicated that the ratio of viable to nonviable PC12 cells was much higher in PSS^--doped PPy-coated VACNFs than Cl^--doped counterparts (Nguyen-Vu et al. 2007). Clearly, adjusting the anionic dopant is an effective method for tailoring the interaction between PC12 cells and PPy-coated VACNF. We note that it is possible to dope the PPy-film with anti-inflammatory compounds, NGF, inhibitors, and pharmaceutical agents for electrically controlled release.

High-resolution images by scanning electron microscopy further elucidate the effect of the PPy-film on the interaction between the PC12 cells and VACNF (Fig. 4). In this study, the PC12 cells were treated with NGF on day 2 of culture to induce neurite formation on the VACNF substrates coated with ECM (Nguyen-Vu et al. 2007). With PPy-coated VACNFs, most cells rest on top of the free-standing VACNF which act like a dense bed of soft high-aspect ratio cylindrical springs. Without PPy coating, the main PC12 cell bodies were found be supported at the

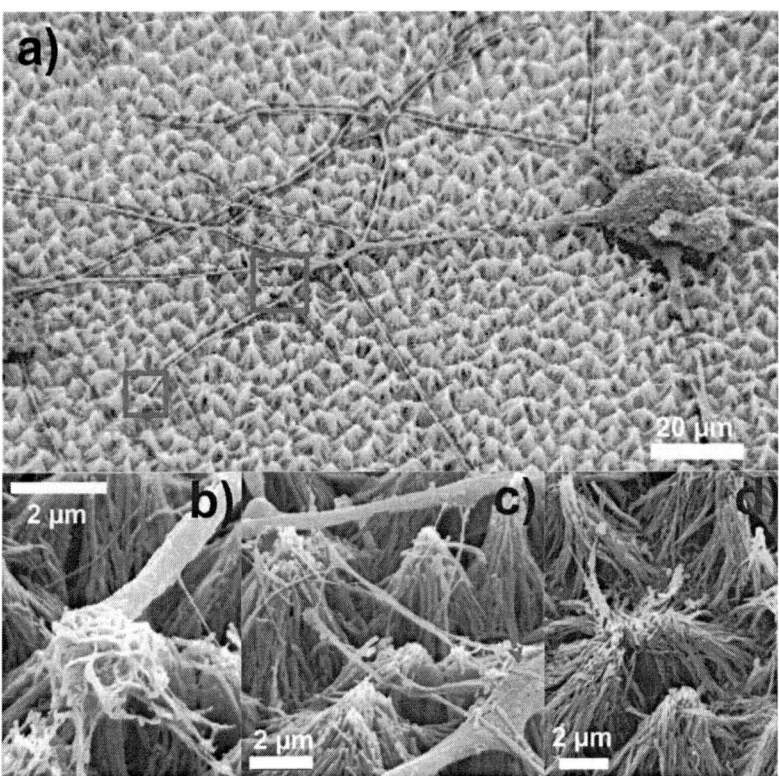

Fig. 4. **PC12 Cells Cultured on As-grown VACNF Microbundles.** SEM images of a) differentiated PC12 cells forming an extensive neural network, b) the fibril filapodia at the extended neurite terminus anchoring on CNF microbundles, c) neural nanofibrils bridging the submicron neurite branches, d) the substrate without subjecting to cell culture.

Color image of this figure appears in the color plate section at the end of the book.

apexes of the collapsed CNF microbundles with the extended neurites of submicron diameters suspended over the valleys between the cell bodies. Many fibril filopodia with the diameter of 0.1–0.2 µm were observed to sprout out from the differentiated neurite terminals strongly anchored on the CNF bundles (Fig. 4b), likely due to the mechanical stress by the stiff CNF bundles. Other neural nanofibrils were entangled with the underlying CNF structure (Fig. 4d). This is similar to the PC12 nanofibrils that extend and retract dynamically in response to electrical stimuli (Mannivannan and Terakawa 1994). Interestingly, most of the cell bodies tended to sit at the tip of the CNF microbundles and did not move to the microbundle sidewalls even though both areas were coated with type IV collagen for cell adhesion. The dense nanofibrils that formed in response to the large continuous mechanical stresses exerted on the PC12 cells indicate that the collapsed CNF microbundles are not mechanically compatible substrate for cell culture.

PPy-coated VACNFs, on the other hand, provide a much softer substrate for PC12 cell culture. Compared to the microbundle CNF substrate, the morphology of PC12 cells is very different. The free-standing PPy-coated VACNFs enable more intimate physical contact between the PC12 cells and individual PPy-coated VACNF (Fig. 5). Cell bodies float at the tips of VACNFs with some CNFs penetrating through the cell membrane while neurites weave through the dense forest of free-standing VACNFs. Far fewer nanofibrils drape across the open brush-like structure of the PPy-coated VACNFs compared to the collapsed CNF microbundles. In some cases, individual PPy-coated VACNFs were subjected to a traction force by the cell body and were clearly bent (Fig. 6a). The PPy-coated VACNF behave like flexible scaffolding which distributed interaction with cell body at multiple nanoscale contact points (i.e., the individual CNF tips) instead of at a single localized point as is the case with the collapsed CNF microbundles. The flexibility of the soft PPy-coated VACNF substrate minimizes the stresses in the PC12 cells, thus reducing neurite formation.

PPy-coated VACNF for Studying Mechanotransduction to Cells in Culture

The free-standing VACNF can also be utilized as a mechanical probe in that, since each individual CNF responds separately to local forces, it is possible to calculate the localized forces acting at different points along the cell body. The bending force of a single CNF can be calculated using Equations (2) and (3):

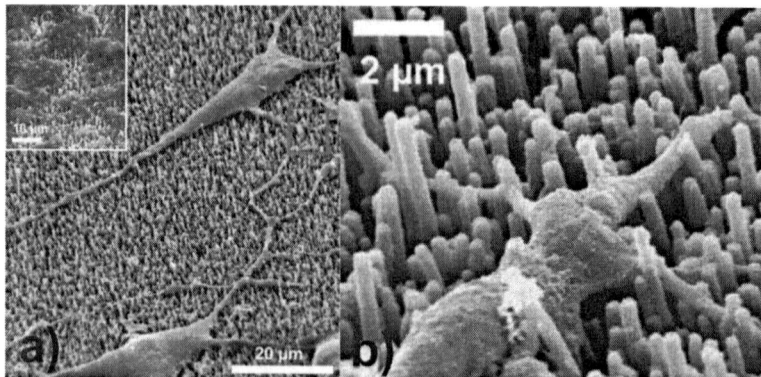

Fig. 5. **PC12 Neural Network Interpenetrating the Freestanding PPy-coated VACNFs.** SEM images of a) NGF-treated PC12 cells adhered and differentiated on the 3-D brush-like matrix of the PPy-coated VACNFs (Inset: cells without NGF) and b) PPy-coated VACNFs supporting a good NEI with the neurites extended through the bed of free-standing CNFs.

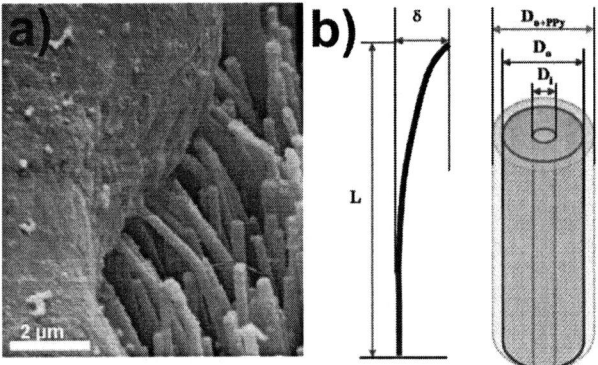

Fig. 6. Cell Adhesion Force Derived from the Bending of VACNFs. a) SEM images of PPy-coated VACNFs bent under the force exerted by the cell body and b) the definition of the parameters of the CNF: L (the length), δ (the lateral deflection of the tip from the original vertical position), D_o (the outer diameter), and D_i (the inner diameter).

Color image of this figure appears in the color plate section at the end of the book.

$$F = \left(\frac{3EI}{L^3}\right)\delta \quad (2)$$

$$I = \left(D_o^4 - D_i^4\right)/64 \quad (3)$$

where F, E, I, L and δ are the bending force, Young's modulus, moment of inertia, length, and lateral deflection of the tip from the original vertical position respectively (Fig. 6b). Here, an individual CNF is modeled as a hollow cylindrical solid with a Young's Modulus E ≈ 40 GPa; thus, the moment of inertia I is given by Equation (3) where D_o and D_i are the outer diameter and inner diameter of the CNF respectively.

From the typical CNF bending displayed in Fig. 6a, the lateral deflection δ varies from 0.5 to 2 μm. For the CNF, the typical L, D_o, and D_i are 7 μm, 150 nm, and 50 nm respectively. Since the Young's modulus of PPy is approximately 0.1 GPa, much lower than that of CNFs, the contribution of the PPy film is negligible. Therefore, the calculated bending force of the CNF is about 4 to 17 nN. This value is comparable to the anomalous value obtained with experiments using elastomeric microposts at focal adhesion area less than 1 μm² (Tan et al. 2003), indicating that we are measuring traction forces at the nanoscale adhesion points instead of normal focal adhesion forces. Equations (2) and (3) indicate that the CNF bending force is proportional to D_o^4/L^3 and, consequently, is a strong function of the CNF aspect-ratio. By controlling the diameter and the length of CNFs, the bending stiffness of the VACNFs can be tuned over several orders of magnitude. Thus, the combination of VACNF with the PPy coating provides a unique solution to study the mechanotransduction on neural function with unprecedented spatial resolution and to dissipate the mechanical stress between the solid electrode and the neural cells.

Electrochemical Properties of PPy-coated VACNF

A good stimulation electrode must be able to deliver the desired current to the contacting tissue with a response time in milliseconds and at an electrode potential ideally less than 1 V to avoid electrolysis of water and biomolecules in the electrolytic medium surrounding the tissue. The current delivery capability of the stimulation electrode is quantified using electrochemical techniques including cyclic voltammetry (CV) and electrochemical impedance spectroscopy (EIS). A high specific capacitance and a low electrode impedance are desired. One approach for enhancing these properties is by increasing the EDL capacitance or incorporating a large pseudocapacitance at the electrode-electrolyte interface. Since VACNFs have a 3D nanostructure, the effective surface area and, as a result, the EDL capacitance of electrodes constructed from VACNFs is greatly increased compared to planar metal electrodes. Moreover, as discussed in Section 5.2, the electroactive PPy film provides a pseudocapacitance, thus increasing the total effective capacitance at the electrode-electrolyte interface which further decreases the electrode impedance.

CV results of as-grown VACNFs and PPy-coated VACNFs indeed demonstrated the increase in the capacitance as expected for the VACNF template and PPy-coating. According to Equation (1), the specific capacitance (C_0) of the bare as-grown VACNF can be calculated as 0.4 mF cm^{-2} using the average Δi value between the charging and discharging currents (Fig. 7a). This value is more than 10 times higher than conventional noble-metal electrodes (20 µF cm^{-2}) but slightly lower than that of metal electrodes covered with a porous hydrous iridium oxide film (0.6 mF cm^{-2}) (Weiland and Anderson 2000). Significantly, coating the VACNF with a Cl$^-$-doped PPy film of about 24 and 147 nm increases Δi by approximately 100 and 250 times, respectively (Fig. 7b). Thus, the specific capacitance is dramatically increased to 40 and 100 mF cm^{-2}. This increase in capacitance at the electrode-electrolyte interface is due to the pseudocapacitive Cl$^-$ flux from the Cl$^-$-doped PPy film upon change in the applied electrode potential. Most importantly, the Cl$^-$ flux causes negligible change to the physiological environment of tissues in contrast to the harmful pH-value changes incurred with the use of Ir electrodes during electrical stimulation in previous studies (Weiland and Anderson 2000).

The change in the electrode impedance is clearly shown in the Nyquist plot of EIS over an AC frequency range from 100 kHz to 1 Hz with the as-grown and PPy-coated VACNFs, respectively (Fig. 7c and d). For the As-grown CNF, the data can be well fitted with a simple serial circuit consisting

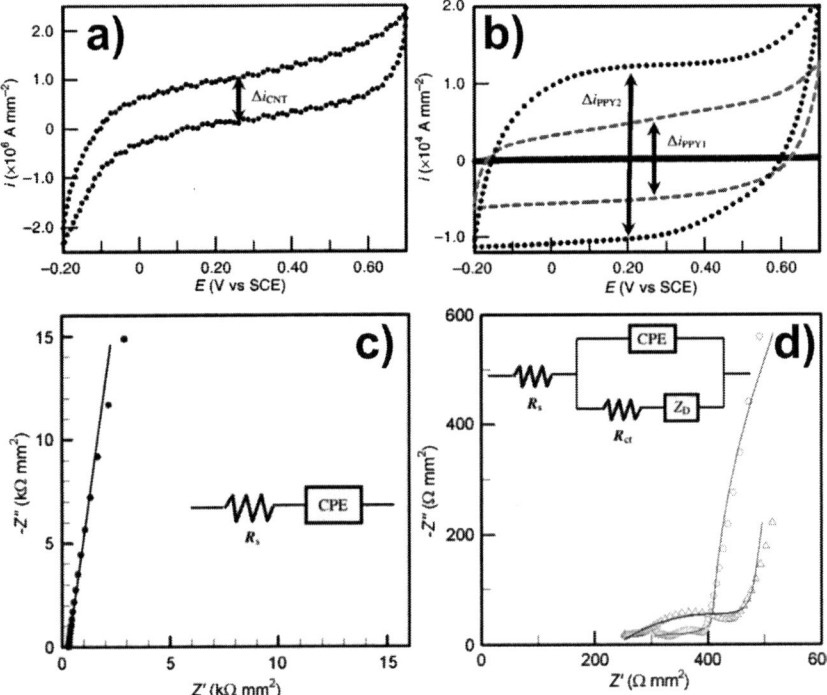

Fig. 7. **Electrochemical Characterization.** CV of a) an as-grown VACNF array and b) two VACNF arrays coated with 24 nm (dashed line) and 127 nm (dotted line) PPy films, respectively, in comparison with the original as-grown VACNF array (solid line). The EIS Nyquist plots of c) an as-grown VACNF array and d) two VACNF arrays coated with 24 nm (circles) and 127 nm (triangles) PPy films.

of a resistor R_s and a constant phase element (CPE) (Nguyen-Vu et al. 2006). The CPE is a reactive element accounting for small deviation from ideal capacitors due to the 3D porous structure and the inhomogeneity of electrochemical activities at different sites of the VACNFs. When the frequency is over 1 kHz (the desired frequency for electrical stimulation), the impedance drops to a constant value of 280 ± 20 Ω mm^{-2}, which is essentially dominated by the ohmic solution resistance between the counter electrode and the VACNF electrodes. As is expected from the CV measurements, coating VACNFs with a thin film of PPy decreases the ac impedance dramatically by more than 20 times. Moreover, the impedance spectrum is dramatically different from that of the as-grown VACNFs. The EIS can be fitted to a circuit model consisting of an additional resistor (R_{ct}) and a Warburg diffusion element (Z_D) in parallel with a CPE. R_{ct} accounts for Faradaic charge transfer at the PPy film and Z_D corresponds to a finite diffusion process of Cl⁻ ions across the ECP films with microstructured pores (Ferloni et al. 1996). Similar to the as-grown VACNFs, the electrode impedance is dominated by the solution resistance (R_s) of $250 \pm 20\ \Omega\ mm^{-2}$ for frequencies over 1 kHz.

The circuit models for the as-grown VACNFs and the PPy-coated VACNFs are relatively simple compared to bare and PPy-coated Au microelectrodes, highlighting a potential advantage of VACNF electrodes over metal microelectrodes for electrophysiology. The well-defined electrical characteristics of the CNF electrodes can simplify data analysis algorithms for FES and neural recording. Furthermore, the extremely low impedance of the PPy-coated VACNFs enables reduction of the dimension of the electrode footprint down to a few µm, much smaller than current metal MEAs (on the order of 10s of µm to several millimeters). This can greatly enhance the spatial resolution of electrophysiological probes so that devices can be potentially fabricated to excite single neurons or even subcellular structures.

Highly-Efficient Extracellular Stimulation with PPy-coated VACNF MBAs

To demonstrate the improved stimulation efficiency of PPy-coated VACNF MBA, we applied a 3x3 PPy-coated VACNF MBA to extracellular stimulation of rat hippocampal brain slices through the hippocampal CA3-SC-CA1 pathway. A PPy-coated VACNF MBA is compared with a pair of tungsten microwires of ~100 µm in diameter, a 3x3 planar Pt MEA, and a 3x3 as-grown VACNF MBA. CA3 and CA1 consist of neuronal cell bodies while SC consists of axonal bundles of CA3 neurons. The axonal bundles comprising SC synapse with the dendrites of CA1 neurons. Stimulation of the neurons in CA3 elicit field potentials in CA1 consisting of excitatory postsynaptic potentials (EPSP) in the CA1 dendrites and somatic action potentials in the CA1 cell body. The somatic action potentials are threshold dependent in that elicited EPSPs must be of a sufficiently large magnitude in order to trigger the somatic action potentials.

The performance of PPy-coated VACNF MBA as an extracellular stimulation electrode was systematically compared to a pair of tungsten wire electrodes, a planar platinum MEA, and an as-grown VACNF MBA utilizing the following procedure. To stimulate the tissue, a variable amplitude monophasic current pulse was applied. Stimulation using W wires was applied at the Schaffer Collateral (SC) (Fig. 8a) by piercing the microwires into this region. For the three electrode arrays, stimulating current was applied to the hippocampal brain slices between electrodes 8 and 9 as the slices laid flat against the electrode array (Fig. 8b). The response was recorded from the striatum pyramidale (SP) dendrites of CA1 with an extracellular glass micropipette electrode connected to signal

Nanostructured NEI 51

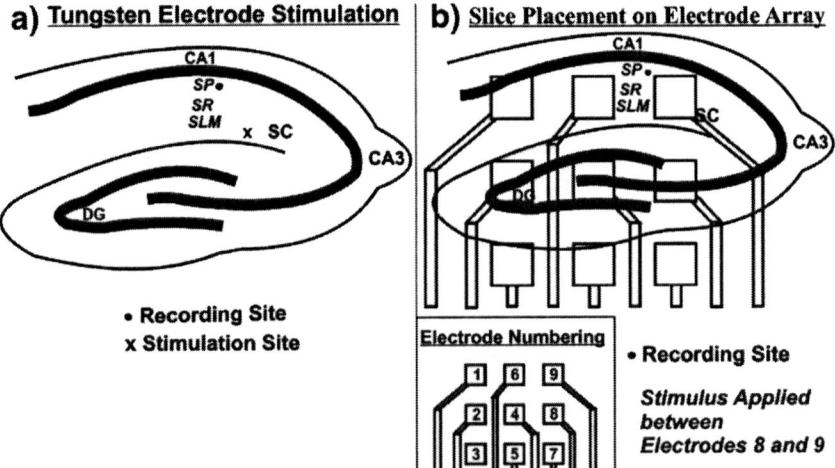

Fig. 8. Schematic of the Electrical Stimulation of Rat Brain Slices. a) the location of stimulation and recording sites on a rat hippocampal slice when stimulating using tungsten wire electrodes and b) the placement of the hippocampal slice on a 3×3 electrode array. Inset: electrode number index.

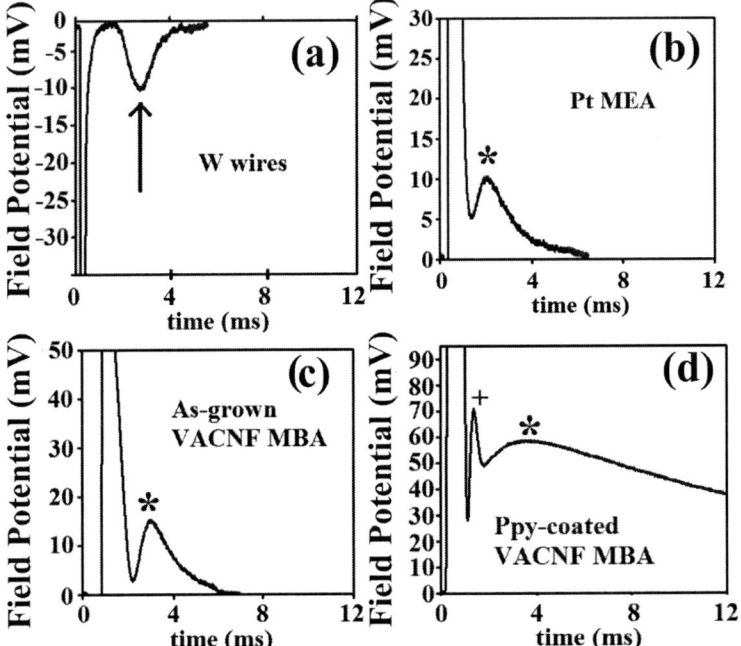

Fig. 9. Waveform of Elicited Field Potentials. Field potentials recorded in response to the 10.5 mA stimulating currents with a) tungsten wire electrodes, b) Pt MEA, c) as-grown VACNF MBA, and d) PPy-coated VACNF MBA.

conditioning circuitry. To characterize the performance of the tungsten wire electrodes, Pt MEA, as-grown VACNF MBA, and PPy-coated VACNF MBA, we measured the amplitude of the field potential recorded at CA1 vs. the effective stimulus outputs (both of voltage and current) across the two stimulation electrode poles as the excitation current pulse amplitude is progressively decreased.

Using the maximum stimulating current pulse amplitude of 10.5 mA, the response voltage recorded at CA1 consisted of the evoked extracellular field potential superimposed upon the stimulus artifacts was observed with all four types of stimulating electrodes (Fig. 9). The typical response to a stimulus applied using W wires consisted of a negative going field EPSP of 3 ms duration and 3 ms latency (delay). The typical observed responses to a stimulus applied using the Pt MEA and as-grown VACNF MBA consisted of a positive going EPSP with a similar duration (4 ms) and latency (2.5–3 ms). However, the response to a stimulus applied using the PPy-coated VACNF MBA is dramatically different, showing two components, i.e., a narrow field potential spike (of 1 ms duration) with a peak latency of 1.3 ms superimposed on a longer field potential wave (20–25 ms duration) with a peak latency of 3.7–4 ms. The narrow field potential spike corresponds to somatic action potentials while the longer potential wave consists of EPSPs. Significantly, a somatic action potential requires a suprathreshold EPSP in CA1 which can not be achieved with the W Wires, the Pt MEA, and the as-grown VACNF MBA. Only the PPy-coated VACNF MBA showed successful elicitation of the somatic action potential. Clearly, this type of electrode is more efficient for electrical stimulation.

The average stimulating electrode voltage and average recorded field potential amplitude versus the stimulation current pulse amplitude for the four types of electrodes are shown in Fig. 10. The W wires, Pt MEA, and as-grown VACNF MBA all failed to evoke cellular responses at stimulating current pulse amplitudes below 1 mA. Only the PPy-coated VACNF MBA can achieve this while, more importantly, the stimulating voltage is below 1.0 V. These results showed that only the 3-D VACNF MBA coated with conformal ECP, i.e., PPy, can indeed stimulate acute hippocampal brain slices at a very low voltage (< 1.0 V) without electrolyzing water. Furthermore, the PPy-coated VACNF MBA stimulates tissue with the highest efficiency as indicated by the much larger field potential amplitude (~30 mV for the fast and narrow wave with the PPy-coated VACNF MBA vs. ~5–15 mV in the broad wave with all other three electrodes) and the ability to elicit a cellular response at stimulation current below 1.0 mA. At higher stimulation current levels, the PPy-coated VACNF MBA was the only probe that elicited large amplitude fast somatic action potentials.

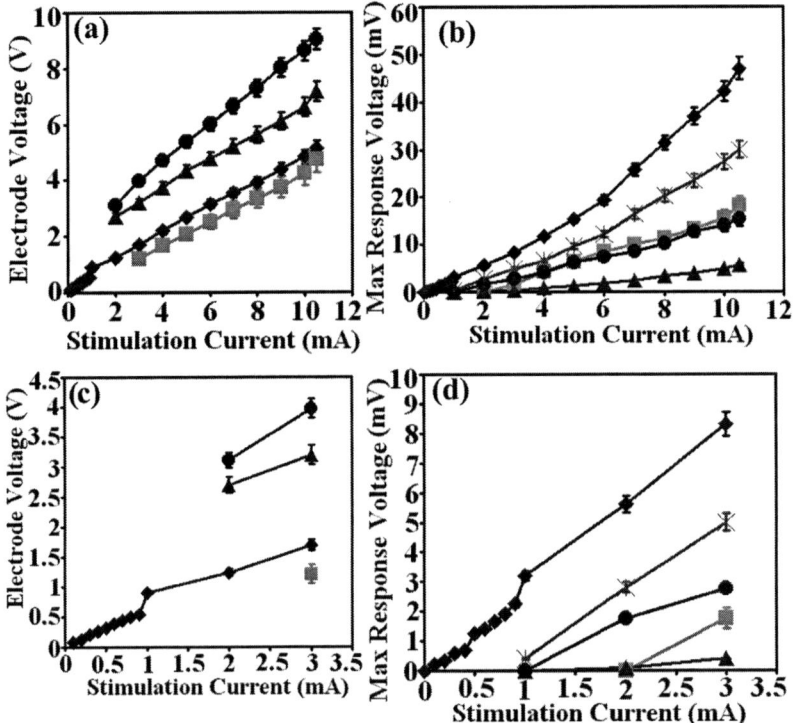

Fig. 10. The Electrode Voltage and Field Potential Amplitude versus the Stimulation Current. a) The electrode voltage and b) the response amplitude of field potential vs. the stimulation current. c) and d) are enlarged plots of a) and b) at low currents. Filled diamonds: PPy-coated VACNF MBA, filled circles: Tungsten wire electrodes, filled square: as-grown VACNF MBA, filled triangle: Pt MEA.

OTHER NEUROPROSTHETIC DEVICES UTILIZING CNT OR CNF ELECTRODES

CNFs and CNTs provide opportunities to fabricate novel nanostructured electrodes for better NEI that can not be obtained with conventional metal electrodes even after modifying with hydrous metal oxides or ECP coatings. Besides using CNFs in vertically aligned brush-like configuration, a linear array of VACNF microbundles (height = 10 μm, base width = 2 μm) separated by 15 μm pitch was also fabricated as ultramicroelectrodes for neural recording (Yu et al. 2007). Extracellular recording of spontaneous and evoked neuroelectrical activity in organotypic hippocampal slice cultures has been demonstrated down to submicron spatial resolution. In addition, these electrodes were also used to stimulate the hippocampal slice with sufficient charge injection capacity (up to 100 μA for 100 μs).

They were able to pass 10 nC of charge without apparent damage to the tissue. In another study, Wang et al. fabricated an array of microelectrodes constructed from pillars of dense entangled multiwalled CNTs to stimulate retinal ganglion cells for potential application in retinal prosthetics (Wang et al. 2006). A high charge injection limit (1–1.6 mC/cm^2) was demonstrated to stimulate hippocampal neurons without faradic reactions. These results suggest that CNTs are capable of providing far safer and more efficacious solutions for neural prostheses than previous metal electrode approaches. CVD synthesized CNT meshes deposited on prepatterned electrode sites have also shown improved signal-to-noise ratio in neural recording (Gabay et al. 2007; Keefer et al. 2008).

Last, nonpatterned materials such as laterally entangled CNT sheets have been explored as neuronal cell culture substrates (Mattson et al. 2000) and prosthetic neural implants (Gheith et al. 2005). With proper chemical functionalization, neurons have shown promoted proliferation due to the more compatible nanostructured topography. In some cases, CNT substrates were found able to boost neuronal electrical signalling (Lovat et al. 2005). Due to the mixed effects as a result of porosity, nanotopography, mechanical compatibility, chemical interaction, and biocompatibility between neural cells/tissues and the CNT or CNF materials, there are still many questions to be answered regarding the nature of the NEI. But there is no doubt that these nanomaterials open a window for many novel properties not available before. Continued research in this area is required to realize these advantages in applicable neural devices.

Key Facts

- PD affects 10–20 people per 100,000 in the U.S. and Internationally (Van Den Eeden et al. 2003).
- DBS using long-term implantable electrodes has been used to treat 50,000 PD patients worldwide (Kuhn and Huff 2010).
- Cochlear Implants have been used to restore hearing in 120,000 patients.
- NEI, Brain-Machine Interface, Brain-Computer Interface, Neural Prosthesis all refer to research in the branch of experimental neuroscience aimed at restoring communication, movement, and sensory function in patients suffering from neurological injury and disease through the use of electrical devices that allow the brain to interact with the outside world (Hatsopoulos and Donoghue 2009).
- The goal of NEI research is the development of a fully implantable neuroprosthetic device that is equipped with electrodes that can record electrical activity and/or transduce those signals into electrical stimuli delivered by stimulating electrodes to nervous system tissue

to execute the sensory, motor, or communication function that has been lost in impaired individuals (Hatsopoulos and Donoghue 2009).
- Efficient stimulating electrodes elicit electrical activity at sufficiently low current levels to minimize damage to tissue, i.e., preclude electrolysis of water, which is toxic to cells.
- Enhancing the integrity of electrical signals captured by recording electrodes entails maximizing the signal-to-noise ratio.
- Present challenges in NEI research include: 1) improving the biocompatibility of the neuroprosthetic device to preclude immune response and minimize damage to tissue and 2) reducing the electrode size to enhance spatial selectivity (i.e., targeted recording and stimulation of single neurons) while maintaining a low impedance which is required for the desired efficiency of the electrical stimuli and integrity of electrical recording (Pancrazio 2008).
- Metal electrodes in contemporary neuroprosthetic devices are on the order of 10s of µm to one mm in diameter, thus precluding recording and stimulation of single neuron units (which are typically less than 20 µm in diameter) (Pancrazio 2008).
- Nanostructured Electrodes such as the PPy-coated VACNFs can be used as an elastic substrate with minimized mechanical stress on cells and reduced impedance to facilitate miniaturization for enhanced spatial resolution while preserving stimulation efficiency.

Definitions

Cyclic Voltammetry (CV): A common electrochemical method for characterizing electron transfer, electrochemical impedance, and capacitance of electrode mateials. A triangular waveform of the electrode potential E is applied between the working electrode and a reference electrode (or sometimes an auxiliary electrode) while the current between the working electrode and the auxiliary electrode is recorded. The difference in current density (Δi) between the positive and negative potential cycles corresponds to the sum of the charging and discharging currents at the electrode/electrolyte interface. The specific capacitance (C_0) which is normalized to the electrode's geometric surface area can be calculated by

$$C_0 = \Delta i / 2\upsilon. \tag{1}$$

Electrical Conductive Polymer (ECP): Macromolecules whose backbone chain consists of repeating conjugate subunits and is thus capable of conducting current under a voltage bias. These polymers are typically involved in electrochemical reactions as the electrode potential is changed, which causes a continuous change in the average oxidation state during CV. As a result, the coating of an ECP at an electrode surface behaves similar to a capacitor (i.e., pseudocapacitor) which can store and release charge by demand.

Electric Double Layer (EDL): A space charge layer appearing across the electrode/electrolyte interface including a close-packed inner layer and a diffusive outer layer consisting of ions and oriented water molecules. The distribution and density of the ions are changed when the electrode potential is varied, generating a current.

Electrochemical Conduction: Mechanisms to convert electronic currents in the solid-state electrodes to ionic currents in liquid electrolytes at the electrode/electrolyte interface, either by electrochemical reaction involving electron transfer between the electrode and electroactive species (faradaic processes) or by redistribution of ions in the EDL (nonfaradaic processes).

Electrode Impedance: An electrical property measuring the capability of an electrode to generate a stimulation current under a voltage bias. The electrode impedance can be derived by electrochemical impedance spectroscopy where the AC impedance of a small AC voltage bias is measured over a wide AC frequency from 1 MHz to 1 mHz. The impedance of the electrode/electrolyte interface is modeled by an equivalent circuit, i.e., the combination of resistors, reactive elements such as capacitors and constant phase elements. The capacitors account for nonfaradaic processes in EDL or pseudocapacitance by fast surface faradaic reaction. The resistors model the faradaic charge transfer involved in the reduction or oxidation (i.e., redox) reaction of electroactive species and the electrolysis of water into H_2 and O_2.

Field Potential: Endogenic bioelectrical potentials emanating from electrically active cells that are recorded by an extracellular electrode. These potentials include the summation of postsynaptic potentials and action potentials from groups of individual cells.

Microelectrode Array (MEA): A two-dimensional array of individually addressable metal electrodes with micron scale footprints on a planar substrate, fabricated using top-down microfabrication techniques.

PC12 Cells: A neuron-like cell line (with average cell size of ~20 μm) derived from a transplantable rat pheochromocytoma.

VACNF Microbrush Array (MBA): An individually addressable microelectrode array in which each microelectrode is constructed with a brush-like structure of VACNFs grown by selective catalytic processes using Plasma Enhanced Chemical Vapor Deposition (PECVD).

Summary Points

- Challenges confronting the development of neuroprostheses include engineering biocompatible neural implants with reliable NEI and reduced size to facilitate intimate interaction with target cells, minimize damage to surrounding tissue, and enable the electrophysiological probing of single neurons.
- The dimension of state-of-the-art neuroprosthetic electrodes presently precludes the selectivity and spatial resolution to reach single neurons.
- Reducing the size of the stimulating electrode causes increase in the electrode impedance and consequently degrades simulation efficiency, lowers the signal-to-noise ratio, and raises the stimulating voltage to elicit electrical activity in neural tissues.
- The implementation of high stimulating voltage (>1.29 V) causes significant water electrolysis and pH value change to levels that are toxic to cells.
- PPy-coated VACNF act like flexible supporting scaffolding, thus minimizing the mechanical stress on cells.
- Compared to metal microelectrodes, PPy-coated VACNFs provide enlarged surface area, increased effective capacitance, and reduced impedance; thus able to enhance the current delivery capability.
- PPy-coated VACNFs achieve higher stimulation efficiency compared to W wires, Pt MEA, and as-grown VACNF MBA, and, as a result, are capable of eliciting larger amplitude electrical signals and stimulating neural tissue at voltage levels that preclude electrolysis of water.
- Electrodes with various configurations using CNFs and CNTs have been successfully demonstrated in neural stimulation and recording with improved efficacy and reduced damage to cells/tissues.

Abbreviations

CNF	:	Carbon Nanofiber
CNT	:	Carbon Nanotube
CA1	:	Cornus Ammonis 1
CA3	:	Cornus Ammonis 3
CV	:	Cyclic Voltammetry
DBS	:	Deep Brain Stimulation
DG	:	Dentate Gyrus
ECP	:	Electrical Conductive Polymer
EIS	:	Electrochemical Impedance Spectroscopy
EPSP	:	Excitatory Postsynaptic Potential

ECM	:	Extracellular Matrix
FES	:	Functional Electrical Stimulation
HD	:	Huntington's Disease
MBA	:	Microbrush Array
MEA	:	Microelectrode Array
NGF	:	Nerve Growth Factor
NEI	:	Neural Electrical Interfaces
OCD	:	Obsessive Compulsive Disorder
PD	:	Parkinson's Disease
PECVD	:	Plasma Enhanced Chemical Vapor Deposition
PPy	:	Polypyrrole
PSS	:	PolystyreneSulphonate
RIE	:	Reactive Ion Etching
SEM	:	Scanning Electron Microscopy
SC	:	Schaffer Collateral
SP	:	Striatum Pyramidale
3D	:	Three-dimensional
VACNFs	:	Vertically Aligned Carbon Nanofibers

References

Cui, X.Y. and D.C. Martin. 2003. Fuzzy gold electrodes for lowering impedance and improving adhesion with electrodeposited conducting polymer films. Sensors and Actuators a-Physical 103: 384–394.

de Asis, E.D. Jr., T.D.B. Nguyen-Vu, P.U. Arumugam, H. Chen, A.M. Cassell, R.J. Andrews, C.Y. Yang and J. Li. 2009. High Efficient Electrical Stimulation of Hippocampal Slices with Vertically Aligned Carbon Nanofiber Microbrush Array. Biomed. Microdevices 11: 801–8.

Ferloni, P., M. Mastragostino and L. Meneghello. 1996. Impedance analysis of electronically conducting polymers. Electrochimica Acta. 41: 27–33.

Gabay, T., M. Ben-David, I. Kalifa, R. Sorkin, Z.R. Abrams, E. Ben-Jacob and Y. Hanein. 2007. Electro-chemical and Biological Properties of Carbon Nanotube Based Multi-electrode Arrays. Nanotechnology 18: 1–6.

Gheith, M.K., V.A. Sinani, J.P. Wicksted, R.L. Matts and N.A. Kotov. 2005. Single-walled carbon nanotube polyelectrolyte multilayers and freestanding films as a biocompatible platform for neuroprosthetic implants. Adv. Mater. 17: 2663–2670.

Hatsopoulos, N.G. and J.P. Donoghue. 2009. The Science of Neural Interface Systems. Annu. Rev. Neurosci. 32: 249–266.

Keefer E.W., B.R. Botterman, M.I. Romero, A.F. Rossi and G.W. Gross. 2008. Carbon Nanotube Coating Improves Neuronal Recordings. Nat. Nanotechnol. 3: 434–439.

Kuhn, J. and W. Huff. 2010. Will Deep Brain Stimulation be as Successful in Major Depression as it has been in Parkinson's Disease. Expert Rev. Neurother. 10: 1363–1365.

Lee, K.H., C.D. Blaha, P.A. Garris, P. Mohseni, A.E. Horne, K.E. Bennet, F. Agnesi, J.M. Bledsoe, D.B. Lester, C. Kimble, H.-K. Min, Y.-B. Kim and Z.-H. Cho. 2009. Evolution of Deep Brain Stimulation: Human Electrometer and Smart Devices Supporting the Next Generation of Therapy. Neuromodulation 12: 85–103.

Lovat, V., D. Pantarotto, L. Lagostena, B. Cacciari, M. Grandolfo, M. Righi, G. Spalluto, M. Prato and L. Ballerini. 2005. Carbon nanotube substrates boost neuronal electrical signaling. Nano Letters 5: 1107–1110.

Manivannan, S. and S. Terakawa. 1994. Rapid sprouting of filopodia in nerve terminals of chromaffin cells, PC12 cells, and dorsal root neurons induced by electrical stimulation. J. Neurosci. 14: 5917–5928.

Mattson, M.P., R.C. Haddon and A.M. Rao. 2000. Molecular functionalization of carbon nanotubes and use as substrates for neuronal growth. J. Molecular Neuroscience 14: 175–182.

Melechko, A.V., V.I. Merkulov, T.E. McKnight, M.A. Guillorn, K.L. Klein, D.H. Lowndes and M.L. Simpson. 2005. Vertically aligned carbon nanofibers and related structures: controlled synthesis and directed assembly. J. Appl. Phys. 97: 41301-1-39.

Nguyen-Vu, T.D.B., H. Chen, A.M. Cassell, R. Andrews, M. Meyyappan and J. Li. 2006. Vertically Aligned Carbon Nanofiber Arrays: An Advance toward Electrical-Neural Interfaces. Small 2: 1, 89–94.

Nguyen-Vu, T.D.B., H. Chen, A.M. Cassell, R. Andrews, M. Meyyappan and J. Li. 2007. Vertically Aligned Carbon Nanofiber Architecture as a Multifunctional 3-D Neural Electrical Interface. IEEE Trans. Biomed. Engr. 54: 1121–1128.

Pancrazio, J.J. 2008. Neural Interfaces at the Nanoscale. Nanomedicine (Lond.). 3: 823–830.

Popovic, M.R., A. Curt, T. Keller and V. Dietz. 2001. Functional Electrical Stimulation for Grasping and Walking: Indications and Limitations. Spinal Cord. 39: 403–412.

Squire, L.R., F.E. Bloom, S.K. McConnell, J.L. Roberts, N.C. Spitzer, M.J. Zigmond. 2003. Fundamental Neuroscience. Academic Press, San Diego, CA. USA.

Tan, J.L., J. Tien, D.M. Pirone, D.S. Gray, K. Bhadriraju and C.S. Chen. 2003. Cells lying on a bed of microneedles: An approach to isolate mechanical force. Proc. Nat. Acad. Sci. USA 100: 1484–1489.

Van Den Eeden, S.K., C.M. Tanner, A.L. Bernstein, R.D. Fross, A. Leimpeter, D.A. Bloch and L.M. Nelson. 2003. Incidence of Parkinson's Disease: Variation by Age, Gender, Race/Ethnicity. Am. J. Epidem. 157: 1015–1022.

Wang, K., H.A. Fishman, H. Dai and J.S. Harris. 2006. Neural Stimulation with a Carbon Nanotube Microelectrode Array. Nano Letters 6: 2043–2048.

Weiland, J.D. and D.A. Anderson. 2000. Chronic Neural Stimulation with Thin-film, Iridium Oxide Electrodes. IEEE Trans. Biomed. Eng. 47: 911–918.

Yu. Z., T.E. McKnight, M.N. Ericson, A.V. Melechko, M.L. Simpson and B. Morrison III. 2007. Vertically Aligned Carbon Nanofiber Arrays Record Electrophysiological Signals from Hippocampal Slices. Nano Letters 7: 2188–2195.

4

Nanoelectrode Arrays for Monitoring and Modulating Nervous System Electrical and Chemical Activity

Russell J. Andrews[1,*] *and Jun Li*[2]

ABSTRACT

The use of neuromodulation to treat disorders of the nervous system will continue to increase exponentially over the next decade. The primary factor limiting the growth of neuromodulation —commonly referred to as deep brain stimulation (DBS)—is the efficacy of the recording and stimulating of the nervous system tissue that is necessary for neuromodulation. Despite the recent laboratory advances in improving the neural-electrical interface (NEI), the electrodes used at present for clinical neuromodulation—DBS—represent technology from the mid-20th century. Additionally, the nervous system communicates chemically

---

[1]NASA Ames Research Center, Moffett Field, CA, USA 94035;
E-mail: rja@russelljandrews.org
[2]Department of Chemistry, Kansas State University, Manhattan, KS 66506;
E-mail: junli@ksu.edu
*Corresponding author

List of abbreviations after the text.

(through neurotransmitters—NTs) as well as electrically (by flux of ions or potentials); thus it is reasonable to expect that incorporating both electrical and chemical nervous system recording and stimulating into neuromodulation devices will enhance efficacy. A further limitation of present-day neuromodulation is the size of the electrodes—hundreds of microns to a millimeter or more in diameter—which is orders of magnitude larger than the neurons and glial cells which compose the nervous system. To neuromodulate in an efficient and precise manner, it is necessary to greatly increase the capability to deliver currents while greatly decreasing the size of the neuromodulatory instrument or device. This chapter reviews the advances in neuromodulation which can result from incorporating nanotechniques into the neuromodulatory device. Nanoelectrode arrays (NEAs) can greatly enhance nervous system recording and stimulating of electrical activity, as well as permit precise recording of changes in neurotransmitter levels in response to this stimulation. NEAs also allow miniaturization of the device to the point where the NEI can actually be at the neuronal or even subneuronal level. Neuromodulation (DBS) can then truly 'modulate' aberrant nervous system activity back to normal—rather than reversibly 'ablate' electrical activity in a region of the brain (which is the effect of macroelectrode DBS at present). Not only will NEAs enhance clinical neuromodulation over the next decade and beyond, but—thanks to the ability of NEAs to probe at the cellular (or even subcellular) level—they will also advance our understanding of the nervous system's intricate electrochemical behavior.

INTRODUCTION

Neuromodulation can be defined as alteration of the activity of the nervous system by external means. It has traditionally been limited to electrical stimulation, either by electrodes placed on the scalp, or electrodes placed surgically either on the brain (or spinal cord) surface or inserted into the brain (or spinal cord). Chemical stimulation (or inhibition) by drugs has not been customarily considered neuromodulation, although as we will see below, as the monitoring of the nervous system's chemical (neurotransmitter—NT) activity becomes accepted, neuromodulation by devices inserted on or into the brain will blur this distinction.

Until the present time, neuromodulation has often been equated with deep brain stimulation (DBS). Clinical DBS entails inserting electrodes approximately one millimeter in diameter into specific regions of the brain in order to alter brain function by providing a controlled electrical current to the region surrounding the electrode. To avoid damage to the brain, the net charge injected into the brain must be zero, and the current delivered sufficiently low to avoid damage to the brain (e.g., by electrolysis—see the preceding chapter in this volume [de Asis et al. 2010]).

DBS was employed with variable success in the 1960s to 1980s primarily for the treatment of patients suffering from chronic pain syndromes. With the discovery by Benabid and colleagues in Grenoble in the late 1980s that a stimulating electrode placed in the thalamus could yield benefits for patients with tremor (both Parkinsonian tremor and tremor of the benign/essential/familial form)—equivalent to a lesion made in the same region of the thalamus—but without ablation of brain tissue (i.e., the effect was reversible if the current was discontinued). This quickly led to the development of totally implantable devices for DBS, with a pulse generator (microprocessor plus battery—similar to a cardiac pacemaker) placed subcutaneously in the upper chest wall (also similar to a cardiac pacemaker) which is connected by a wire beneath the skin of the neck and scalp to the electrode placed stereotactically into the thalamus through a small hole in the skull [Fig. 1].

DBS rapidly became an accepted treatment for patients with movement disorders such as advanced Parkinson's disease and disabling essential or familial tremor. DBS for tremor (thalamic stimulation) received the CE Mark for use in the European Union in 1993, and for stimulation of other regions (notably the subthalamic nucleus and globus pallidus) for Parkinson's disease in 1997. In the US, Food and Drug Administration (FDA) approval for DBS was received four years later (in 1997 and 2002, respectively). At present, upwards of 100,000 patients worldwide have been implanted with DBS systems, the majority for advanced Parkinson's disease and essential/familial tremor—but increasing numbers for dystonia as well as experimentally for refractory epilepsy, major depression and other mood disorders, chronic headache, morbid obesity, etc. The results for the indications apart from movement disorders have been inconsistent and generally much less favorable.

There appears to be no shortage of both patients with extremely debilitating and costly neurological disorders, and neurosurgeons with a desire to treat these unfortunate individuals. Although disorders such as Parkinson's disease and epilepsy affect roughly one to two million people in North America (and approximately the same number in Europe), disorders such as chronic headache and major depression affect perhaps 10 times that number. For the promise of neuromodulation to offer a treatment for those suffering from these common neurological disorders

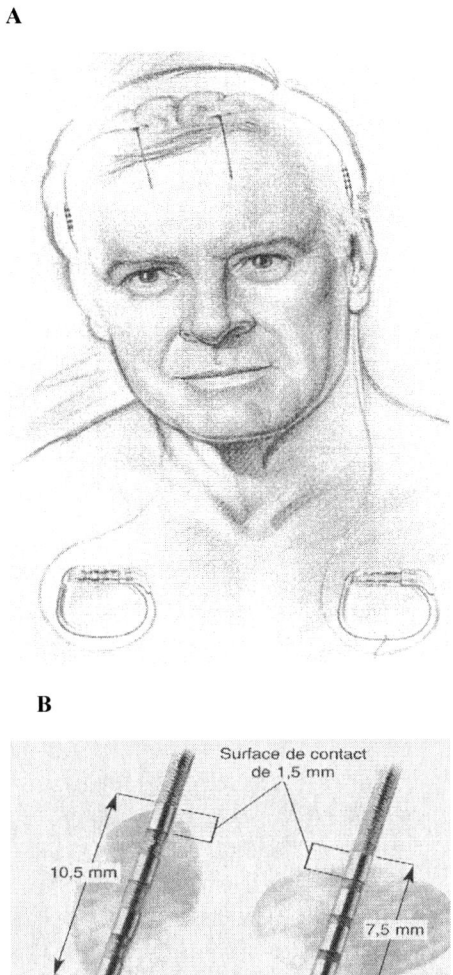

Fig. 1. Typical DBS system. (courtesy of Medtronic, Inc).

who have failed other modalities (usually pharmacologic therapy), it will be necessary for DBS to improve dramatically in efficacy. NEAs can offer that improvement in efficacy.

NANOELECTRODE ARRAYS FOR NERVOUS SYSTEM ELECTRICAL ACTIVITY

DBS to date has used high frequency (> 100 Hz) constant electrical stimulation to modulate brain function. While this has worked remarkably well for Parkinson's disease and tremor (although certainly not ideally, since incomplete effectiveness and side effects are common), for similar efficacy in epilepsy and other disorders it appears that closed-loop feedback will be advantageous. Using the brain's electrical activity (and changes in focal electrical activity that signal abnormal brain function) to guide DBS has motivated one company to develop a DBS system for refractory epilepsy which incorporates electrode(s) that continuously record brain electrical activity in order to optimize the stimulation for preventing seizures [Fig. 2].

The advantages of NEAs for improving the NEI and thus the efficacy of DBS are considered in detail in the preceding chapter in this volume [de Asis et al. 2010]. Data supporting the benefits of NEAs for recording and stimulating the nervous system will be considered here.

The improvements in impedance and capacitance gained with carbon nanofiber (CNF) NEAs reported in the preceding chapter have been demonstrated with similar CNTs [Keefer et al. 2008]. This group electrochemically deposited an aqueous suspension of multiwalled CNTs and potassium-gold cyanide on indium-tin oxide MEAs. Although the CNTs appear to provide a very irregular surface [Fig. 3], marked improvement in the NEI resulted: 23-fold decrease in impedance (from 748 kΩ to 32 kΩ), and 45-fold increase in capacitance (225 pF to 10,187 pF). Coating the CNTs with the conducting polymer polypyrrole also improved charge transfer. Their experiments included cell culture stimulation and recording as well as *in vivo* recording in rats and monkeys.

Although DBS is the main focus of this chapter, some of the most elegant NEI work has been done with microarrays for both cochlear and retinal implants. Findings from groups working in these fields are relevant to improving DBS. One study used MEAs (with individual electrodes ~ 10 μm in diameter) to determine where is the region of low-threshold for electrical stimulation of retinal ganglion cells in the peripheral macaque retina [Sekirnjak et al. 2008]. The low-threshold for stimulation region was found to be on the proximal axon. This concurs with a later study that also found the low-threshold region for stimulation in retinal ganglion cells to be on the proximal axon—but also that the low-threshold region is coextensive with the high-density sodium-channel region (i.e., also on the proximal axon) [Fried et al. 2009]. This is illustrated in Fig. 4 A & B.

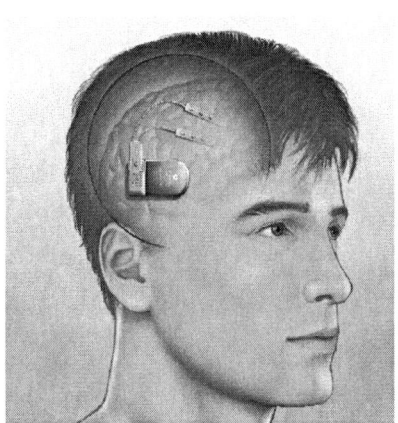

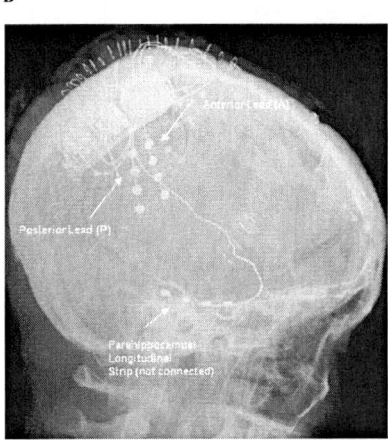

Fig. 2. A closed-loop DBS system for the treatment of refractory epilepsy. (courtesy of NeuroPace, Inc).

It is likely that the initial application of nanotechniques to improving the NEI for DBS will be in coating present-day macro- or micro-electrodes with CNFs or CNTs. Reducing the electrical power needs by an order of magnitude or more would allow the pulse generator battery to be reduced dramatically in size—likely allowing the pulse generator to be implanted in the skull. This would eliminate a lengthy subcutaneous connecting lead along the neck, a source of significant surgical morbidity. However, the major contribution of nanotechniques to DBS will be in the more precise

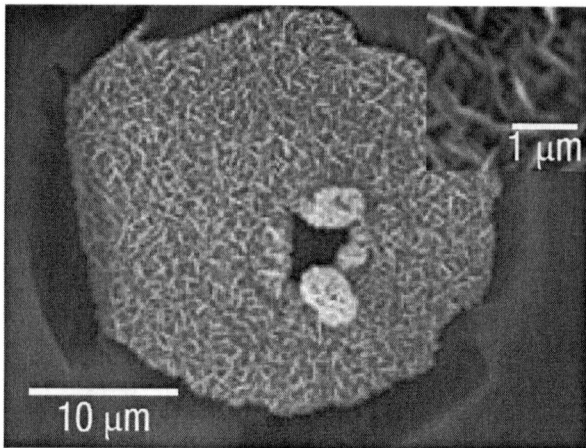

Fig. 3. Scanning electron micrograph image of CNT-coated electrode. Inset shows the porous nature of the CNT coating. (from Keefer et al. 2008, with permission).

recording and stimulation of the nervous system (down to the single cell level, if need be) that NEAs will permit. This will allow DBS to advance from 'reversible neural ablation' to true 'neuromodulation' of errant electrical activity.

NANOELECTRODE ARRAYS FOR NERVOUS SYSTEM CHEMICAL ACTIVITY

Many nervous system disorders involve neurotransmitters (NTs). A loss of dopamine underlies Parkinson's disease, many mood disorders involves NTs such as dopamine and serotonin, and an imbalance in the excitatory NT glutamate and the inhibitory NT gamma-amino-butyric-acid (GABA) appears relevant in many forms of epilepsy. One of the drawbacks of pharmacologic (drug) treatments for mood disorders, for example, is that drugs given orally tend to affect all regions of the brain [Schlaepfer et al. 2008]. However, it has been demonstrated in rodent models that NT levels (e.g., dopamine) are heterogeneous within the nucleus accumbens—suggesting that there are subpopulations of dopamine neurons within this structure which is likely involved in many mood disorders [Wightman et al. 2007]. Neuromodulation which incorporates precise recording and stimulating (or inhibiting) of NT levels—in addition to precise recording and stimulating of electrical activity—is likely to greatly enhance DBS efficacy.

Electrochemical Nanoelectrode Arrays 67

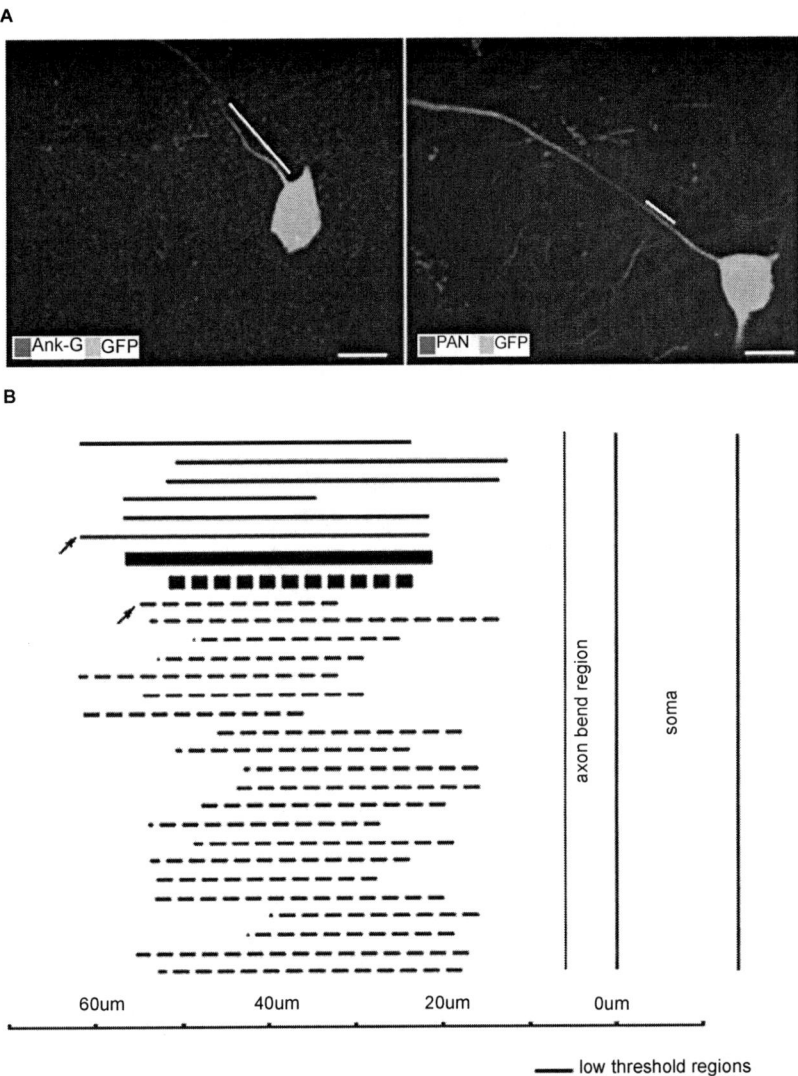

Fig. 4. A. Examples of sodium-channel bands from two unidentified ganglion cell types. Ank-G: ankyrin G (protein associated with high-density sodium channels). GFP: green fluorescent protein. PAN: pan sodium antibody. Scale bar: 25 μm. B. The regions of high-density sodium-channels and low-threshold (to electrical stimulation) are coextensive. Each thin line corresponds to the length and position of a low-threshold region (solid lines) or high-density sodium-channel region (dashed lines). Each thick line represents the mean of the low-threshold regions (thick solid line) or the high-density sodium channel regions (thick dashed line). (from Fried et al. 2009, with permission).

Color image of this figure appears in the color plate section at the end of the book.

A second consideration regarding NTs in CNS function is the role of glial cells (which compose roughly 90% of the CNS, compared with 10% neurons). A review article has summarized the many ways in which the NT glutamate signals between neurons and glial cells (astrocytes) [Fig. 5] [Ni et al. 2007]. One group has recently studied the effect of *in vitro* DBS-like high frequency electrical stimulation on astrocytic glutamate and adenosine release—and the possible relevance of this NT release to the efficacy of DBS [Tawfik et al. 2010].

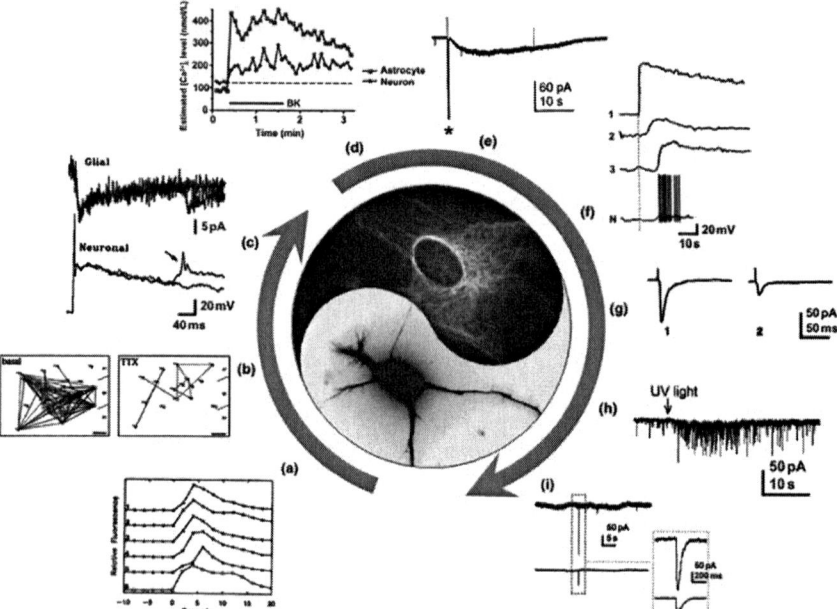

Fig. 5. Examples of interactions between neurons and astrocytes mediated by glutamate (details given in the source publication). (from Ni et al. 2007, with permission).

Members of the same group have developed—and continue to refine—an elegant system for real-time monitoring of focal NT levels *in vivo* [Agnesi et al. 2009; Bledsoe et al. 2009]. The system involves an implanted carbon microfiber electrode which uses cyclic voltammetry to measure one or more NTs in a free-roaming animal—and which transmits data continuously via a wireless (Bluetooth) connection [Fig. 6]. This group has demonstrated in animal models the focal NT response to focal electrical stimulation of the brain with a DBS-type electrode. NTs monitored have included dopamine, adenosine, serotonin, and glutamate; the feasibility of monitoring two NTs simultaneously has also been shown [Shon et al. 2010; Griessenauer et al. 2010].

Electrochemical Nanoelectrode Arrays 69

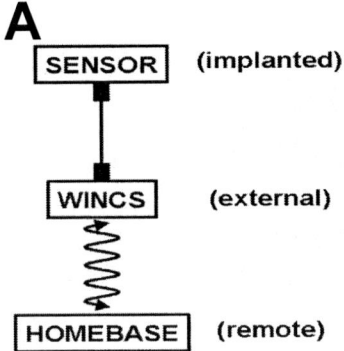

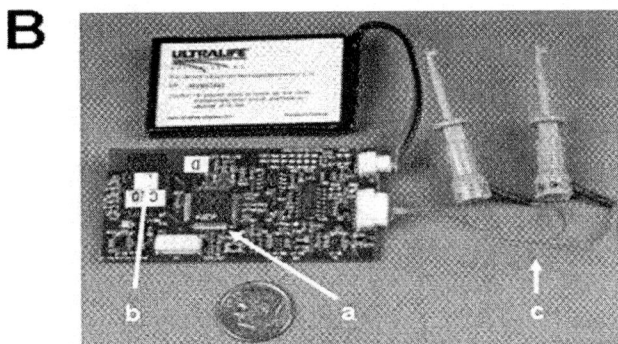

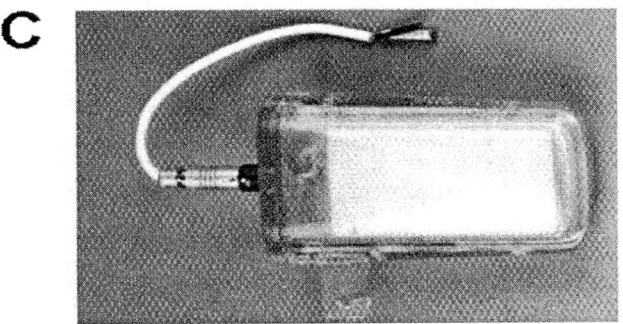

Fig. 6. Schematic and photographs of the WINCS (Wireless Instantaneous Neurotransmitter Concentration System) under development at the Mayo Clinic. (from Agnesi et al. 2009, with permission).

The spatial and temporal resolution of the electrodes used for cyclic voltammetry NT detection can be improved upon using NEAs fabricated with vertically aligned CNFs or CNTs rather than carbon microfiber electrodes. One method for fabricating such NEAs is described in our previous publications, as shown in Fig. 7 [Li et al. 2003; Li et al. 2005]. A vertically aligned CNF array is grown on a prepatterned MEA and then encapsulated with an insulating material such as SiO_2. A mechanical polishing or reactive ion etching procedure is followed to removed excess SiO_2 and expose only the very end of some CNF tips. This forms an array of disk-like nanoelectrodes. Figure 8 shows unpublished data demonstrating the greater sensitivity of a CNF NEA than carbon microfiber electrodes for the detection of dopamine. Preliminary data using CNF NEAs with the wireless NT monitoring system described in the preceding paragraph suggest that CNF NEAs will enhance the real-time monitoring of one or more NTs in clinical DBS/neuromodulation devices in the near future.

Fig. 7. SEM images of (a) a 3x3 microelectrode array, (b) an array of vertically aligned CNFs on one of the microelectrode pads, (c) an array of vertically aligned CNFs at e-beam patterned Ni spots, and (d) the surface of a polished CNF array grown on 200 nm spots, forming a well-defined nanoelectrode array. Panels (a-c) are 45° perspective views and panels (d) are top views. The scale bars are 200, 50, 5, and 2 µm, respectively. (from Li et al. 2003, with permission).

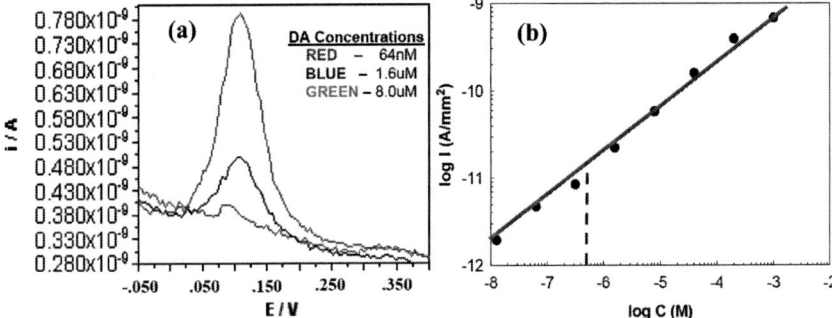

Fig. 8. (a) The signal of differential pulse voltammetry for detection of dopamine with a CNF NEA in 64 nM, 1.6 µM, and 8.0 µM dopamine in PBS solutions. (b) The calibration curve of the peak current in differential pulse voltammetry vs. the concentration of dopamine from 13 nm to 1.0 mM, with the blue vertical line indicating the detection limit of a typical carbon microelectrode at ~500 nM. (unpublished data, NASA Ames Research Center–J.L.).

Color image of this figure appears in the color plate section at the end of the book.

Applications to other Areas of Health and Disease

Neuromodulation/DBS is currently a standard of care intervention only for advanced movement disorders such as Parkinson's disease. It is under intense investigation for refractory epilepsy, mood disorders not responsive to pharmacologic interventions, and other disorders such as chronic headaches and morbid obesity. The incorporation of NT monitoring (and potentially modulating) raises the possibility of enhancing diagnosis and treatment of neuroendocrine disorders (e.g., hypothalamic and pituitary dysfunction and/or tumors). The ability to monitor and modulate precisely brain electrical and chemical activity will open new avenues in both the understanding and the treatment of a wide range of CNS disorders.

Key Facts

- Neuromodulation/DBS, using electrical stimulation alone, has already proven to be a remarkably successful treatment for patients with advanced movement disorders such as Parkinson's disease—but side effects and lack of efficacy in specific patients are significant problems.
- Brain tissue is composed of neurons (10%) and glial cells (90%), and these cells communicate both chemically (NTs) and electrically.
- The incorporation of monitoring and modulating brain chemical (NT) activity as well as electrical activity will dramatically enhance the effectiveness of neuromodulation.

- Nanoelectrode arrays (NEAs) will improve both the spatial and temporal precision of neuromodulation (due to the much smaller size of NEAs than macro or microelectrodes).
- NEAs will improve the sensitivity of recording and stimulating both electrical and chemical activity in the brain—enhancing efficacy.

Definitions

Deep brain stimulation: stimulation of brain activity at the subcortical level, customarily using electrodes placed into the brain through a small hole made in the skull. It is contrasted with 'cortical brain stimulation' through electrodes placed on the surface of the brain. At present it is limited to electrical stimulation with electrodes hundred of microns to a millimeter or more in diameter.

Neuromodulation: alteration (modulation) of the activity of the nervous system, particularly with electrical stimulation. Customarily it has been limited to electrical stimulation of the brain, spinal cord, or peripheral nerves with either surface or depth electrodes. Pharmacologic modulation of the nervous system (e.g., psychotropic drugs) has not be considered to be a form of neuromodulation in this limited definition. More recently, other techniques for neuromodulation have been developed, e.g., transcranial magnetic stimulation (TMS).

Neurotransmitter: a chemical (most commonly a small amino acid or monoamine) which transmits signals from a neuron to a target cell, typically across a synapse. The most common neurotransmitters are glutamate (excitatory) and gamma-amino-butyric-acid (GABA—inhibitory); dopamine is another neurotransmitter which is especially important in Parkinson's disease and many mood disorders.

Summary Points

- Neuromodulation/DBS has proven in the past 20 years to be a remarkably effective treatment in many patients with advanced Parkinson's disease and other movement disorders.
- Neuromodulation/DBS is limited by the inefficient nature of recording and stimulating the electrical activity of the CNS using present-day electrode technology.
- Neuromodulation/DBS is limited by the lack of recording and stimulating the chemical (NT) activity of the CNS using present-day electrode technology.
- The large size of present-day DBS electrodes (greater than one millimeter in diameter) limits the specificity (anatomic and physiologic precision) of recording and stimulating.

- Nanoelectrodes arrays (NEAs) composed of carbon nanofibers (CNFs) or carbon nanotubes (CNTs) have been shown to greatly improve charge transfer (neural-electrical interface—NEI).
- Nanoelectrodes arrays (NEAs) composed of carbon nanofibers (CNFs) or carbon nanotubes (CNTs) have been shown to be more efficient than traditional electrodes at detecting NTs such as dopamine.
- NEAs will allow simultaneous monitoring and modulating of brain electrical and chemical (NT) activity, improving both our understanding of CNS disorders and our ability to correct the electrochemical imbalances underlying those disorders.

Abbreviations

CNF	:	carbon nanofiber
CNT	:	carbon nanotube
CNS	:	central nervous system
DBS	:	deep brain stimulation
MEA	:	microelectrode array
NEA	:	nanoelectrode array
NEI	:	neural-electrical interface
NT	:	neurotransmitter

References

Agnesi, F., S.J. Tye, J.M. Bledsoe, C.J. Griessenauer, C.J. Kimble, G.C. Sieck, K.E. Bennet, P.A. Garris, C.D. Blaha and K.H. Lee. 2009. Wireless Instantaneous Neurotransmitter Concentration System-based amperometric detection of dopamine, adenosine, and glutamate for intraoperative neurochemical monitoring. J. Neurosurg. 111: 701–711.

Bledsoe, J.M., C.J. Kimble, D.P. Covey, C.D. Blaha, F. Agnesi, P. Mohseni, S. Whitlock, C.M. Johnson, A. Horne, K.E. Bennet, K.H. Lee and P.A. Garris. 2009. Development of the Wireless Instantaneous Neurotransmitter Concentration System for intraoperative neurochemical monitoring using fast-scan cyclic voltammetry. J. Neurosurg. 111: 712–723.

de Asis, E., R.J. Andrews and J. Li. 2012. Electrical stimulation of brain tissue with carbon nanofiber microbrush arrays. In: C. Martin et al. (Eds.)—Nanomedicine and the Nervous System, Science Publishers, Enfield NH. pp

Fried, S.I., A.C.W. Lasker, N.J. Desai, D.K. Eddington and J.F. Rizzo. 2009. Axonal sodium-channel bands shape the response to electrical stimulation in retinal ganglion cells. J. Neurophysiol. 101: 1972–1987.

Griessenauer, C.J., S.Y. Chang, S.J. Tye, C.J. Kimble, K.E. Bennet, P.A. Garris and K.H. Lee. 2010. Wireless Instantaneous Neurotransmitter Concentration System: electrochemical monitoring of serotonin using fast-scan cyclic voltammetry—a proof-of-principle study. J. Neurosurg. 113: 656–665.

Keefer, E.W., B.R. Botterman, M.I. Romero, A.F. Rossi and G.W. Gross. 2008. Carbon nanotube coating improves neuronal recordings. Nat. Nanotech. 3: 434–439.

Li, J., H.T. Ng, A. Cassell, W. Fan, H. Chen, Q. Ye, J. Koehne, J. Han and M. Meyyappan. 2003. Carbon Nanotube Nanoelectrode Array for Ultrasensitive DNA Detection. Nano Lett. 3: 597–602.

Li, J., J.E. Koehne, A.M. Cassell, H. Chen, H.T. Ng, Q. Ye, W. Fan, J. Han and M. Meyyappan. 2005. Inlaid multi-walled carbon nanotube nanoelectrode arrays for electroanalysis. Electroanalysis 17: 15–27.

Ni, Y., E.B. Malarkey and V. Parpura. 2007. Vesicular release of glutamate mediates bidirectional signaling between astrocytes and neurons. J. Neurochem. 103: 1273–1284.

Schlaepfer, T.E., M.X. Cohen, C. Frick, M. Kosel, D. Brodesser, N. Axmacher, A.Y. Joe, M. Kreft, D. Lenartz and V. Sturm. 2008. Deep brain stimulation to reward circuitry alleviates anhedonia in refractory major depression. Neuropsychopharmacology 33: 368–377.

Sekirnjak, C., P. Hottowy, A. Sher, W. Dabrowski, A.M. Litke and E.J. Chichilnisky. 2008. High-resolution electrical stimulation of primate retina for epiretinal implant design. J. Neurosci. 28: 4446–4456.

Shon, Y.M., S.Y. Chang, S.J. Tye, C.J. Kimble, K.E. Bennet, C.D. Blaha and K.H. Lee. 2010. Comonitoring of adenosine and dopamine using the Wireless Instantaneous Neurotransmitter Concentration System: proof of principle. J. Neurosurg. 112: 539–548.

Tawfik, V.L., S.Y. Chang, F.L. Hitti, D.W. Roberts, J.C. Leiter, S. Jovanovic and K.H. Lee. 2010. Deep brain stimulation results in local glutamate and adenosine release: investigation into the role of astrocytes. Neurosurgery 67: 376–375.

Wightman, R.M., M.L.A.V. Heien, K.M. Wassum, L.A. Sombers, B.J. Aragona, A.S. Khan, J.L. Ariansen, J.F. Cheer, P.E.M. Phillips and R.M. Carelli. 2007. Dopamine release is heterogeneous within microenvironments of the rat nucleus accumbens. Euro. J. Neurosci. 262046–2054.

5

Probing Protein Nanoassemblies in Neurodegenerative Disease

Min S. Wang[1] and Michael R. Sierks [2,]*

ABSTRACT

Atomic force microscopy (AFM) has been utilized in many biological applications and has evolved into a useful tool for structural characterization and for direct measurement of intermolecular forces at the nanoscale level. AFM offers unique advantages compared to other analytical tools, including the ability to perform label-free and fluid phase imaging which allows researchers to visualize, probe, and analyze intricate structures of biological molecules in their native environments. In addition, binding and intermolecular interactions between two biomolecules can be studied using the AFM, making it a very powerful tool for investigating the biomolecular nano-assemblies. This chapter focuses on the application of AFM in neurodegenerative disease studies. We show how AFM can be used to characterize protein

---

[1]Department of Chemistry, University of Colorado Denver, Campus box 194, PO Box 173364, Denver, CO 80217-3364 USA; E-mail: min.wang@ucdenver.edu
[2]Department of Chemical Engineering, Arizona State University, Tempe, AZ 85287-6006 USA; E-mail: sierks@asu.edu
*Corresponding author

List of abbreviations after the text.

structures at various aggregation stages; to detect changes in conformations of proteins and protein–antibody complexes, to determine the binding interactions between an antibody and antigen and to quantify binding specificity of an antibody to its antigen. In addition, we highlight the development of novel AFM techniques to facilitate the study of protein morphologies and interactions in neuroscience research. We hope that this chapter offers new insights and appreciation to the reader on the versatility of the AFM in nanotechnology research, particularly nanobioscience.

INTRODUCTION TO PROTEIN MISFOLDING AND NEURODEGENERATIVE DISEASES

Protein folding is one of the fundamental processes in living organism where linear proteins self assemble into their correct three-dimensional or "native" structure in order to carry out essential biological functions (Dobson 2003). Under normal circumstances, a protein will fold efficiently and spontaneously into its intended structure often with the help of molecular chaperones. However, proteins that do not fold properly tend to misfold or aggregate, thus rendering the protein inactive and sometimes dysfunctional (Dobson 2003; Agorogiannis et al. 2004). Many natively unfolded proteins such as beta amyloid (A-beta) and alpha-synuclein (a-syn) lack secondary structure and have a higher propensity to misfold and aggregate into fibrillar amyloid structures known as plaques. Aggregation and accumulation of plaques can disrupt other cellular components which may impair cellular function. Misfolded protein aggregates are a central component in many neurodegenerative diseases including Alzheimer's disease (AD), Parkinson's disease (PD), Huntington's disease (HD), Creutzfeldt-Jakob diseases (CJD), and transmissible spongiform encephalopathies (TSEs).

The presence of extracellular or intracellular aggregates of amyloidogenic proteins is a characteristic of all the recognized amyloid diseases. While the respective misfolded proteins differ in their pathogenesis and amino acid sequences (Lipfert et al. 2005), they share a common 'cross-β' structured motif and form insoluble amyloid fibrils or plaques. AD is characterized by the presence of A-beta plaques and neurofibrillary tangles (NFT) in the brain (Fig. 1) (Lee et al. 1991); Lewy bodies (LBs), comprised predominantly of aggregated and insoluble fibrils of a-syn, are associated with PD (Volles et al. 2003); while aggregates of huntingtin protein comprising of expanded polyglutamine (poly Q) repeats are a characteristic hallmark of HD (Bhattacharyya et al. 2005).

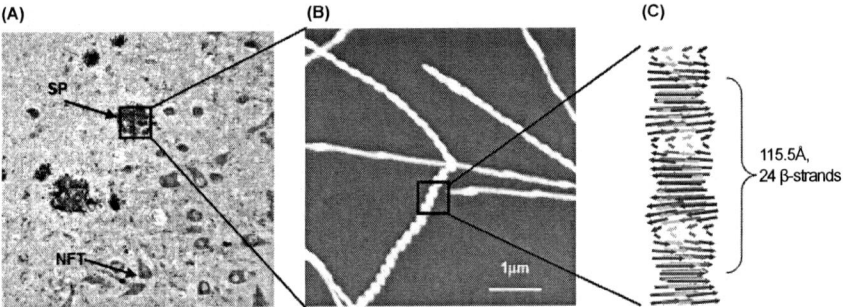

Fig. 1. Aggregated protein and plaque deposits in AD. (A) Accumulation of amyloid senile plaques (SPs) and neurofibrillary tangles (NFTs) found in AD brain (Image courtesy of Millipore, Anti Amyloid-beta clone W0-2, Catalog No. MABN10). (B) Atomic force microscopy image of an amyloid fibril. (C) Cross β–sheets run parallel to the axis of the fibril that give rise to helical repeats of 115.5Å containing 24 β-strands. (Figure adapted from Sunde, J. Mol. Biol. 1997.)

KINETICS OF PROTEIN AGGREGATION

Protein aggregation and fibrillization follows a seeded nucleation polymerization mechanism (Agorogiannis et al. 2004). In this process unfolded monomeric proteins self-aggregate into oligomeric nuclei in the lag-phase known as the nucleation step, followed by a rapid growth phase in which larger oligomers and protofibrils are formed and finally reaching a plateau phase where insoluble amyloid fibrils are formed (Fig. 2).

In many amyloidogenesis studies, the formation of a variety of toxic oligomeric intermediates at early stages of protein aggregation are linked to pathogenesis and are therefore responsible for initiating a cascade of events leading to neurodegeneration and neuronal cell death (Harper et al. 1999). The mechanism of amyloid fibrillogenesis is obviously critical to the pathogenesis of protein misfolding disorders. Understanding the relationship between amyloid formation and disease onset and progression is key to facilitate the development of new therapeutic strategies aimed at targeting the toxic intermediate aggregate species that are present during fibril formation. Therefore, well characterized reagents and assays are needed to probe and analyze these intermediate protein morphologies to gain further insight into the onset and progression of these neurodegenerative diseases.

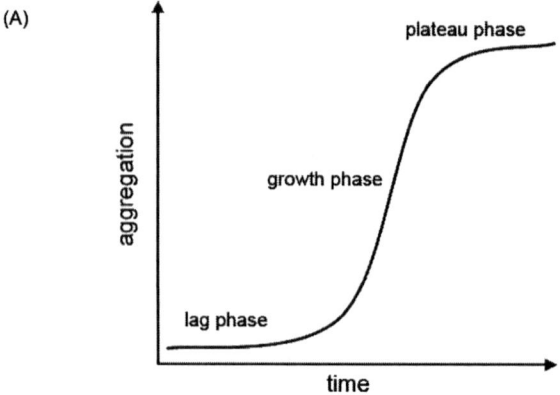

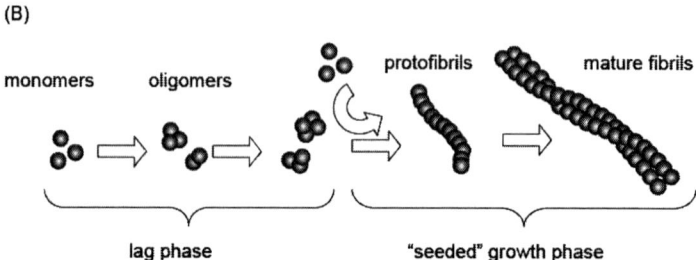

Fig. 2. Kinetics of protein aggregation. (A) Protein aggregation follows a seeded nucleation polymerization process. (B) Aggregation mechanism and formation of protein aggregates. (Figure illustrated by M.S. Wang.)

ATOMIC FORCE MICROSCOPY AS A TOOL TO STUDY NEURODEGENERATIVE DISEASE

Because of its superior imaging capability and ease of sample preparation, the atomic force microscope (AFM) is a very powerful tool to study protein aggregation in neurodegenerative disease. AFM can be used to elucidate the kinetic mechanism of protein aggregation; to provide detailed structural information of the transient oligomeric forms of A-beta and a-syn morphologies; to aid in the isolation and design of antibodies with improved binding affinity and specificity and to provide a basis for improved therapeutics that target underlying causes of neurodegenerative disease. Here we will describe how AFM can be used in the study of neurodegenerative diseases and how morphology-specific antibodies can be selected using the AFM.

INTRODUCTION TO ATOMIC FORCE MICROSCOPY

The AFM was first introduced by Binnig, Quate and Geber in 1986 to provide high quality topographic images of a sample surface and to detect intermolecular forces with atomic scale resolution (Binnig et al. 1986). The main components of an AFM are the piezoelectric scanner, detector and a cantilever probe (Fig. 3A). AFM cantilevers are nanofabricated using either silicon or silicon nitride with typical tip radii of 10 nm or less. Most

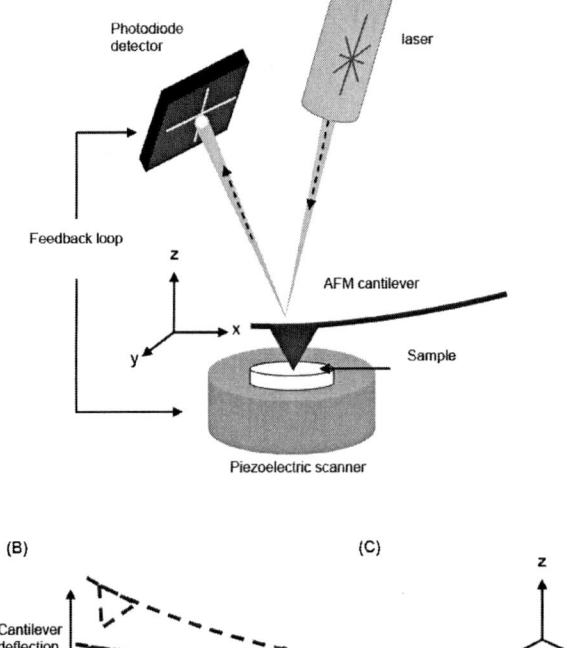

Fig. 3. Schematic of an Atomic Force Microscope setup. (A) The laser is focused on the backside of the tip and the beam is reflected to a photodiode detector. The detector sends signal to the piezoelectric scanner via a feedback loop, such that the position of the sample relative to the tip is held at either constant voltage or height. (B) Topographical image of the sample is acquired as tip scans the sample area (x and y- axes), while the deflection of the tip in the z-axis in gives the 3rd dimension information (e.g., height). (C) Height profile of the sample. (Figure illustrated by M.S. Wang.)

AFM operate by reflecting a laser beam off the back of the AFM cantilever and into a photodiode detector (Fig. 3A). A sample is mounted on the piezoelectric scanner and the distance between the AFM tip and sample is held constant by the detector and piezoelectric scanner via a closed-loop feedback mechanism. During operation, the detector senses subtle changes in deflection experienced by the tip as it moves across the sample and sends a signal to the piezoelectric scanner to adjust the tip-to-sample distance (z-axis) to prevent tip and sample damage (Fig. 3B). An AFM image is acquired by the movement of the scanner in the z-direction as the tip rasters in the x- and y-direction over the sample (Fig. 3B), thus creating a "topographic map" of the sample surface (Fig. 3C). Since AFM can operate under fluid and physiological conditions while still provide image resolution that rivals electron microscopy, such as TEM (Nag et al. 1999), the AFM has been used to resolve intricate surface structures of individual molecules or molecular assemblies, including DNA (Lyubchenko et al. 1992), proteins (Browning-Kelley et al. 1997) and cells (Rossetto et al. 2007).

MONITORING PROTEIN AGGREGATION WITH AFM

Traditionally, protein aggregation such as formation of amyloid aggregates is monitored by spectroscopic methods using thioflavin T (ThT) fluorescence assay, turbidity measurements, circular dichroism (CD) and light scattering techniques (Murphy et al. 2000; Zameer et al. 2008). ThT associates specifically with beta-sheet structures such as amyloid fibrils, when upon incorporation with A-beta oligomers or fibrils, causes an increase in fluorescence emission at 482 nm (Zameer et al. 2008). Turbidity assay measures the aggregation absorbance peak at 405 nm (Zameer et al. 2008), whereas CD monitors the conformational changes of protein from α-helical to β-sheet at far-UV (222 nm) and light scattering techniques quantify the extent of protein aggregation by measuring the changes of protein size. These methods rely heavily on the α-helix to β-sheet conformational changes of the protein but do not provide detailed information at early aggregation stages. However with the AFM, we can monitor protein aggregation at different stages, and provide complementary information both qualitatively and quantitatively by measuring the height changes in protein aggregates (Fig. 4). The AFM has been used to characterize the aggregation mechanism and to describe the morphological details of various amyloid-like proteins including a-syn (Conway et al. 2000), A-beta (Stine et al. 1996) and huntingtin protein (Dahlgren et al. 2005).

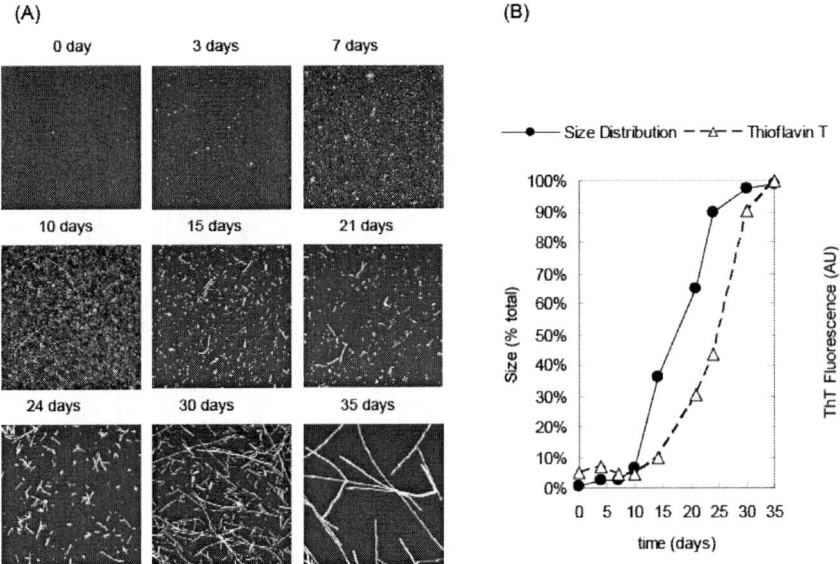

Fig. 4. Monitoring aggregation of beta-amyloid with AFM. (A) Time-course AFM images of beta-amyloid aggregation. (B) Particle size distribution of A-beta aggregates measured as % of total particle sizes of the AFM image (compared with ThT data). (Wang, unpublished data.)

CHARACTERIZING BINDING INTERACTIONS WITH AFM

AFM is also a powerful tool to probe intermolecular interactions between biomolecules, for example the interactions between two identical proteins or between an antibody and its target antigen. Typically, one biomolecule is tethered to the tip via covalent linkage, while the second protein of interest is immobilized on the substrate surface (Fig. 5A) (Yu et al. 2009). Lyubchenko's group has developed an AFM-based methodology to quantify the inter-protein interactions between two A-beta peptides (McAllister et al. 2005) and between dimers of a-syn (Yu et al. 2009). Using AFM dynamic force spectroscopy (DFS), Lyubchenko and co-workers quantified the strength of intermolecular complexes between proteins immobilized on the AFM tip and substrate by performing multiple approach-retraction cycles (Fig. 5C) (Yu et al. 2009). This pioneering work by Lyubchenko's group has provided evidence linking protein misfolding to an increase in interprotein interactions, which subsequently leads to aggregation (McAllister et al. 2005). They concluded that under conditions that favor protein aggregation, proteins adopt misfolded conformations

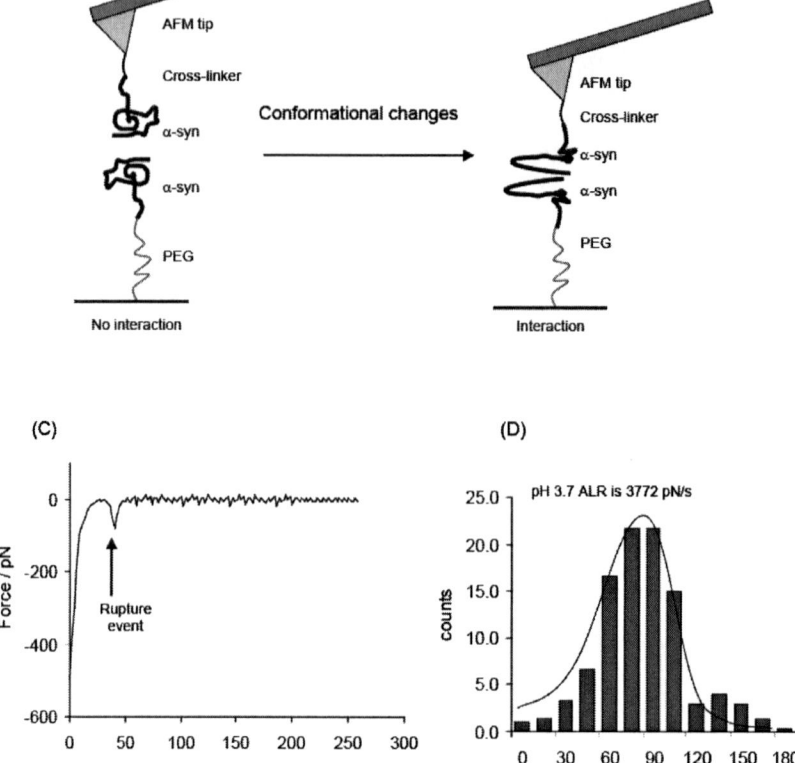

Fig. 5. Probing for the conformational change of a-syn and interprotein force measurement with dynamic force spectroscopy. (A) Alpha-synuclein was covalently immobilized on the AFM tip and mica surface using a PEG linker. (A) No interaction between natively unfolded a-syn proteins. (B) Strong interaction between misfolded dimers of a-syn. (Figure adapted from Yu, Nanomedicine, 2010). (C) A representative force curve for the interaction between a-syn measured at pH 3.7. The rupture event is indicated with an arrow. (D) The a-syn-a-syn dimer rupture force distribution at an apparent loading rate of 3772 pN/s. (Figure adapted from Yu, J. Mol. Biol. 2008.)

that are responsible for the increase in interprotein interaction and promote formation of additional protein aggregates (McAllister et al. 2005; Yu et al. 2008; Yu et al. 2009). Using this technique, they were also able to show that a-syn dimers are stable over a relatively long lifetime compared to that of the monomeric form (Yu et al. 2008; Yu et al. 2009) suggesting that the dimeric a-syn form may serve as a nuclei or "seed" to shift the protein aggregation kinetics towards the formation of larger aggregates. Therefore,

DFS can be used to predict which protein morphologies are responsible for initiating the aggregation process and appropriate strategies can be employed to prevent the formation of these morphologies. By combining AFM imaging and DFS, efficient therapeutics for protein misfolding neurodegenerative diseases can be developed.

CHARACTERIZING ANTIBODY SPECIFICITY WITH AFM

We have taken advantage of the fast and label-free surface characterization capabilities of the AFM for direct detection and quantification of binding interactions between single chain antibody variable domain fragments (scFvs) and specific protein morphologies associated with neurodegenerative diseases (Wang et al. 2009). Calculation of height distribution data with the AFM enables precise characterization of scFv binding to different protein aggregate morphologies (Fig. 6). The combination of height distribution and mean particle height analyses provides a simple and accurate means to characterize protein binding specificity (Wang et al. 2009). We isolated antibody fragments to a number of different protein morphologies of a-syn including an anti-oligomeric scFv (Emadi et al. 2007), a pan-specific scFv (Zhou et al. 2004) and anti-fibrillar scFv (Barkhordarian et al. 2006). Since every particle or pixel on the AFM image are accounted for regardless of whether or not an scFv is bound to that particle, two important advantages of this approach are: 1) the results are not biased by analysis of individual particles selected because of a desired binding event, and 2) heterogeneous samples containing multiple morphologies and sizes, such as those used here, can be analyzed.

This protocol has several other distinct advantages over more traditional analytical methods such as enzyme linked immunosorbant assay (ELISA) and Western blotting, including minimal amounts of antibody or target proteins, minimal handling and modification of the target proteins, thus allowing them to be studied in their native state, and the insignificant need of a highly purified target antigen. This enables characterization of binding specificity of samples that are either difficult to purify or available in limited quantities. In addition, different protein morphologies and sizes can easily be visualized and characterized using the AFM which otherwise cannot be distinguished by traditional ELISA or Western blots. Height distribution data along with visual image characterization can provide valuable information about various aggregated protein samples and how other proteins such as antibodies interact with them. Furthermore, characterization of scFvs that target specific protein morphologies will provide valuable tools for studying protein aggregation diseases such as

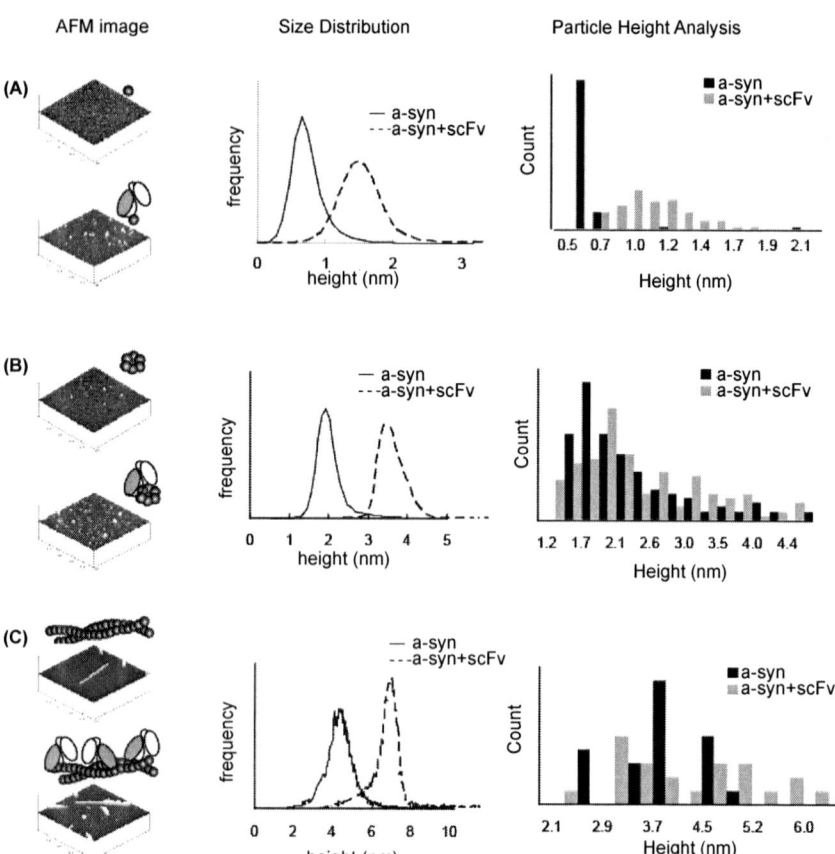

Fig. 6. Binding of scFvs to various a-syn morphologies analyzed by AFM. AFM image, particle size distribution and height analysis of (A) monomeric a-syn and a-syn+scFv, (B) oligomeric a-syn and a-syn+scFv, and (C) fibrillar a-syn and a-syn+scFv. (Wang, Langmuir 2009.)

AD, PD and HD as these morphology specific scFvs can be used as probes to detect or block specific protein morphologies in cell cultures, animal models and tissue samples.

AFM BASED BIOPANNING

Protein misfolding diseases including AD, PD, HD, Amyotrophic Latereral Sclerosis, and diabetes are associated with that can self-assemble into various aggregate morphologies including small soluble oligomers, ring-like structures, thin protofibrils and long fibrils (Dobson 2003). Since proteins with multiple morphologies are common in protein misfolding diseases, antibodies that can specifically identify or control the assembly of these different structures could serve as extremely valuable tools for studying the role of various morphologies in the progression of each of these diseases. One method to isolate antibody fragments to specific morphologies of the target antigen, is to combine the imaging capabilities of AFM with the recognition diversity of phage display antibody libraries (Barkhordarian et al. 2006).

In the AFM-based biopanning protocol (Barkhordarian et al. 2006), a mica substrate is used to immobilize the target antigen (Fig. 7). Since the negatively charged mica surface preferentially binds the positively charged target antigen but does not bind phage particles, the target antigen or antigen/phage complexes can be isolated without any modification to the target antigen. After immobilization on mica, the target antigen or antigen/phage complex can be imaged by AFM to confirm not only that the desired antigen morphology is present, but also that the phage is binding to its target (Fig. 7). The efficiency of each step in the biopanning process, such as binding, washing, or elution, can be monitored separately and optimized using AFM, thus, facilitating the selection and recovery of positive binders while reducing the recovery of non-specific bound phage. The combination of phage display technology and AFM thus provides a very powerful method to isolate reagents that recognize a specific protein morphology. This AFM-based biopanning method has been used by our group to successfully isolate antibody fragments that specifically recognize

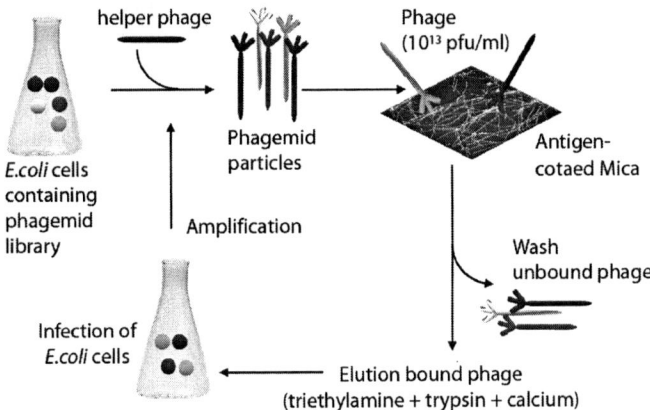

Fig. 7. Schematic of biopanning protocol. Target antigen was adsorbed on mica surface and selection of phage was performed using four rounds of biopanning. Unbound phage was subsequently washed and removed, and bound phage was eluted from the mica surface. Eluted phage were amplified, purified and used for subsequent rounds of panning. (Figure illustrated by P. Schulz.)

specific morphologies of target proteins such as different oligomeric (Emadi et al. 2009) and fibrillar (Barkhordarian et al. 2006) forms of a-syn and oligomeric (Zameer et al. 2008; Kasturirangan et al. 2010) and fibrillar (Marcus et al. 2008) forms of A-beta.

AFM TOPOGRAPHY AND FORCE RECOGNITION

In addition to providing a means to generate reagents that recognize specific protein forms, AFM also provides unique capabilities to analyze the interactions between the antibody reagents and different protein morphologies. A recent advancement in AFM imaging technology is the development of a simultaneous topography and recognition (TREC) imaging technique by Lindsay and co-workers (Stroh et al. 2004). TREC imaging has the capability of acquiring two image modes simultaneously, i.e., one, a topographic image of the sample, and the other, a lateral force recognition map of the corresponding binding sites (Stroh et al. 2004).

In TREC imaging, the antibody is tethered to the AFM tip via a poly(ethylene glycol) (PEG) linker (Fig. 8). The PEG linker increases the flexibility of the antibody which reduces the occurrence of non-specific

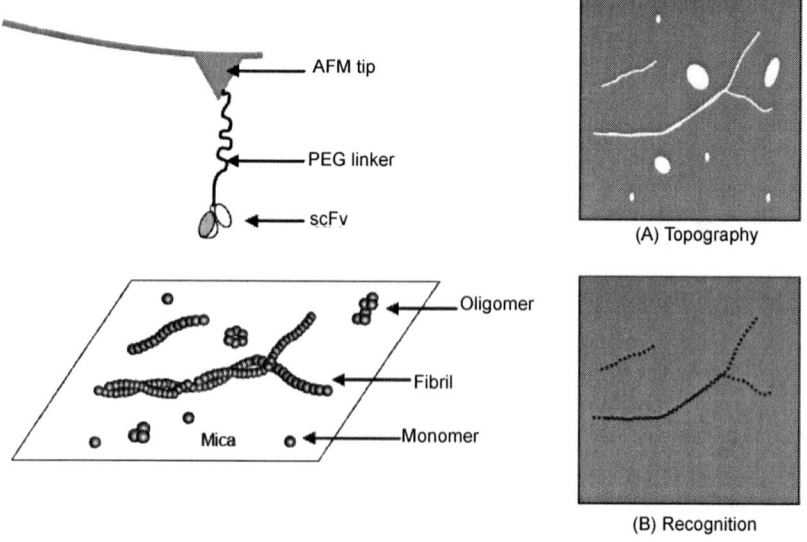

Fig. 8. AFM Topography Recognition Imaging. A fibril specific scFv is attached to the AFM tip via a PEG linker and is subjected to a mica substrate coated with a mixed morphologies of monomers, oligomers and fibrils. (A) The AFM topography image shows different morphologies present on the substrate. (B) The recognition image reveals the specific interaction between the scFv and fibrils (dotted lines), but does not recognize the other types of morphologies on the substrate. (Figure illustrated by M.S. Wang.)

target binding while still enabling accurate localization of the target antigen site. TREC imaging detects specific antigen-antibody binding through small changes in the absolute deflection voltage (DC) experienced by the AFM cantilever (Stroh et al. 2004). The changes in these small DC signals are caused by the reduction of the oscillation amplitude due to specific binding during lateral scan. To overcome the sudden reduction in oscillation amplitude, the microscope servo pulls the tip away from the surface to restore the amplitude to its original set point. The result is a downshift in the peak amplitude which corresponds to the reduction in amplitude (Stroh et al. 2004). Using a custom image analysis program, a map of these DC changes is superimposed on the topographic images to determine the location of the target antigen on the image.

Using this imaging technique, a monoclonal scFv specific for fibrillar amyloid aggregates was used to demonstrate that specific morphological regions occur at regular periodic intervals in the A-beta fibrillar structure (Marcus et al. 2008). Since TREC imaging can detect periodic binding to a particular site, anti-fibrillar antibodies can be used to further probe the structure of amyloid fibrils. This work has outlined a protocol for studying protein aggregation using TREC imaging and it can be a valuable tool for probing amyloid structure (Marcus et al. 2008). Because of its capability to distinguish specific components of the target molecule, recognition imaging has great potential for use in a wide range of biological applications. The combination of TREC imaging and AFM biopanning techniques offer an improved AFM imaging and phage display technology where topography and recognition images of specific target sites can be acquired simultaneously at the molecular level without sacrificing the topographic imaging qualities. This method can be used as a generalized protocol for biopanning with any type of target since the surface of mica can be chemically modified in order to immobilize any desired antigen.

AFM NANOMANIPULATION FOR SINGLE-MOLECULE PHAGE RECOVERY

Isolation of scFvs using traditional panning protocols generally involves several rounds of panning to enrich the number of positive clones present in the eluted mixture, and requires significant amounts of purified target for panning and screening. An alternative panning protocol utilizes the precision and sensitivity of the AFM to "fish" out the desired scFv directly from the mica surface using only minimal amounts of unpurified sample. Toward this goal, a single-molecule antibody selection technique using AFM was developed, based on the principle of AFM biopanning (Shlyakhtenko et al. 2007). This AFM technique enables both the visualization of protein morphologies, and the ability to recover an antibody that binds specifically to a target protein morphology of interest, for example, oligomeric a-syn (Fig. 9). The single phage recovery process begins with the imaging of the target protein-phage displayed antibody complex. Once the protein-antibody complex of interest is determined, the AFM tip is manipulated by nanolithography software over the target where the scanning force is increased in order to "pick up" the specific phage particle displaying the antibody associated with the target. The DNA of the scFv is recovered from the single phage particle picked up by the tip using polymerase chain reaction (PCR). The advantages of this technology include: (i) the ability to select morphology-specific antibodies from a mixture of different antigens; (ii) the ability to both detect and recover antibodies that form transiently stable protein-antibody complexes; and (iii) the capability to recover antibodies to different regions of the same protein aggregate, e.g., the backbone, sides or ends of a fibril. The ability of this AFM technology to detect and recover transient intermediate species of misfolded proteins opens exciting possibilities to detect the *in vivo* presence of these protein structures involved in many neurodegenerative diseases. This single-molecule immunoselection protocol can provide a very powerful tool to generate affinity reagents to a variety of biomolecules that may be difficult to isolate by conventional immunological screening methods.

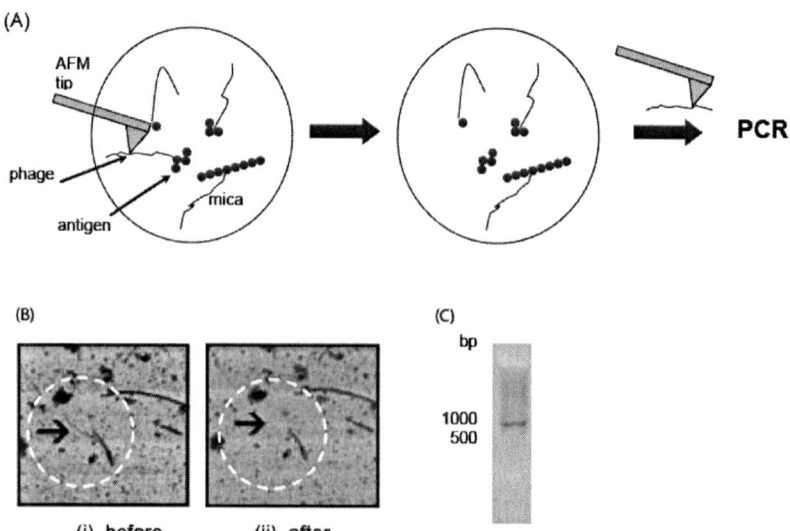

Fig. 9. Single phage recovery using AFM. (A) Schematic of an AFM phage recovery technique. (B) AFM images of a before (i) and after (ii) phage removal by AFM tip. (C) PCR amplification of phage DNA on AFM tip. (Wang, unpublished data.)

Conclusion

Accumulation of misfolded protein aggregates is the characteristic hallmark of many neurodegenerative diseases, including AD, PD and HD. Although the pathogenesis of these diseases is linked to the presence of insoluble amyloid plaques, emerging evidence has implicated various intermediate oligomeric species as the likely neurotoxic species. In order to facilitate the development of new therapeutic strategies, a link between amyloid formation and disease onset and progression is crucial in the understanding of toxicity and morphologies of intermediate species that are present prior to fibril formation. Therefore, detailed and precise characterization methods are needed to probe and analyze these intermediate protein morphologies in order to gain further insight into disease formation and progression.

AFM offers unique advantages over many other tools for the study of biological processes at the nanometer scale. AFM offers a label-free imaging technology that allows researchers to visualize, probe, and analyze intricate structures of biological molecules in their native environments with nanometer scale resolution. In addition, intermolecular forces between two biomolecules can be studied using the AFM, and this unique feature has made the AFM an important tool for studying the nanomechanical properties and interactions of biological samples. In this chapter, we have highlighted the use of AFM and the development of novel AFM techniques to facilitate the study of protein morphologies and interactions in neuroscience research. Several different applications of AFM to study protein aggregation were discussed.

Monitoring protein aggregation with AFM. Protein aggregation has traditionally been monitored by spectroscopic methods using fluorescent dyes or the intrinsic fluorescence of the protein. One of the major drawbacks of spectroscopy is the need of a label probe and a large sample volume at high concentrations. Moreover, these methods do not usually provide detailed information at early aggregation stages because they rely on conformational changes of the protein. Using the imaging capabilities of AFM, we can monitor protein aggregation at various stages of aggregation and provide qualitative and quantitative information about the protein state by measuring changes in protein heights utilizing minimal sample volume and concentration. Therefore, AFM is a powerful tool that can help characterize the protein aggregation mechanism and to describe the morphological details of different morphologies of various amyloid-like proteins including a-syn, A-beta, insulin and huntingtin protein.

Characterizing binding interactions with AFM. Lyubchenko's group has developed an AFM-based methodology to quantify the inter-protein interactions between two amyloid peptides. Using DFS, the strength of intermolecular complexes between proteins immobilized on the AFM tip and substrate were quantified. By combining AFM imaging and DFS, efficient therapeutics for protein misfolding neurodegenerative diseases can be developed.

Characterizing antibody specificity with AFM. The combination of height distribution and mean particle height analyses provides a simple and accurate means to characterize protein binding specificity. Calculation of height distribution with the AFM enables precise characterization of scFv binding to different protein aggregate morphologies. The availability of well-characterized reagents that target specific protein morphologies will aid in the development of useful probes to detect or inhibit specific protein morphologies in cell cultures, animal models and tissue samples.

AFM Based Biopanning. The combination of AFM imaging and phage display technology has allowed us to successfully isolate antibody fragments to specific morphologies of the target antigen. The efficiency of each step in the biopanning process can be monitored separately and optimized using AFM, thus, facilitating the selection and recovery of positive binders while reducing the recovery of non-specific bound phage. This novel biopanning technique provides a very powerful method to isolate reagents that recognize a specific protein morphology and has been used to isolate antibody fragments that selectively recognize specific morphologies of target proteins such as different oligomeric and fibrillar forms of a-syn and A-beta.

AFM Topography and Force Recognition. AFM TREC imaging detects specific antigen-antibody binding force through small changes in the absolute voltage experienced by the AFM cantilever. TREC imaging has the capability to simultaneously acquire a topographic image and a lateral force recognition map of the corresponding binding sites. TREC imaging is used to quantify and map specific biomolecular interactions between a variety of biological molecules of scientific interest and physiological importance. Combinatory topographical and force mapping of the antibody-antigen interaction allows researchers to locate the exact position on the antigen which experiences the strongest interaction with the antibody. This will allow for better antibody design because the binding site can be elucidated with TREC imaging.

Single-molecule selection of antibody using AFM. Single-molecule antibody selection technique using AFM was developed, based on the principles of AFM biopanning. This AFM technique allows for both the visualization of protein morphologies and the selection and recovery of an antibody that binds specifically to the target protein of interest. Single-molecule nanomanipulation takes advantage of the precision of the AFM to pick up the antibody of interest carefully from a pool of mixed-morphologies sample, essentially revolutionizing the antibody screening and selection process.

Overall, the AFM is a very useful tool to elucidate the kinetic mechanism of protein aggregation; to provide detailed structural information of the transient oligomeric forms of proteins such as A-beta and a-syn; to aid in the isolation and design of reagents with improved binding affinity and specificity and to provide a basis for better therapeutic design that targets the underlying cause of neurodegenerative diseases.

Key Facts of the Applications of Atomic Force Microscopy For the Study of Protein Aggregation and Neurodegenerative Diseases

- Protein misfolding occurs when proteins do not fold into their intended functional 3D structure.
- Protein aggregation is a consequence of protein misfolding which leads to the formation protein aggregates called amyloid plaques. It is a common hallmark in many neurodegenerative diseases including Alzheimer's, Parkinson's and Huntington's diseases.
- Alzheimer's disease is the most prevalent form of neurodegenerative disease affecting 26 million people worldwide. In the US, 5.1 million were affected by the disease with an annual estimated cost of $91 billion.
- Alzheimer's disease was named after Dr. Alois Alzheimer, who discovered the accumulation of plaques and stringy tangles in the otherwise empty spaces between nerve cells in 1906.
- Parkinson's disease is the second most common form of neurodegenerative disease affecting 1.5 million people in the U.S. and 4.1 million worldwide. The annual estimated cost is $25 billion.
- Parkinson's disease is named after Dr. James Parkinson, who first gave a detailed description of the disease in 1817.
- While fibrillar aggregates are diagnostic features in the brain for neurodegenerative disease, smaller intermediate species are thought to be the primary cause of neurodegeneration and toxicity to brain cells.
- Antibodies against specific toxic intermediate protein species are very helpful as diagnostic and therapeutic agents.
- The atomic force microscope is a powerful technique to look at the small protein structures and to help distinguish toxic species. It also allows for the measurement of specific interactions between antibody and protein morphologies associated with neurodegenerative diseases.

Applications to other Areas of Health and Disease

Owing to its superior imaging capability and ease of sample preparation, the AFM is a very useful tool in nanomedical research to elucidate intricate protein structures. The various AFM applications described here are not limited to the study of neurodegenerative diseases; they can be readily adapted to study mechanisms of other diseases that are associated with changes in native protein structure, including prion disease, diabetes and cancer. In addition to imaging, the AFM can also be used to quantify specific biomolecular interactions. Therefore, the methods described here have a broad range of applications in other areas of health and medical research, and can be applied to study interactions between any protein targets or antibody-antigen interactions.

Summary Points

- Accumulation of misfolded protein aggregates is the characteristic hallmark of many neurodegenerative diseases, including Alzheimer's, Parkinson's and Huntington's diseases.
- Since emerging evidence has implicated intermediate oligomeric species as the likely neurotoxic species, detailed and precise characterization methods are needed to probe and analyze these intermediate protein morphologies in order to gain further insight into disease onset and progression.
- The AFM offers a label-free imaging technology that allows researchers to visualize, probe, and analyze intricate structures of biological molecules in their native environments with nanometer scale resolution.
- Protein aggregation at various stages of aggregation can be monitored using the AFM to provide qualitative and quantitative information about the protein.
- Specific antibody-antigen interactions can be quantified using the AFM.
- Antibody fragments to specific morphologies of the target antigen can be and have been successfully isolated using a combination of AFM imaging and phage display technology.
- Specific binding sites on the target antigen can be elucidated using AFM-TREC imaging to allow for better antibody design.
- Single-molecule antibody selection and recovery can be made using the AFM to specifically select an antibody that binds selectively to the target protein of interest.

Definitions

Alpha-synuclein (a-syn): a-syn is a small, natively unfolded protein of 140 amino acids found in the cytosol and expressed primarily at pre-synaptic terminals in the central nervous system.

Alzheimer's disease (AD): AD is a progressive neurodegenerative disease that is characterized clinically by the loss of memory and language skills, damaged cognitive function, and altered behavior and is the most common form of dementia. The major neuropathological hallmarks of AD include the presence of neurofibrillary tangles and amyloid plaques.

Atomic force microscopy (AFM): AFM is a high resolution imaging technique that generates topographical images of surfaces using a physical probe that scans the specimen.

AFM cantilever/tip: The AFM consists of a cantilever with a sharp tip (probe) with nanoscale radius of curvature at its end that is used to scan the specimen surface.

Beta amyloid (A-beta): A-beta is a 4 kDa amphiphilic peptide that consists of 39–43 amino acids and is the major component of the senile plaques found in AD patients.

Biopanning: Biopanning is an affinity selection technique which selects for reagents that bind to a given target. Successive rounds of panning are generally required to select and screen for peptides with the highest affinity.

Neurodegenerative disease: A class of disease affecting the nervous system that is characterized by the progressive loss of structure or function of neurons, including neuronal cell death.

Parkinson's disease (PD): PD is a progressive neurodegenerative disease that is characterized clinically by resting tremor, muscular rigidity, and bradykinesia resulting from the progressive loss of dopaminergic neurons in the brain. The pathological hallmark of PD is the cytoplasm inclusions of insoluble protein aggregates (i.e., a-syn) in dopaminergic neurons.

Phage display technology: Phage display technology is a technique that describes the display of peptides and proteins on the surface of bacteriophage (virus) that is widely used in biotechnological applications including selection of antibodies, identifying enzyme inhibitors, discovery of new therapeutic targets, mapping of epitopes on antigens and others. This technology relies on the utilization of phage display libraries in a screening process known as biopanning.

Protein misfolding/aggregation: Protein misfolding is a physical process where proteins adopt an improperly folded 3D-conformation. Misfolded proteins can aggregate to generate toxic species such as those correlated with various neurodegenerative diseases.

Single chain antibody fragment (scFv): ScFv is a fusion protein that contains only one light chain (V_L) and one heavy chain (H_L) of immunoglobulins and is connected via a peptide linker.

Abbreviations

AFM	:	Atomic force microscope/microscopy
A-beta	:	Beta-amyloid
AD	:	Alzheimer's disease
a-syn	:	Alpha-synuclein
CD	:	Circular dichroism
CJD	:	Creutzfeldt-Jakob disease
DC	:	Deflection voltage
DFS	:	Dynamic force spectroscopy
DNA	:	Deoxyribonucleic acid
ELISA	:	Enzyme linked immunosorbant assay
HD	:	Huntington's disease
LBs	:	Lewy bodies
NFT	:	Neurofibrillary tangles
PD	:	Parkinson's disease
PEG	:	Poly(ethylene glycol)
scFv	:	Single chain antibody fragment
TEM	:	Transimission electron microscopy
ThT	:	Thioflavin T
TREC	:	Topography and recognition
TSE	:	Transmissible spongiform encephalopathy
UV	:	Ultraviolet

Acknowledgments

We thank Mr. Philip Schulz for his assistance with the graphics for this chapter.

References

Agorogiannis, E.I., G.I. Agorogiannis, A. Papadimitriou and G.M. Hadjigeorgiou. 2004. Protein misfolding in neurodegenerative diseases. Neuropath Appl. Neuro. 30: 215–224.

Barkhordarian, H., S. Emadi, P. Schulz and M. Sierks. 2006. Isolating recombinant antibodies against specific protein morphologies using atomic force microscopy and phage display technologies. Prot. Eng. Des. Sel. 19: 497–502.

Bhattacharyya, A.M., A.K. Thakur and R. Wetzel. 2005. Polyglutamine aggregation nucleation: Thermodynamics of a highly unfavorable protein folding reaction. Proc. Natl. Acad. Sci. USA 102: 15400–15405.

Binnig, G., C.F. Quate and C. Gerber. 1986. Atomic Force Microscope. Phys. Rev. Lett. 56: 930–933.

Browning-Kelley, M.E., K. WaduMesthrige, V. Hari and G.Y. Liu. 1997. Atomic force microscopic study of specific antigen/antibody binding. Langmuir 13: 343–350.

Conway, K.A., S.J. Lee, J.C. Rochet, T.T. Ding, R.E. Williamson and P.T. Lansbury. 2000. Acceleration of oligomerization, not fibrillization, is a shared property of both alpha-synuclein mutations linked to early-onset Parkinson's disease: Implications for pathogenesis and therapy. Proc. Natl. Acad. Sci. USA 97: 571–576.

Dahlgren, P.R., M.A. Karymov, J. Bankston, T. Holden, P. Thumfort, V.M. Ingram and Y. L. Lyubchenko. 2005. Atomic force microscopy analysis of the Huntington protein nanofibril formation. Nanomedicine 1: 52–57.

Dobson, C.M. 2003. Protein folding and misfolding. Nature 426: 884–90.

Emadi, S., H. Barkhordarian, M. Wang, P. Schulz and M. Sierks. 2007. Isolation of a human single chain antibody fragment against oligomeric alpha-synuclein that inhibits aggregation and prevents alpha-synuclein-induced toxicity. J. Mol. Biol. 368: 1132–1144.

Emadi, S., S. Kasturirangan, M.S. Wang, P. Schulz and M.R. Sierks. 2009. Detecting morphologically distinct oligomeric forms of alpha-synuclein. J. Biol. Chem. 284: 11048–58.

Harper, J.D., S.S. Wong, C.M. Lieber and P.T. Lansbury, Jr. 1999. Assembly of A beta amyloid protofibrils: an *in vitro* model for a possible early event in Alzheimer's disease. Biochemistry 38: 8972–80.

Kasturirangan, S., L. Lin, S. Emadi, S. Boddapati, P. Schulz and M.R. Sierks. 2010. Nanobody specific for oligomeric beta-amyloid stabilizes non-toxic form. Neurobiol Aging In press.

Lee, V.M.Y., B.J. Balin, L. Otvos and J. Q. Trojanowski. 1991. A68—a Major Subunit of Paired Helical Filaments and Derivatized Forms of Normal-Tau. Science 251: 675–678.

Lipfert, J., J. Franklin, F. Wu and S. Doniach. 2005. Protein misfolding and amyloid formation for the peptide GNNQQNY from yeast prion protein Sup35: Simulation by reaction path annealing. J. Mol. Biol. 349: 648–658.

Lyubchenko, Y.L., A.A. Gall, L.S. Shlyakhtenko, R.E. Harrington, B.L. Jacobs, P.I. Oden and S.M. Lindsay. 1992. Atomic Force Microscopy Imaging of Double-Stranded DNA and Rna. J. Biomol. Struct. Dyn. 10: 589–606.

Marcus, W.D., H. Wang, S.M. Lindsay and M.R. Sierks. 2008. Characterization of an antibody scFv that recognizes fibrillar insulin and beta-amyloid using atomic force microscopy. Nanomedicine 4: 1–7.

McAllister, C., M.A. Karymov, Y. Kawano, A.Y. Lushnikov, A. Mikheikin, V.N. Uversky and Y.L. Lyubchenko. 2005. Protein interactions and misfolding analyzed by AFM force spectroscopy. Neurobiol. Aging. 354: 1028–1042.

Murphy, R.M. and M.R. Pallitto. 2000. Probing the kinetics of beta-amyloid self-association. J. Struct. Biol. 130: 109–122.

Nag, K., J.G. Munro, S.A. Hearn, J. Rasmusson, N.O. Petersen and F. Possmayer. 1999. Correlated atomic force and transmission electron microscopy of nanotubular structures in pulmonary surfactant. J. Struct. Biol. 126: 1–15.

Rossetto, G., P. Bergese, P. Colombi, L.E. Depero, A. Giuliani, S.F. Nicoletto and G. Pirri. 2007. Atomic force microscopy evaluation of the effects of a novel antimicrobial multimeric peptide on Pseudomonas aeruginosa. Nanomed-Nanotechnol. 3: 198–207.

Shlyakhtenko, L.S., B. Yuan, S. Emadi, Y.L. Lyubchenko and M.R. Sierks. 2007. Single-molecule selection and recovery of structure-specific antibodies using atomic force microscopy. Nanomed-Nanotechnol. 3: 192–197.

Stine, W.B., S.W. Snyder, U.S. Ladror, W.S. Wade, M.F. Miller, T.J. Perun, T.F. Holzman and G.A. Krafft. 1996. The nanometer-scale structure of amyloid-beta visualized by atomic force microscopy. J. Protein. Chem. 15: 193–203.

Stroh, C., H. Wang, R. Bash, B. Ashcroft, J. Nelson, H. Gruber, D. Lohr, S.M. Lindsay and P. Hinterdorfer. 2004. Single-molecule recognition imaging microscopy. Proc. Natl. Acad. Sci. USA 101: 12503–7.

Volles, M.J. and P.T. Lansbury. 2003. Zeroing in on the pathogenic form of alpha-synuclein and its mechanism of neurotoxicity in Parkinson's disease. Biochemistry 42: 7871–7878.

Wang, M.S., A. Zameer, S. Emadi and M.R. Sierks. 2009. Characterizing antibody specificity to different protein morphologies by AFM. Langmuir 25: 912–8.

Yu, J.P., and Y.L. Lyubchenko. 2009. Early Stages for Parkinson's Development: alpha-Synuclein Misfolding and Aggregation. J. Neuroimmune Pharm. 4: 10–16.

Yu, J.P., S. Malkova and Y.L. Lyubchenko. 2008. alpha-Synuclein Misfolding: Single Molecule AFM Force Spectroscopy Study. Neurobiol. Aging. 384: 992–1001.

Zameer, A., S. Kasturirangan, S. Emadi, S.V. Nimmagadda and M.R. Sierks. 2008. Anti-oligomeric Abeta single-chain variable domain antibody blocks Abeta-induced toxicity against human neuroblastoma cells. J. Mol. Biol. 384: 917–28.

Zhou, C., S. Emadi, M.R. Sierks and A. Messer. 2004. A human single-chain Fv intrabody blocks aberrant cellular effects of overexpressed alpha-synuclein. Mol. Ther. 10: 1023–1031.

6

Nanoparticle-assisted Laser Desorption/ionization (nano-PALDI)-based Imaging Mass Spectrometry (IMS) and its Application to Brain Sciences

Saira Hameed,[1,a] Yuki Sugiura,[2] Yoshishige Kimura,[1,b] Kalmesh Kumar Shrivas[3] and Mitsutoshi Setou[1,c,]*

ABSTRACT

Imaging mass spectrometry (IMS; also referred to as mass spectrometry imaging [MSI]) is an emerging mass-spectrometry-based imaging technique that enables visualization of the distribution of various biomolecules in biological tissue sections.

---

[1]Department of Cell Biology and Anatomy, Hamamatsu University School of Medicine, 1-20-1 Handayama, Hamamatsu, Shizuoka 431-3192, Japan.
[a]E-mail: saira101@hama-med.ac.jp
[b]E-mail: ykimura@hama-med.ac.jp
[c]E-mail: setou@hama-med.ac.jp.
[2]Department of Biochemistry, School of medicine, Keio University, PRESTO, Japan Science and Technology Agency (JST).
E-mail: yuki.sgi@z6.keio.jp
[3]Department of Chemsitry, Guru Ghasidash University, Bilaspur, CG, India.
E-mail: kshrivas@gmail.com
*Corresponding author

List of abbreviations after the text.

This technique, which can be used for a variety of tissues having vast structures, was initially developed as a tool for protein imaging. However, because of the general versatility of IMS and the lack of established imaging technology for small organic molecules, the number of studies reporting IMS of small molecules has recently increased. In fact, IMS is an effective technique for the visualization of endogenous small metabolites, especially lipids, facilitated by the unique advantages of mass-spectrometry-based molecular detection. For IMS, the choice of a proper analyte ionization technique is critical. Matrix-assisted laser desorption/ionization (MALDI) has been regarded as the most effective analyte ionization method and has been applied to the analyses of brain disorders, such as Alzheimer's and Parkinson's diseases. Despite the promising capability of MALDI-based IMS for imaging of small metabolites, this technique suffers from several critical drawbacks, especially with regard to spatial resolution. One of the critical limitations of the spatial resolution of MALDI-IMS is the size of the organic matrix crystal and analyte migration during the matrix-crystallization process. To overcome these problems, we report herein a nanoparticle (NP)-assisted laser desorption/ionization (nano-PALDI)-based IMS technique, in which NPs are used as the ionization-enhancing reagent and the organic matrix crystallization process is eliminated. Another important advantage of the use of NPs for IMS comes from the recently increasing availability of various NPs with different core-metals, surface modifications, and particle diameters, which has expanded the range of molecular species that can be analyzed by means of this technique, to include species that cannot be ionized by MALDI-IMS. Hence, we believe that this new approach will lead to a better understanding of physiological processes as well as the diagnosis and pathophysiology of complex biological process, especially in the brain. This chapter summarizes the recent technological developments in the field of IMS and also describes the utilization of nano-PALDI in IMS as an attractive alternative to traditional MALDI-IMS.

INTRODUCTION

In the last decade, the practical use of nanoparticles (NPs) in the field of biomedicine, particularly as nanomachines, molecular imaging probes, biosensors, diagnostic tools, and drug-delivery systems, has been reported extensively. With the goal of improving the therapeutic efficacy of drug-delivery systems, NPs, which are small-sized particles having diameters in the range of 100–1000 nm, have been frequently exploited as carriers for macromolecules [e.g., plasmid DNA, siRNA, peptides, and genes] and small molecules [e.g., a corticosteroid and alkaloids]. The details of these applications have been described in a recent review by Taira et al. (2009). NPs have also found practical application in mass spectrometry (MS) research.

Historically, one of the currently employed "*standard*" soft ionization techniques in MS, matrix-assisted laser desorption/ionization (MALDI), was initiated by Tanaka's study, which was awarded the Nobel Prize in Chemistry; in this technique, a suspension of an inorganic NP such as cobalt, silicon, or titanium nitride was utilized as an analyte ionization enhancing reagent, referred to as a matrix (Tanaka et al. 1988). Although subsequent studies have successively developed various series of organic matrix compounds, which are currently the most frequently used matrices (Karas and Hillenkamp 1988), NPs have again begun to attract increasing attention owing to the progress in technology, which has resulted in the production of a wide variety of NPs. The increasing availability of various kinds of NPs, especially with variations in sizes, core materials, and surface coating/chemical modifications, has allowed researchers the option of choice from a selection of NPs to suit the chemical/physical requirements of their research purposes. In fact, the use of NPs as an alternative to organic matrices in MS research has been widely studied, particularly in the analyses of small molecules. In this regard, the use of NPs offers a number of attractive advantages including the elimination of noise generated by the organic matrix compounds. In addition, based on the proper choice of core metal material and surface modification, analyte molecules that are difficult to ionize using conventional organic matrices can be very efficiently ionized using NP matrices (Su and Tseng 2007).

This chapter describes the application of NPs to the emerging MS methodology known as imaging mass spectrometry (IMS). In IMS, which is an MS based molecular imaging technique, distributions of analyte molecules are visualized from the mass spectra obtained from thousands of data points collected from thin biological tissue sections as well as inorganic samples. Owing to the MS-based detection principle, IMS has now opened up a new frontier, particularly in the imaging of a variety of small organic molecules, such as endogenous metabolites, and the *in vivo* monitoring of administered drugs (within the animal/human body). The most frequently used ionization techniques are MALDI or SIMS (secondary ion mass spectrometry) (Yang et al. 2010) and recently, the application of NP-assisted laser desorption/ionization (nano-PALDI) has been initiated.

IMAGING MASS SPECTROMETRY (IMS)

Principles of MS and IMS

MS is an analytical technique that measures the mass-to-charge (m/z) ratio of charged atoms, molecules, and molecular clusters/fragments. MS is one of the fastest and most reliable methods for the high-accuracy determination of the mass of analyte particles. A typical mass spectrometer is composed of several functionally distinct components—an ion source, a mass analyzer, and a detector. Within the ion source, the analyte atoms or molecules are ionized by means of various ionization techniques; the analyzer then separates the ions on the basis of their m/z values and the detector outputs electrical signals in response to the reception of separated ions. The mass spectra can then be constructed from these signals (Fig. 1). The development of analyte ionization techniques, including recent progress in the soft-ionization technique, particularly MALDI and electrospray ionization (ESI), has opened up the possibility for the analysis of quite a wide range of molecules using a simple procedure, especially for large biopolymers such as proteins, nucleotides, and polysaccharides. MALDI, in particular, permits the analysis of solid phase samples and has therefore has been adopted in IMS from early studies mainly for biological tissue samples (Stoeckli et al. 2001). The details of this application are further discussed in a subsequent section. The MALDI process is triggered by a laser beam. An organic matrix compound is used to protect the large

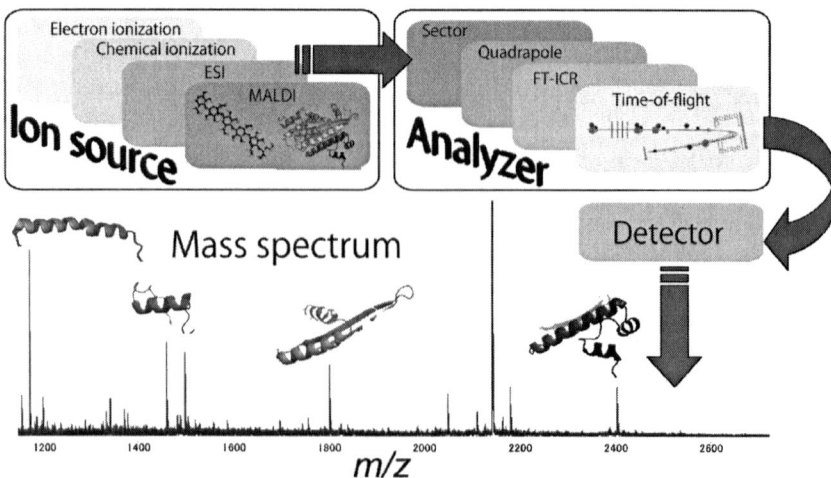

Fig. 1. **Principle schematic of mass spectrometer.** A typical mass spectrometer consists of three separate components—an ion source, a mass analyzer, and a detector. Numerous variations have been developed for each component based on different principles.

biomolecules from being destroyed by direct contact with the laser beam and to facilitate vaporization and ionization. The currently derived, general versatility of MALDI was established by a momentous effort directed at the development of novel matrix compounds suitable for the ionization of various molecules of interest. This effort has, therefore, facilitated the utilization of MALDI-IMS for the visualization of a variety of molecules (Sugiura and Setou 2010a).

Figure 2 illustrates the general workflow of MALDI-IMS. The basic technique involves the mounting of thin tissue slices on conductive glass slides and application of a suitable MALDI matrix to the tissue section. The slide is then inserted into a mass spectrometer and a focused laser beam is directed at predetermined positions of the tissue slice. The mass spectrometer records the spatial distribution of the molecular species (typically with a 10–200 μm scan pitch). Automated data collection takes 2–6 h, depending on the number of points assayed. Appropriate image processing software is required to import data from the mass spectrometer in order to allow visualization of the ion distribution images and comparison with the histological images of the sample. The unique advantages of MALDI imaging that facilitate the versatility of IMS as a molecular imaging technique are summarized as follows. (1) IMS does not

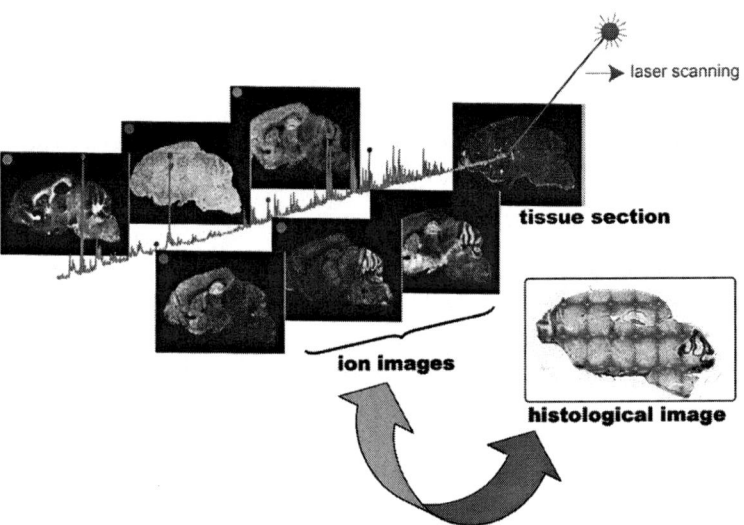

Fig. 2. Schematic representation of matrix-assisted laser desorption/ionization imaging mass spectrometry (MALDI-IMS) procedures. The MALDI laser scans through a set of preselected locations on the tissue (10–200 μm scan pitch), and the mass spectrometer records the spatial distribution of molecular species. Suitable image processing software can be used to import data from the mass spectrometer to allow visualization and comparison with the histological image of the sample.

require any specific chemical labels or probes. (2) IMS is a "non-targeted" imaging method. (3) The simultaneous imaging of multiple types of molecular species is possible. With the unique and powerful detection principle facilitated by MS, the MALDI-imaging mass spectrometry (IMS) can be used for the visualization of the distribution of a large number of biomolecules in cells and tissues, ranging from small metabolite molecules (Khatib-Shahidi et al. 2006) to much larger proteins (Chaurand et al. 2006).

IMS Applications to Areas of Health and Diseases

Current MALDI-IMS applications can be subdivided into two major categories, namely, IMS analysis of large proteins/peptides and IMS analysis of small organic molecules. In the early days, most of the reports on MALDI-IMS were geared at the detection and imaging of proteins or peptides. This area of application particularly targeted the detection of biomarker proteins, which localize specifically in lesions, by utilizing the capacity of IMS for the direct and simultaneous detection of multiple proteins in the tissues. Figure 3a shows one of the earliest medical application studies, which reports the detection and imaging of cancer-specific proteins in a mouse glioma model (Chaurand et al. 2004). As discussed previously, one of the significant advantages of IMS is that a number of cancer/normal specific protein distribution images can be acquired in a non-targeted manner from a single measurement. In addition, disease diagnosis by distinguishing between normal and cancerous biopsy specimens has been attempted by statistical evaluation of such multiple protein expression levels. Figure 3b shows that by applying hierarchical clustering analyses of IMS datasets obtained from human lung specimens, Yanagisawa et al. (2003) achieved successful classification of not only normal and cancerous biopsy samples but also of different cancer types, i.e., primary and non-primary, non-small-cell lung cancer. Such IMS-based molecular diagnosis studies continue to attract growing attention.

Another major medical application of IMS to the analysis of small molecules is conducted in the pharmaceutical field, e.g., for pharmacokinetic monitoring, pharmacotoxicology, and pharmacometabolomics (Fig. 4). For example, an important phase of drug discovery is determining how a drug-candidate compound is distributed and metabolized within the body. The application of IMS to the monitoring of drug delivery has also attracted much interest. Compared to traditional whole-body autoradiography (WBA) using radio labeled compounds, IMS offers many advantages in the determination of drug distribution. First, IMS allows for the simultaneous and discriminate monitoring not only of the intact drug molecules, but also of their metabolites (Khatib-Shahidi et al. 2006), whereas WBA cannot

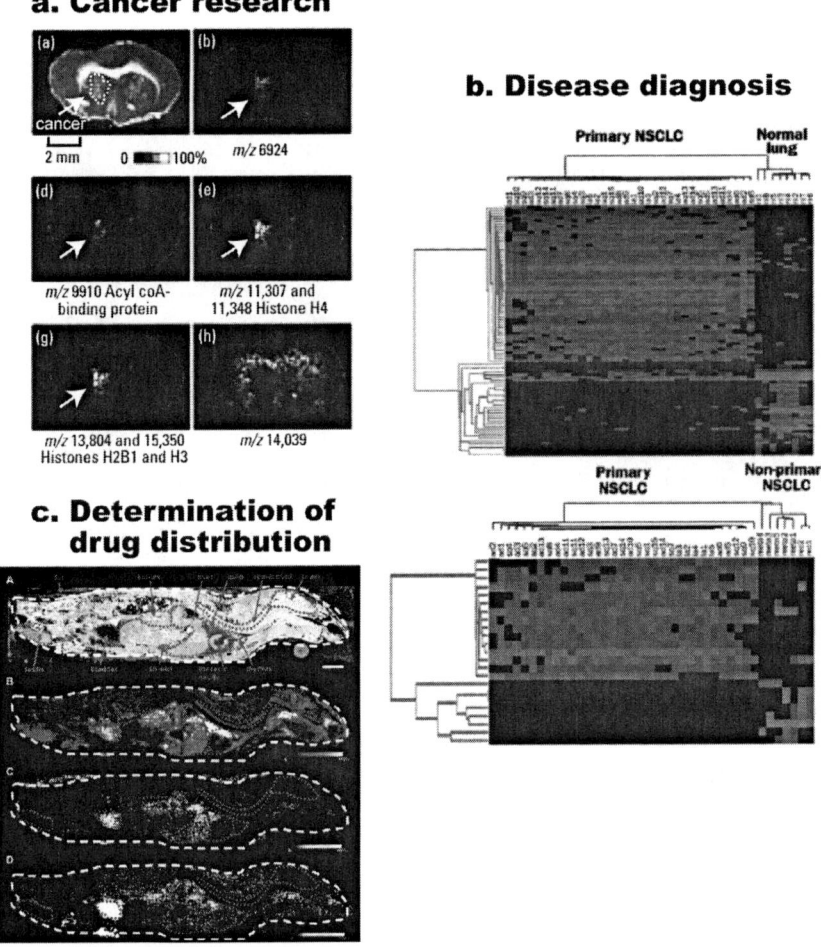

Fig. 3. **Representative imaging mass spectrometry (IMS) applications in areas of health and diseases.** Shown are representative examples of IMS application to brain cancer research using the mouse glioma model (a); human disease diagnosis of healthy and different types of lung cancers (b); and determination of pharmacokinetics in the whole animal body (c). NSCLC = non-small cell lung cancer. Reprinted from Chaurand et al. 2004 (a); Yanagisawa et al. 2003 (b); and Khatib-Shahidi et al. 2006 (c) with permission of ACS Publications (a and b) and Elsevier, Ltd. (c).

distinguish these molecules. Thus, IMS can be used to determine whether or not medicinally intact drugs have reached the target organs. Secondly, IMS can be used to visualize the distribution of drugs at a lower cost and in a much shorter time than with detection using isotopes. Figure 3c shows the detection of drugs that have been delivered orally to mice. In this study, the distribution of the antipsychotic drug olanzapine and its metabolites in the sagittal section of an intact (whole) rat was successfully investigated 6 h after administration. This study clearly showed the distinct distribution of intact drugs and their metabolites; the intact drug reached the target organ (the brain), whereas its metabolites were localized in the bladder.

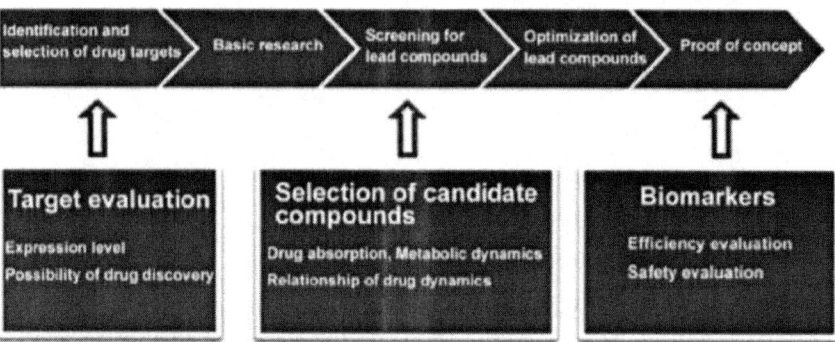

Fig. 4. Processes of new drug development and areas of application of imaging mass spectrometry (IMS) in each step. New drug development can be divided into several processes. IMS can be widely applied to many of the steps along the way through the ability to visualize various molecules.

APPLICATION OF NANOPARTICLES TO IMS

Utilization of a Variety of NPs in Basic MS Research

As described previously, MALDI had its genesis as a soft-ionization technique employing NPs (Tanaka et al. 1988). Since then, rapid technical progress over the ensuing couple of decades has made NP handling much easier, and studies on the utilization of NPs as ionization assisting reagents have been intensively reported. The main motivation behind the use of NP-assisted ionization is to overcome the limitations of MALDI, particularly for the detection of molecules in the low mass range, without interference from matrix-derived ions. The low m/z region of a MALDI spectrum contains a large population of ions from biological metabolites as well as matrix-related adduct clusters and fragments, which are dominantly observed in the MALDI mass spectrum. This high density of ions increases the risk of sharing of the same mass window by matrix ions and analyte molecules. The utilization of NPs as a matrix is one of the effective methods for prevention of this problem, because the NPs produce few background ions. Figure 5 shows an example in which reserpine, an indole alkaloid antipsychotic and antihypertensive drug, was analyzed using both fNP and α-cyano-4-hydroxycinnamic acid (CHCA) as respective matrices. The spectra presented clearly demonstrate the complete elimination of background ions by use of NPs (Sahashi et al. 2010). Furthermore, the

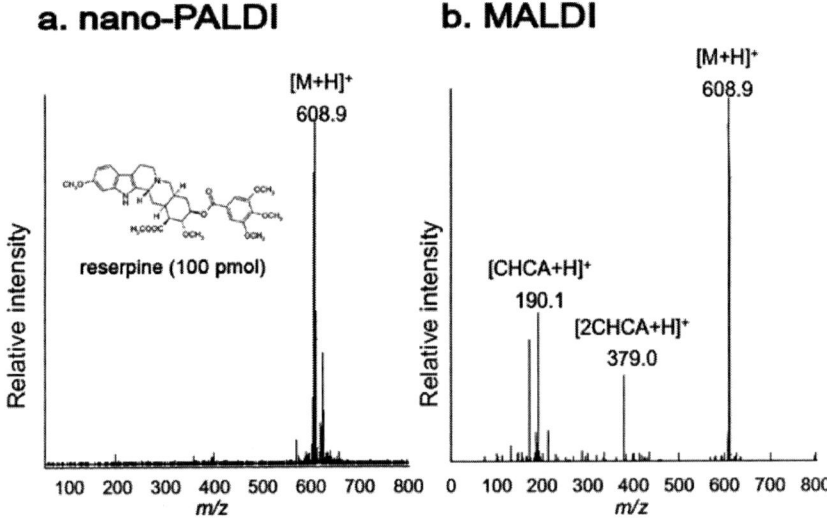

Fig. 5. Comparison between mass spectra obtained by nanoparticle-assisted laser desorption/ionization (nano-PALDI) and matrix-assisted laser desorption/ionization (MALDI). Mass spectra of reserpine, an indole alkaloid antipsychotic and antihypertensive drug, obtained by both nano-PALDI and MALDI-mass spectrometry in which functional nanoparticle (fNP) and α-cyano-4-hydroxycinnamic acid (CHCA) were used as matrices, respectively. This example clearly demonstrates the complete elimination of background ions by use of NPs. Reprinted from Sahashi et al. 2010 with the permission of Elsevier, Ltd.

use of NPs has improved the homogeneity of analyte distribution in the samples on the target plate, which could increase sample-to-sample reproducibility. In comparison, the crystallization process in MALDI inevitably causes artificial localization of analyte molecules within the sample; therefore, researchers have had to locate a *hot spot* where the analytes were concentrated by moving the laser irradiation spots. This is a time-consuming process and one of the major reasons for the lowered quantitative performance of MALDI.

In addition to these general advantages of NP-assisted ionization, a variety of characteristic NPs have been used for analyte-specific MS research (as summarized in Table 1). For example, graphite NPs have been utilized for the detection of relatively large molecules such as peptides and proteins from liquid solutions, with sensitivities in the pico- to nanomole range (Sunner et al. 1995). Bifunctional magnetic iron oxide particles immobilized on silane have also been used for the analysis of proteins and peptides (Chen and Chen 2006). Recently, citrate capped gold nanoparticles (AuNPs) have been utilized for the analyses of important signaling biomolecules—progesterone, testosterone, and cortisol (Wu et al. 2009), which are difficult to detect using conventional organic matrices. Silver

Table 1. Representative nanoparticle assisted laser desorption/ionization technique and its applications. Listed are representative mass spectrometry studies in which various nanoparticles were utilized as ionization enhancing reagents; applied to the measurement of biological molecules.

Name of NP	Core metal	Diameter	Analyte	Reference	Imaging
AgNPs	Ag	3.84±0.45nm	Fatty acids	Hayasaka et al. 2010	yes
		34±3nm	Estrogen	Chiu et al. 2008	yes yes
MnNPs	MnO_2, Mn_2O_3	5.4±0.2nm	Ginsenosides	Sahashi et al. 2010	yes
fNP	Fe_2O_3	3.7nm	Phospholipids	Moritake et al. 2009	yes
			Sulfatide	Ageta et al. 2009	yes
			Lipids, peptides	Taira et al. 2008	yes
AuNPs	Au	4.3±0.7nm	Glycosphingolipids	Goto-Inoue et al. 2010	yes
		13.2±1.2nm	Progesterone, Cortisol, Testesteron	Wu et al. 2009	no
TiO_2 NPs	TiO_2	<0.05μm	Trypsinogen	Watanabe et al. 2009	no

Ageta et al. 2009. Med. Mol. Morphol. 42: 16–23.
Chiu et al. 2008. J. Am. Soc. Mass. Spectrom. 19: 1343–1346.
Goto-Inoue et al. 2010. J. Am. Soc. Mass. Spectrom. 21: 1940–1943.
Hayasaka et al. 2010. J. Am. Soc. Mass. Spectrom. 21: 1446–1454.
Moritake et al. 2009. J. Nanosci. Nanotechnol. 9: 169–176.
Sahashi et al. 2010. Food Chem. 123: 865–871.
Taira, S. et al. 2008. Anal. Chem. 80: 4761–4766.
Watanabe et al. 2009. J. Mass Spectrom. 44: 1443–1451.
Wu et al. 2009. J. Am. Soc. Mass. Spectrom. 20: 875–882.

nanoparticles (AgNPs) capped with several types of functional groups have been used for the detection of sulfur drugs and biothiols (Shrivas and Wu 2008). It has also been demonstrated that AgNPs can selectively ionize cholesterol, phosphatidylcholine, and carotenoids (Sherrod et al. 2008). Moreover, AgNPs have also been used for the determination of small molecular hormones, such as estrone, estradiol, and estriol (Chiu et al. 2008). In another study, titanium dioxide (TiO_2) NPs modified with urea have been shown to increase the ionization efficiency of analytes owing to the photocatalytic effect of TiO_2, which was easily activated by UV irradiation. Furthermore, the modified TiO_2 NPs could also be applied to the detection of large proteins of sizes greater than 20 kDa, such as trypsinogen (Watanabe et al. 2009).

Nano-PALDI for IMS

The usefulness of MALDI-based IMS was briefly reviewed in the preceding sections; however, utilization of the NP-assisted ionization technique is an attractive alternative to MALDI. In this context, utilization of NPs in IMS is expected to be a quite useful tool, especially for the imaging of small molecule distribution, and therefore, this state-of-the-art imaging technology, nano-PALDI-based IMS, and its future perspectives are discussed in the following sections.

Current Limitations of MALDI-IMS and Nano-PALDI-IMS as a Solution

Despite the promising capability of MALDI-IMS, this technique still has several critical limitations. An important challenge is the improvement in spatial resolution toward ion imaging within cellular organelles, which requires resolution at the sub-micrometer level. However, when MALDI is employed as an ionization technique for IMS, the nature of the MALDI process requires the formation of analyte-matrix co-crystals on the tissue section. The typical size of these co-crystals is >50 μm; they function to protect the analyte molecules from direct laser irradiation, i.e., act as a "cushion" and eventually enhance the soft ionization of biomolecules. Unfortunately, this crystal size effectively limits the spatial resolution of IMS to as large as the crystal size. In this regard, imaging with SIMS, a matrix-free ionization technique, has already achieved submicron spatial resolution. In SIMS, the use of a tightly focused ion beam for ionization offers a resolution at several tens of nanometers and has been successfully used to visualize sub-cellular structures in biological samples (Monroe et al. 2005; Ostrowski et al. 2004). However, SIMS is a much "harder" ionization method than MALDI, and consequently, it is not the best choice for intact ionization of various biomolecules because heavier molecules (<1000 Da) and molecules with easily fragmented groups cannot be ionized in their intact from using SIMS (Kraft et al. 2006). In order to overcome these issues, the current authors have reported an NP-assisted laser desorption/ionization (nano-PALDI)-based IMS technique, in which the organic matrix is replaced with NPs, and therefore, the matrix crystallization process is eliminated (Taira et al. 2008). Figure 6 presents a simple illustration of the relationship among MALDI, SIMS, and nano-PALDI.

This novel nano-PALDI-IMS technique affords high-resolution imaging of complex biological specimens (Taira et al. 2008). For this purpose, functional nanoparticles (fNPs) with a diameter of 3.7 ± 0.1 nm were also developed. The inset of Fig. 7 shows the structure of the developed fNPs. Surface-positioned silicon dioxide (SiO_2) groups could be used for attaching various chemical groups, and in the study, hydroxyl

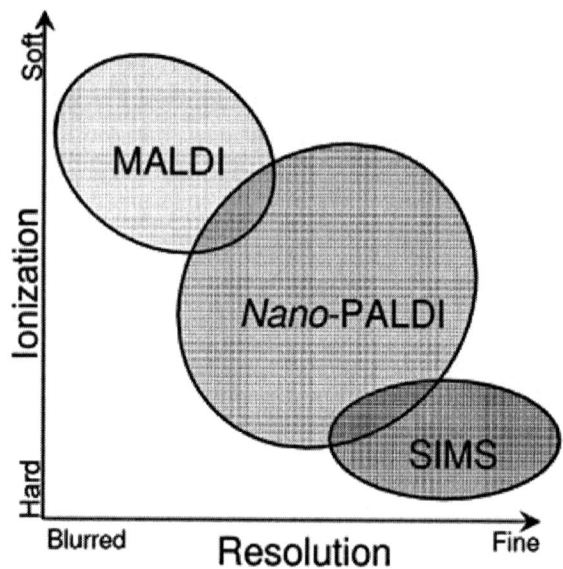

Fig. 6. Applicable areas of matrix-assisted laser desorption/ionization (MALDI), nanoparticle-assisted laser desorption/ionization (nano-PALDI) and secondary ion mass spectrometry (SIMS) techniques in IMS. Three different IMS techniques are compared based on the resolution and the severity of ionization.

and amino groups were linked, resulting in improved ionization efficiency for the small molecules. The improved ionization efficiency presumably results from the capture of analyte molecules close to the fNP surface, which facilitates efficient transfer of laser energy from the NP to the analyte molecules (Fig. 8h). Figure 7 shows representative mass spectra of mouse brain sections obtained using fNP and DHB as respective matrices. Comparison of the spectra shows that in the mass range of $700 < m/z < 900$, signals derived from phospholipids and glycolipids were detected at almost the same intensity using either technique. On the other hand, in the lower mass range, i.e., $100 < m/z < 500$, a larger number of mass peaks with higher intensities were detected using fNP as matrix, whereas when DHB was used as a matrix, most of the intense peaks detected in the same mass region were derived from DHB-originated ions. This example clearly demonstrates the highly effective nature of nano-PALDI for small molecule imaging.

Another important advantage: spraying fNP on the tissue surface did not alter the optical image of the biological tissue surface (Fig. 8). In contrast, when the mouse brain sections were sprayed with a DHB solution, non-homogeneous DHB crystals were formed on the section, which obscured the optical view of the sample surface (Fig. 8a–b). This obscurity resulting

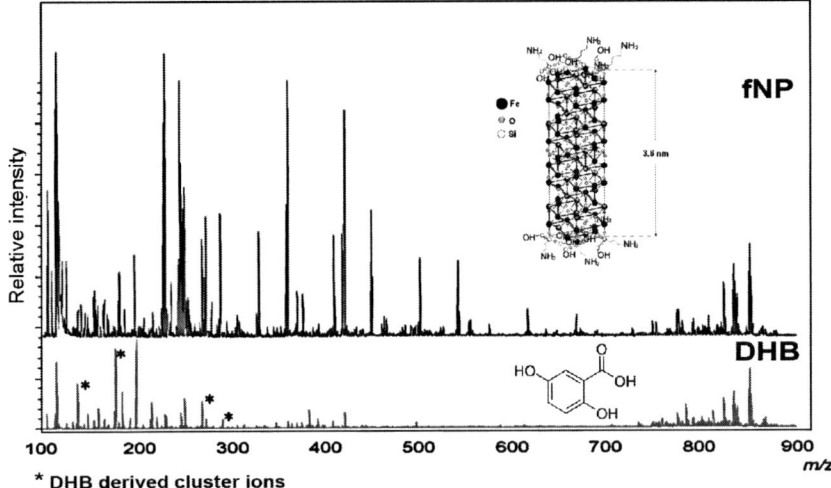

Fig. 7. Representative mass spectra of mouse brain section obtained using functional nanoparticle (fNP) and DHB as respective matrices. Either fNP or 2,5-dihydroxybenzoic acid (DHB) was applied as a matrix on a mouse brain section, and mass spectra were obtained from each applied spot. These mass spectra clearly demonstrate the elimination of DHB-derived background signals as well as increased detection sensitivity for small molecules within the $100 < m/z < 600$ range using fNP as a matrix. Reprinted from Sahashi et al. 2010 with the permission of Elsevier, Ltd.

from crystal formation with DHB makes it difficult to predefine the tissue region of interest before conducting MS or IMS measurements. Owing to the quite large number of MS measurements performed during the IMS experiment, which is equal to the number of pixels of the resulting image, there is a practical requirement that the measurement area be limited (generally, ten thousand MS measurements per single IMS analysis is the upper limit mainly because of the huge size of the data set). In addition, SEM observation of the DHB coated tissue surface revealed inhomogeneous needle like crystals having typical lengths of > 50 μm (Fig. 8c), which limits the spatial resolution of MALDI-IMS, as described above. In contrast, because of the extremely small particle size of fNPs, spraying these fNPs onto the tissue surface did not alter the optical image of the tissue structure (Fig. 7d–e), thereby allowing the researcher to perform MS and IMS measurements with concomitant optical observation of the tissue structure (using a CCD camera with which MALDI-MS instruments are generally equipped). The SEM images demonstrate that the fNPs which were sprayed onto the tissue surface were distributed in a manner similar to the as-synthesized particles (Fig. 8f–g).

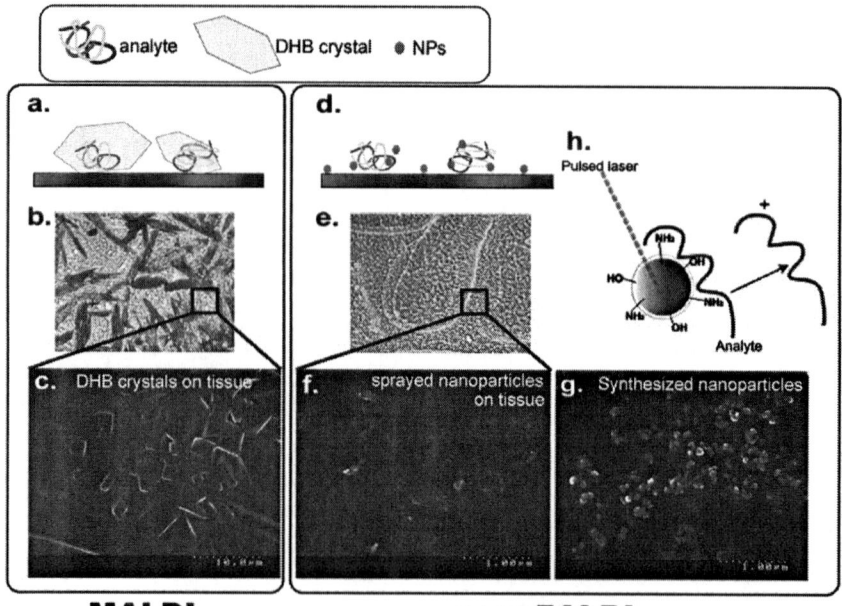

Fig. 8. Principle of nanoparticle-assisted laser desorption/ionization (nano-PALDI) and matrix-assisted laser desorption/ionization based imaging mass spectrometry (MALDI-IMS). Shown are schematic, light microscopic, and scanning electron microscopy images of 2,5-dihydroxybenzoic acid (DHB)/nanoparticles (NPs) applied to biological tissue section. MALDI requires formation of analyte-matrix co-crystals with typical sizes of <50 μm on the tissue section, which limits the spatial resolution of MALDI-IMS (left), whereas in nano-PALDI, the use of NPs as the ionization enhancing reagent eliminates formation of such crystals (right) and enables clear observation of tissue surface even during mass spectrometry measurement. Modified from Taira et al. 2008 with the permission of ACS Publications.

The considerable advantage of nano-PALDI-based IMS in the low m/z region is clearly demonstrated in Fig. 9, which shows the results of the feasibility study using IMS measurements performed with both nano-PALDI and MALDI. Tissue samples were obtained from rat cerebella, and NP fluid was sprayed on a thin tissue section of the cerebellum (Fig. 9b, d, f), whereas successive tissue sections were treated by application of DHB (Fig. 9c, e, g). Each IMS measurement was performed at spatial resolution of 15 μm, using a MALDI-TOF/TOF-type instrument. Consistent with the aforementioned features of fNPs, the comparative analysis showed that the use of fNP produced a unique ion distribution image that could not be detected when DHB was used (Fig. 9e and g), and a much finer ion distribution image (Fig. 9d and f), without background noise. On the other hand, ion images with DHB showed crystal-shaped ion localization patterns (especially in Fig. 9g), suggesting that analyte molecules could

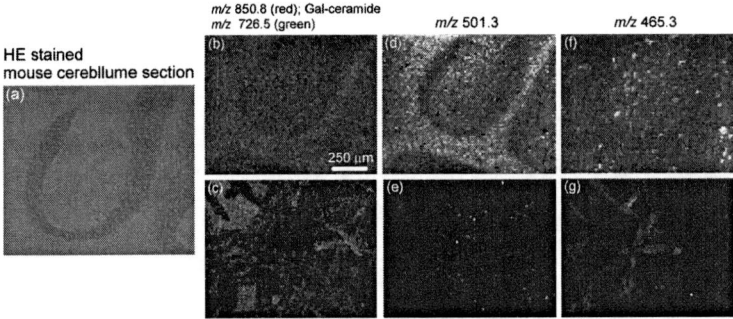

Fig. 9. Nanoparticle-assisted laser desorption/ionization based imaging mass spectrometry (nano-PALDI-IMS) of low molecular weight compounds showing improved ion distribution image quality. Optical images of rat cerebellum tissue before spraying with nanoparticles (NPs) (a) and 2,5-dihydroxybenzoic acid (DHB) solution, and ion images obtained with NPs (b, d, f) and DHB (c, e, g) are shown. Visualized ions were identified as galactosylceramide (C24h:0) and phosphotidylcholine (PC) (diacyl-34:2) by tandem mass spectrometry on both DHB and NP coated sections. Reprinted from Taira et al. 2008 with the permission of ACS Publications.

Color image of this figure appears in the color plate section at the end of the book.

only be ionized from the analyte-matrix co-crystals. Even though the results are only representative, this example clearly demonstrates that the organic matrix crystals severely limit the IMS spatial resolution.

Analysis of Molecular Distribution of Sulfatide

As frequently described, lipids constitute half of the dry weight in the brain, and play important roles especially as fundamental building components of cell structures and as signaling molecules with strong bioactivity (Bosio et al. 1998). Owing to the technical difficulty presented by the insolubility of lipids in aqueous media, the study of lipids has been mainly performed using traditional biochemical techniques. The emergence of IMS as a tool for the imaging of lipids has had significant impact, because there was previously no established technique for two-dimensional mapping of lipids, whereas transcripts can be visualized with oligonucleotide probes using *in situ* hybridization and proteins can be visualized using immunohistochemistry with appropriate antibodies. In this context, lipid imaging by IMS, particularly utilizing nano-PALDI based IMS should prove critical to the interpretation of the role of lipids in brain research.

Nano-PALDI-IMS has also been applied to the mapping of sulfatide distribution. Sulfatides are important lipid components of the myelin sheath. A direct correlation between sulfatide deficiency and neurological disorders, such as Alzheimer's disease (Han et al. 2002) has been reported. Furthermore, one decomposition pathway of sulfatides is catalyzed by arylsulfatase A (ASA) and the functional deficiency of ASA results in metachromatic leukodystrophy (MLD), which causes the accumulation of sulfatide in lysozymal storage deposits, and eventually, demyelination in the peripheral and central nervous systems (PNS and CNS) (Krivit 2004). In addition, structural variations of sulfatide arise from hydroxylation of the fatty acid moiety as shown by the arrow in Fig. 10. The hydroxylation is catalyzed by fatty acid 2-hydroxylase (FA2H), which also causes leukodystrophy with spastic paraparesis and dystonia (Kruer et al. 2010).

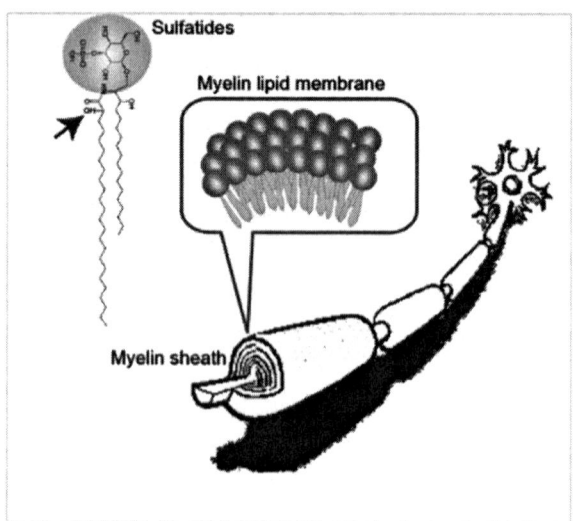

Fig. 10. Schematic representation of myelin sheath and one of its major lipid components, sulfatides. Nanoparticle-assisted laser desorption/ionization imaging mass spectrometry (nano-PALDI-IMS) was applied to the mapping of sulfatide distribution. Sulfatides are important lipid components of the myelin sheath. The black arrow indicates the hydroxylation site on the fatty acid moiety that causes structural variation.

The distribution pattern of sulfatide in the brain has been assessed by using the anti-sulfatide antibody to track the distribution of sulfatide in the CNS and PNS of rodent brains (Pernber et al. 2002). However, immunostaining with this antibody did not allow for discrimination of the fatty acid moiety of this lipid specie or of the presence/absence of the abovementioned hydroxylation. In comparison, based on the MS detection principle, distinct images of these species were obtained using IMS (Ageta et al. 2009). Figure 11 shows different regions of the dentate gyrus of rat hippocampus, such as the granular cell layer (GCL), inner molecular layer (IML), and middle molecular layer (MML) (left column), as well as the intensity of the different mass peaks in these regions (right column). It was found that the intensity of the peaks with m/z of 906.3,

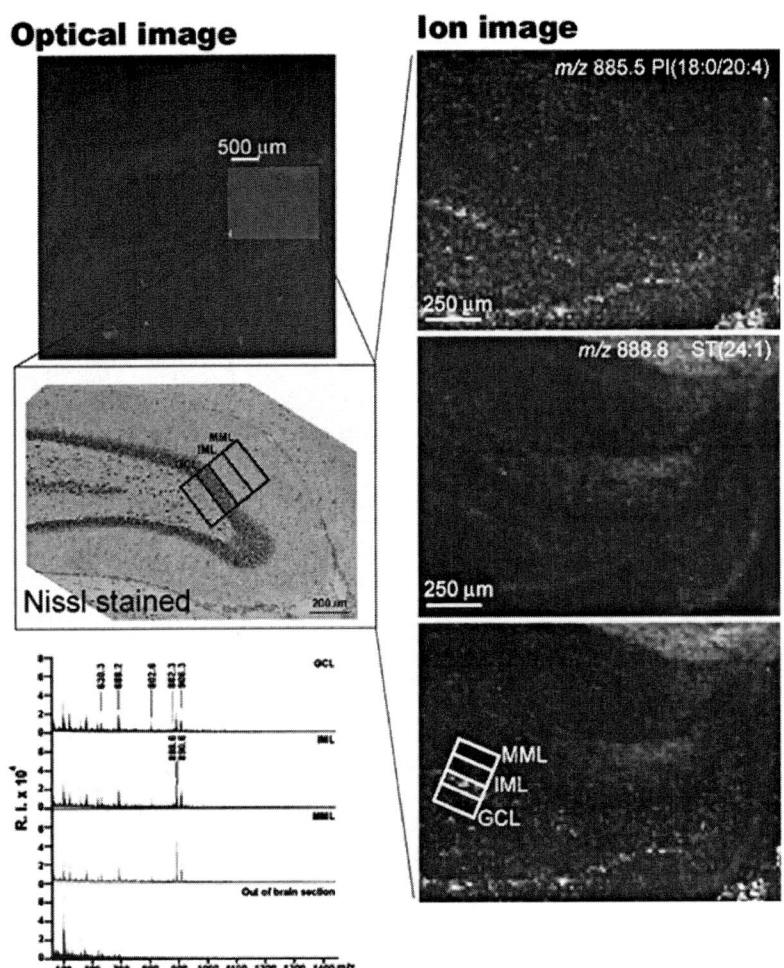

Fig. 11. Nanoparticle-assisted laser desorption/ionization improves spatial resolution in imaging mass spectrometry (IMS). Optical image of rat hippocampus indicating measurement area for nanoparticle-based IMS is shown along with Nissl-stained section indicating fine layer structure of rat hippocampus (left panels). Ion images, which reveal hippocampal layer specific distribution of phosphatidylinositol (18:0/20:4) at m/z 885.5 and sulfatide (24:1) at m/z 888.8, are presented (right panels). GCL = granular cell layer, IML = inner molecular layers, and MML = middle molecular layers. Reprinted from Ageta et al. 2009 (left column) and Sugiura and Setou 2010b (right column) with the permission of Springer.

Color image of this figure appears in the color plate section at the end of the book.

904.6, 890.6, and 888.6 was higher in the MML region than in the GCL and IML regions, whereas the mass peaks with m/z of 862.3, 802.6, 688.2, and 630.3 were detected in the GCL, IML, and MML regions only, indicating their biological origin. Therefore, this technique can potentially be used to explore physiological processes for better understanding of diagnostics and the pathology of neurological disorders, such as leukodystrophy and Alzheimer's disease (Ageta et al. 2009).

Key Facts

- A key fact about MALDI-imaging mass spectrometry (IMS) is that it is an MS-based molecular imaging technique with the following unique advantages. (1) MALDI-IMS can be used for the visualization of the distribution of large numbers of biomolecules in cells and tissues, ranging from small metabolite molecules to much larger proteins. (2) MALDI-IMS does not require any specific chemical labels or probes. (3) MALDI-IMS is a "non-targeted" imaging method. (4) MALDI-IMS enables the simultaneous imaging of multiple types of molecular species.
- A key fact about nanoparticle (NP)-assisted laser desorption/ionization (nano-PALDI)-based IMS is that it is an IMS technique, in which the organic matrix is replaced with NPs such as gold, silver, and titanium. Recently, researchers have begun using NPs for MS research owing to the progress in the development of NP technology, which has made various kinds of NPs available. In MS research, the use of NPs has been studied particularly in the analyses of small molecules, because of the following advantages. (1) The use of nano-PALDI may eliminate background noise generated by organic matrix compounds. (2) Judicious choice of NPs allows for the efficient analysis of analyte molecules that are otherwise difficult to ionize using conventional organic matrices.

Summary

- Imaging mass spectrometry (IMS) enables visualization of the distribution of various biomolecules in biological tissue sections.
- IMS has several unique advantages—(1) It does not require any specific chemical labels or probes. (2) It is a "non-targeted" imaging method. (3) The simultaneous imaging of multiple types of molecular species is available.
- MALDI-IMS in particular, which is a soft ionization method using an organic "matrix," can be used for a large number of biomolecules ranging from small metabolite molecules to much larger proteins.

- Although MALDI-based IMS has promising capacity for the imaging of small metabolites, this technique has a critical problem in spatial resolution derived from matrix crystallization.
- In nanoparticle (NP)-assisted laser desorption/ionization (nano-PALDI)-based IMS, the organic matrix is replaced with NPs, and therefore, the matrix crystallization process is eliminated; therefore, it can be used to overcome the resolution problems of MALDI-IMS.
- Recent development of NPs with different core metals, surface modifications, and particle diameters has expanded the measurable range of analytes as well as the application of the analyses to physiological processes and the diagnosis and pathophysiology of complex biological process, especially in the brain.

Abbreviations

ASA	:	Arylsulfatase A
AuNP	:	Gold Nanoparticle
AgNP	:	Silver Nanoparticle
CCD	:	Charge-Coupled Device
CHCA	:	α-Cyano-4-Hydroxycinnamic Acid
CNS	:	Central Nervous System
DHB	:	Dihydroxy Benzoic Acid
DNA	:	Deoxyribonucleic Acid
ESI	:	Electrospray Ionization
FA2H	:	Fatty Acid 2-Hydroxylase
fNP	:	Functional Nanoparticle
GCL	:	Granular Cell Layer
HE	:	Hematoxilin-Eosin
IML	:	Inner Molecular Layer
IMS	:	Imaging Mass Spectrometry
ITO	:	Indium Tin Oxide
MALDI	:	Matrix-Assisted Laser Desorption/Ionization
MLD	:	Metachromatic Leukodystrophy
MML	:	Middle Molecular Layer
MS	:	Mass Spectrometry
m/z	:	Mass-to-Charge Ratio
nano-PALDI	:	Nanoparticle-Assisted Laser Desorption/Ionization
NP	:	Nanoparticle
NSCLC	:	Non-Small Cell Lung Cancer
PC	:	Phosphotidylcholine
PNS	:	Peripheral Nervous System
SEM	:	Scanning Electron Microscopy
SIMS	:	Secondary Ion Mass Spectrometry

SiO$_2$:   Silicon dioxide
siRNA   :   Short-Interfering Ribonucleic Acid
TiO$_2$:   Titanium Dioxide
TOF   :   Time-of-Flight
WBA   :   Whole-Body Autoradiography

Key Terms

- MS (Mass spectrometry): Analytical technique that measures the mass (m)-to-charge (z) ratio of charged atoms, molecules, and molecular clusters/fragments.
- IMS (Imaging MS): Molecular imaging technique based on MS.
- MALDI (Matrix-assisted laser desorption/ionization): A soft ionization method used for MS using an analyte ionization-enhancing reagent, called a matrix, which allows the analysis of biomolecules and large organic molecules.
- NP (nanoparticle): A particle whose diameter is 100~1000 nm.
- nano-PALDI (NP-assisted laser desorption/ionization): An ionization method used for MS using NP as an analyte ionization-enhancing reagent.

References

Ageta, H., S. Asai, Y. Sugiura, N. Goto-Inoue, N. Zaima and M. Setou. 2009. Layer-specific sulfatide localization in rat hippocampus middle molecular layer is revealed by nanoparticle-assisted laser desorption/ionization imaging mass spectrometry. Med. Mol. Morphol. 42: 16–23.

Bosio, A., E. Binczek, W.F. Haupt and W. Stoffel. 1998. Composition and biophysical properties of myelin lipid define the neurological defects in galactocerebroside- and sulfatide-deficient mice. J. Neurochem. 70: 308–315.

Chaurand, P., J.L. Norris, D.S. Cornett, J.A. Mobley and R.M. Caprioli. 2006. New developments in profiling and imaging of proteins from tissue sections by MALDI mass spectrometry. J. Proteome Res. 5: 2889–2900.

Chaurand, P., S.A. Schwartz and R.M. Caprioli. 2004. Profiling and Imaging Proteins in Tissue Sections by MS. Anal. Chem. 76: 86–93.

Chen, W.Y. and Y.C. Chen. 2006. Affinity-based mass spectrometry using magnetic iron oxide particles as the matrix and concentrating probes for SALDI MS analysis of peptides and proteins. Anal. Bioanal. Chem. 386: 699–704.

Chiu, T.C., L.C. Chang, C.K. Chiang and H.T. Chang. 2008. Determining estrogens using surface-assisted laser desorption/ionization mass spectrometry with silver nanoparticles as the matrix. J. Am. Soc. Mass. Spectrom. 19: 1343–1346.

Han, X., D.M. Holtzman, D.W. McKeel Jr., J. Kelley and J.C. Morris. 2002. Substantial sulfatide deficiency and ceramide elevation in very early Alzheimer's disease: potential role in disease pathogenesis. J. Neurochem. 82: 809–818.

Karas, M. and F. Hillenkamp. 1988. Laser desorption ionization of proteins with molecular masses exceeding 10,000 daltons. Anal. Chem. 60: 2299–2301.

Khatib-Shahidi, S., M. Andersson, J.L. Herman, T.A. Gillespie and R.M. Caprioli. 2006. Direct molecular analysis of whole-body animal tissue sections by imaging MALDI mass spectrometry. Anal. Chem. 78: 6448–6456.

Kraft, M.L., P.K. Weber, M.L. Longo, I.D. Hutcheon and S.G. Boxer. 2006. Phase separation of lipid membranes analyzed with high-resolution secondary ion mass spectrometry. Science 313: 1948–1951.

Krivit, W. 2004. Allogeneic stem cell transplantation for the treatment of lysosomal and peroxisomal metabolic diseases. Springer Semin. Immunopathol. 26: 119–132.

Kruer, M.C., C. Paisan-Ruiz, N. Boddaert, M.Y. Yoon, H. Hama, A. Gregory, A. Malandrini, R.L. Woltjer, A. Munnich, S. Gobin, et al. 2010. Defective FA2H leads to a novel form of neurodegeneration with brain iron accumulation (NBIA). Ann. Neurol. 68: 611–618.

Monroe, E.B., J.C. Jurchen, J. Lee, S.S. Rubakhin and J.V. Sweedler. 2005. Vitamin E imaging and localization in the neuronal membrane. J. Am. Chem. Soc. 127: 12152–12153.

Ostrowski, S.G., C.T. Van Bell, N. Winograd and A.G. Ewing. 2004. Mass spectrometric imaging of highly curved membranes during Tetrahymena mating. Science 305: 71–73.

Pernber, Z., M. Molander-Melin, C.H. Berthold, E. Hansson and P. Fredman. 2002. Expression of the myelin and oligodendrocyte progenitor marker sulfatide in neurons and astrocytes of adult rat brain. J. Neurosci. Res. 69: 86–93.

Sahashi, Y., I. Osaka and S. Taira. 2010. Nutrition analysis by nanoparticle-assisted laser desorption/ionisation mass spectrometry. Food Chem. 123: 865–871.

Sherrod, S.D., A.J. Diaz, W.K. Russell, P.S. Cremer and D.H. Russell. 2008. Silver nanoparticles as selective ionization probes for analysis of olefins by mass spectrometry. Anal. Chem. 80: 6796–6799.

Shrivas, K. and H.F. Wu. 2008. Applications of silver nanoparticles capped with different functional groups as the matrix and affinity probes in surface-assisted laser desorption/ionization time-of-flight and atmospheric pressure matrix-assisted laser desorption/ionization ion trap mass spectrometry for rapid analysis of sulfur drugs and biothiols in human urine. Rapid Commun. Mass Spectrom. 22: 2863–2872.

Stoeckli, M., P. Chaurand, D.E. Hallahan and R.M. Caprioli. 2001. Imaging mass spectrometry: a new technology for the analysis of protein expression in mammalian tissues. Nat. Med. 7: 493–496.

Su, C.L. and W.L. Tseng. 2007. Gold nanoparticles as assisted matrix for determining neutral small carbohydrates through laser desorption/ionization time-of-flight mass spectrometry. Anal. Chem. 79: 1626–1633.

Sugiura, Y. and M. Setou. 2010a. Imaging mass spectrometry for visualization of drug and endogenous metabolite distribution: toward in situ pharmacometabolomes. J Neuroimmune Pharmacol. 5: 31–43.

Sugiura, Y. and M. Setou. 2010b. Matrix-Assisted Laser Desorption/Ionization and Nanoparticle-Based Imaging Mass Spectrometry for Small Metabolites: A Practical Protocol. pp. 173–195. In: S.S.S. Rubakhin and J.V.V. Sweedler. [eds.] Mass Spectrometry Imaging: Principles and Protocols, Methods Mol. Biol. 656. Springer. New York.

Sunner, J., E. Dratz and Y.C. Chen. 1995. Graphite surface-assisted laser desorption/ionization time-of-flight mass spectrometry of peptides and proteins from liquid solutions. Anal. Chem. 67: 4335–4342.

Taira, S., S. Moritake, T. Hatanaka, Y. Ichyanagi and M. Setou. 2009. Functionalized Magnetic Nanoparticles as an *In Vivo* Delivery System pp. 571–587. *In:* J.W. Lee and R.S. Foote. [eds.] Micro and Nano Technologies in Bioanalysis: Methods and Protocols, Methods Mol. Biol. 544. Humana press. New York.

Taira, S., Y. Sugiura, S. Moritake, S. Shimma, Y. Ichiyanagi and M. Setou. 2008. Nanoparticle-assisted laser desorption/ionization based mass imaging with cellular resolution. Anal. Chem. 80: 4761–4766.

Tanaka, K., H. Waki, Y. Ido, S. Akita, Y. Yoshida, T. Yoshida and T. Matsuo. 1988. Protein and polymer analyses up to m/z 100 000 by laser ionization time-of-flight mass spectrometry. Rapid Commun. Mass Spectrom. 2: 151–153.

Watanabe, T., K. Okumura, H. Kawasaki and R. Arakawa. 2009. Effect of urea surface modification and photocatalytic cleaning on surface-assisted laser desorption ionization mass spectrometry with amorphous TiO_2 nanoparticles. J. Mass Spectrom. 44: 1443–1451.

Wu, H.P., C.J. Yu, C.Y. Lin, Y.H. Lin and W.L. Tseng. 2009. Gold nanoparticles as assisted matrices for the detection of biomolecules in a high-salt solution through laser desorption/ionization mass spectrometry. J. Am. Soc. Mass. Spectrom. 20: 875–882.

Yanagisawa, K., Y. Shyr, B.J. Xu, P.P. Massion, P.H. Larsen, B.C. White, J.R. Roberts, M. Edgerton, A. Gonzalez and S. Nadaf. 2003. Proteomic patterns of tumour subsets in non-small-cell lung cancer. Lancet 362: 433–439.

Yang, H.J., I. Ishizaki, N. Sanada, N. Zaima, Y. Sugiura, I. Yao, K. Ikegami and M. Setou. 2010. Detection of characteristic distributions of phospholipid head groups and fatty acids on neurite surface by time-of-flight secondary ion mass spectrometry. Med. Mol. Morphol. 43: 158–164.

7

Nanomaterials and Ion Channels: Observed Effects and Possible Mechanisms

Lorin M. Jakubek and *Robert H. Hurt*

ABSTRACT

Voltage-gated ion channels are essential to all functions of the nervous system, from maintaining membrane potential and synaptic transmission, to endocrine functions. Recent research has shown altered channel function and consequently altered nervous tissue response with the addition of certain nanomaterials. This chapter provides an overview of voltage-gated sodium, potassium and calcium ion channels, their structure, function and a review of current literature regarding channel interactions with carbon nanomaterials, nanometals and quantum dots. Based on this literature we postulate potential mechanisms of interaction, namely: release of metal ions, adsorption of target ions onto nanomaterial surfaces, direct nanomaterial interaction and oxidative stress. There are many health issues such as migraines, vertigo and arrhythmias that arise from disregulation of ion

---

[1]Box D, Brown University, Providence, RI 02912 USA.
[a]E-mail: Lorin_Jakubek@Brown.edu
[b]E-mail: Robert_Hurt@Brown.edu

List of abbreviations after the text.

channel homeostasis. Additionally, pharmaceuticals target ion channels to treat chronic pain, provide anesthetics, and control blood pressure. Given the alterations of ion-channel homeostasis observed for the small number of nanomaterials tested to date, we can anticipate both the potential for nanomaterial-based therapies that target channel activity and the risk of toxicological effects associated with undesired nanomaterial-channel interactions. To exploit the potential and avoid the risk, the field needs a much better understanding of the mechanisms underlying channel response to nanomaterials and the mechanisms determining the dose and mode of exposure of ion channels to nanoparticles in tissue systems. This chapter attempts to take a first step in the systematic exploration of this new area.

INTRODUCTION

The importance of ion homeostasis to the proper function of nervous tissue has been widely known and accepted for over a century. The early work of Sidney Ringer in the late 1800s showed the extracellular solution perfusing a frog heart must contain defined concentrations of sodium, potassium and calcium ions in order for the heart to beat normally (Hille 2001). Today, the cellular roles and mechanisms of entry of these ions are well established and applied to the development of pharmaceuticals targeting conditions such as anxiety, epilepsy and pain. Therefore, the development of novel biomaterials for the nervous system must consider the material interaction with the rudiments of cellular excitability—maintenance of membrane potential through ion channels. Quantum Dots (QD) (Tang et al. 2009), nanometals (Gramowski et al. 2010) and carbon nanomaterials (Cellot et al. 2008) have all been shown to alter neural network activity. This chapter summarizes the available literature regarding the interactions of nanomaterials with VGICs and postulates potential mechanisms underlying these effects.

ION CHANNELS—TYPES AND FUNCTIONS

The fluid in and around excitable cells of the nervous system is composed of numerous ionic species, the most abundant being sodium, calcium, potassium, and chloride. Finite ionic gradients define an electrochemical potential across the cellular membrane of excitable cells. At rest, the flow of ions through ion channels and pumps maintains the membrane potential. Ion channels are membrane-bound, pore-forming proteins that enable and facilitate the transfer of ions across the cellular membrane (Fig. 1B). The response of ion channels to chemical signals, electrochemical gradients and mechanical tension enables efficient cellular signaling and control of cellular homeostasis. Ion channels are essential to all forms of life, from humans to single-celled organisms, and many pathological conditions derive from, or are exacerbated by, abnormal ion channel function.

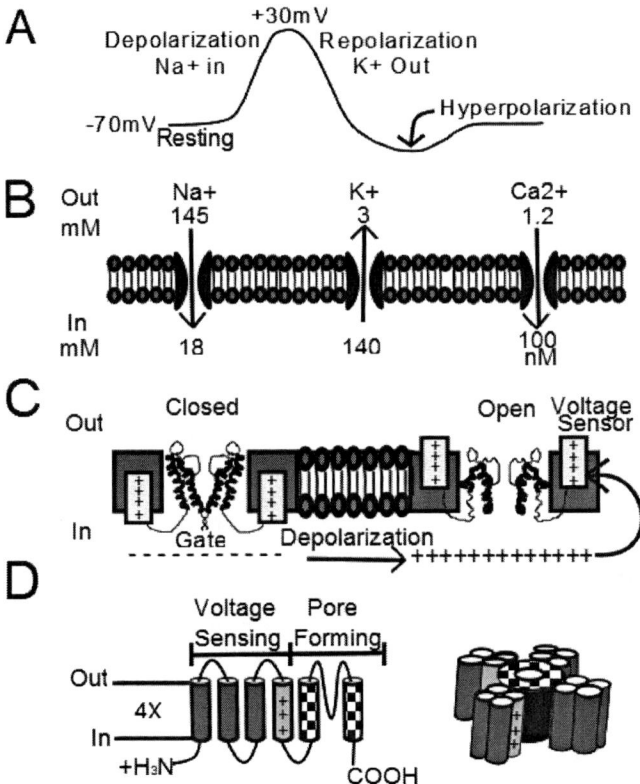

Fig. 1. **Summary of Ion Channel Structure and Fuction.** A. Action potential diagram. At rest, membrane potential is –70mV. When an action potential is fired, VGSCs open and the membrane depolarizes. Depolarization opens VGKCs which pass potassium ions to the extracellular space. VGKCs are responsible for repolarizing the membrane. (Unpublished Material). B. Direction of movement of target ions of voltage-gated ion channels. Relative concentration gradients for ions are provided. (Unpublished Material) C. VGIC mechanics. At rest, the membrane potential is negative and the voltage sensor is at the interior of the membrane. Depolarization causes an influx of positive charge that electrostatically repulses the voltage sensor and consequently opens the channel enabling the passage of current. With permission from (Elinder et al. 2007). D. Simplified schematic of six transmembrane regions comprising one domain of the VGKCs α-subunit. Four domains form the channel shown on the right. Grey transmembrane region is the voltage sensor. With permission from (Börjesson and Elinder 2008).

Under excitatory conditions, voltage-gated ion channels (VGICs) successively open and close in response to alterations in membrane potential, enabling the flow of ionic current and the propagation of the action potential down the axon. This process is the hallmark of neuronal communication (Fig. 1A). This chapter will focus on sodium, calcium and potassium channel members of the superfamily of VGICs. Members of this family are structurally related (Hille 2001) and have three complimentary aspects, namely, ion conductance, pore gating and regulation (Yu et al. 2005). For all members, the pore motif is formed by the four homologous domains consisting of six transmembrane regions (Fig. 1D) comprising the α subunit. For voltage-gated calcium channels (VGCCs) and voltage-gated sodium channels (VGSCs), the four domains of the α subunit are linked together. In voltage-gated potassium channels (VGKCs) they are unlinked. Each domain contains a voltage-sensing transmembrane region which contains repeated positively charged amino acid residues (Hille 2001). Upon depolarization, the change in membrane potential moves this region towards the extracellular space causing a conformational shift in the protein that opens the channel to allow the flow of current (Elinder et al. 2007) (Fig. 1C). Each member of the superfamily has a unique selectivity filter that extends into the pore to preferentially coordinate the throughput of the respective target ions.

VGSCs are responsible for the upstroke of the action potential (Fig. 1A) in nerve, muscle and endocrine cells and in some cases contribute to pacemaker and subthreshold potentials that underlie the decision to fire an action potential. Given the higher intracellular potassium concentration, VGKCs shuttle potassium ions out of the cell and serve to dampen excitation. In this way, VGKCs repolarize the membrane after depolarization events, time the interspike interval during repetitive firing, and lower the effectiveness of excitatory inputs (Hille 2001) (Fig. 1A).

VGCCs convert membrane depolarization events into an inward calcium flux essential for muscle contraction, neurotransmission, hormone secretion, gene expression and neurite outgrowth (Catterall and Few 2008). VGCCs are classified into three families: Ca_v1, Ca_v2 and Ca_v3. Members of the Ca_v1s are abundant in the retina, skeletal and cardiac muscles. Ca_v2 members are distributed throughout the nervous system and cluster near presynaptic terminals. The entry of calcium through some Ca_v2 members enables vesicle fusion and release of neurotransmitter across the synapse. Ca_v3 members activate near resting potential and are responsible for establishing oscillating firing patterns and rebound bursts (Catterall 2000).

NANOMATERIALS AND ION CHANNELS—KNOWN EFFECTS

Studies on Carbon Nanomaterials

Carbon nanotubes are electrically conductive and on the same scale as growing neurons. Consequently, nanotubes have been considered in a number of studies as potential novel biomaterials for neuronal growth (Ni et al. 2005) and stimulation (Cellot et al. 2008). Neurons grown in the presence of water soluble, arc-synthesized SWNT functionalized with PEG, demonstrated altered neurite outgrowth patterns consistent with alterations in calcium homeostasis. Using calcium sensitive dyes, Ni et al. showed that upon exposure to SWNT-PEG, the intracellular calcium levels were not being restored, suggesting SWNT-PEG was an inhibitor of depolarization-dependent calcium influx (Ni et al. 2005). A later study focused on the interaction of arc-synthesized SWNTs on the cultures of primary neurons and glia cells (Belyanskaya et al. 2009). Whole-cell patch clamp recordings of dorsal root ganglia cells exposed to bundled SWNTs revealed diminished inward conductivity and a more positive resting potential. The authors postulate that their results could be explained if VGCCs were affected by the nanotubes (Belyanskaya et al. 2009). Each of these studies utilizes arc-synthesized SWNTs manufactured with nickel/yttrium catalytic particles and suggests that the SWNTs are affecting the conductance of VGCCs. A more recent study showed that physiological solutions containing arc-synthesized SWNTs (with a Ni/Y catalyst) and surprisingly SWNT-free supernatant, inhibit neuronal VGCCs in a dose-dependent and SWNT-sample-dependent manner (Fig. 2). The inhibitory activity involved very low concentrations of soluble yttrium released from the catalytic particle. Yttrium potently inhibits calcium ion channel function with an inhibitory efficacy, IC50, of 0.07 ppm w/w (Jakubek et al. 2009). As a result of this potency, unpurified and even some "purified" SWNT samples contain sufficient bioavailable yttrium to potentially inhibit channel function.

There is evidence that the effect of carbon nanotubes on ion channels is variable as exemplified by the interaction of MWNTs and VGKCs. Early research showed that unmodified SWNTs of varying diameter could inhibit potassium channel function in Chinese hamster ovary cells, whereas MWNTs did not show appreciable inhibition. The authors suggest direct pore occlusion as the mode of action (Park et al. 2003). In contrast, later work suggests that MWNTs are antagonists to VGKCs expressed in the pheochromocytoma (PC12) cell line (Xu et al. 2009a). The mode of action of the inhibition is presently unknown, as is the relation between wall number, surface chemistry, and metal impurities in the nanotubes.

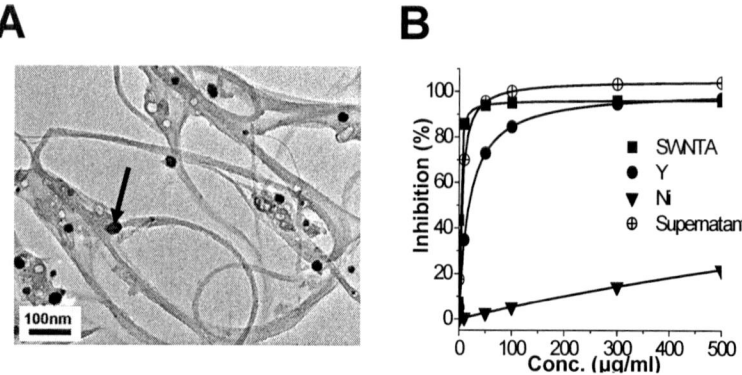

Fig. 2. **Metal catalysts from purified samples can affect calcium channels.** A. Scanning electron microscope image of a vendor purified, aryl-sulfonated SWNT sample. Arrow points to residual catalyst nanoparticle remaining after vendor purification B. Summary diagram showing the effects of SWNTs on VGCC. SWNT-free supernatant has similar inhibition curve to nanotube containing solution. Combining inhibition studies conducted with nickel and yttrium salts (not shown) and metal bioavailability studies from SWNTs (not shown), the amount of inhibition due to yttrium and nickel can be determined. While yttrium is the minor component, it is responsible for a majority of the VGCC inhibition caused by the SWNT sample. With permission from (Jakubek et al. 2009).

Although these studies demonstrate an inhibition of ion channel conductance, other studies suggest enhancement of channel activity. Single-cell electrophysiology techniques, which have been used to assess the ability of carbon nanotube mats to stimulate networks of neurons, recorded after potential depolarization events in stimulated cells. These events were inhibited by calcium channel blockers suggesting the nanotubes were in fact enhancing calcium uptake through alterations in the location or function of VGCCs (Cellot et al. 2008).

Studies on Nanometals

Low particle concentrations of nanometals in solution have been shown to disrupt electrical activity of neural networks grown on multielectrode array neurochips (Gramowski et al. 2010). In a series of studies highlighting the effects of ZnO, CuO and Ag nanoparticles on VGKC and VGSC currents in rat hipocampal neurons, the following contrasting effects were observed. ZnO nanoparticles increase the transient outward potassium current (I_A) and delayed rectifier potassium current (I_K) while increasing the overshoot and diminishing the peak half-width of VGKCs (Zhao et al. 2009). Contrastingly, CuO nanoparticles have no effects on I_A, but inhibited I_K. Furthermore, CuO nanoparticles did not shift the steady-state activation curve of I_K and I_A, but the inactivation curve of I_K was shifted negatively (Xu et al. 2009b).

Relatedly, ZnO nanoparticles increase the peak amplitudes of the VGSC while the inactivation and the recovery from inactivation of the sodium current are promoted by ZnO (Xu et al. 2009b). Silver nanoparticles reduce the amplitude of the sodium current, produce a hyperpolarizing shift in the activation–voltage curve of the sodium current and delay the recovery after inactivation. Furthermore, peak amplitude and overshoot of the evoked single action potential are decreased and half-width is increased with Ag nanoparticles (Liu et al. 2009). If combined, these results suggest that nanosilver decreases peak amplitude and nano-zinc oxide increases peak amplitude (Fig. 3). More research is necessary in order to understand the mechanism underlying each effect.

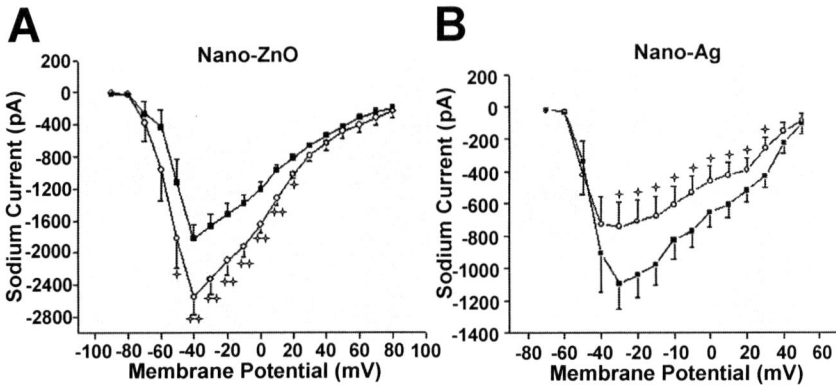

Fig. 3. Comparison of the effect of nanosilver and nanoscale zinc oxide on peak sodium currents in hippocampal pyramidal neurons. Current-voltage graphs plot the entry current of sodium as a function of membrane potential. A. Nanoscale zinc oxide (open circles); Control (closed square) Nanoscale ZnO increases sodium peak amplitude. B. Nanosilver (open circle) decreases sodium peak amplitude as compared to control (closed square). With permission from (Yang et al. 2010).

Interaction of Quantum Dots with Ion Channels

Quantum dots (QDs) are 2–100 nm fluorescent semiconducting nanocrystals of interest to biologists as imaging agents due to their high quantum yield, high resistance to photobleaching, broad absorption and narrow emission spectra. In a study of rat primary cultured hippocampal neurons exposed to CdSe QDs, the QDs were shown to increase cytoplasmic calcium levels through the influx of calcium from both extracellular and intracellular (mostly endoplasmic reticulum) stores. Concurrently, the CdSe QDs enhanced activation and inactivation, slowed recovery, reduced the percent of available VGSCs and prolonged the time course of activation. Interestingly, the effect is observed with QDs 10 nm and above, however below 10nm the effect is not seen (Tang et al. 2008b). The authors suggest that some of these channel effects were similar to the effects of β-scorpion toxin and may in part be due to the binding of the QDs or QD degradation particles to the S3-S4 loop affecting the function of the VGSC voltage sensor. However, the shift in activation to more depolarizing potentials is not consistent with this hypothesis and further research suggested a more elaborate mechanism.

In studying the mechanisms underlying the elevation of intracellular calcium levels, Tang et al. (2008a) observed that the addition of T-type calcium channel blocker, mibefradil, and L-type calcium channel blocker, verapamil, did not block the elevation of intracellular calcium observed with CdSe QD exposure. However, the addition of N-type antagonist, ω-conotoxin, partially blocked the calcium influx. Surprisingly, the inhibition of VGSCs with TTX abolished the elevation of intracellular calcium associated with CdSe QD exposure. Interestingly, the group observed that under exposure to CdSe QDs, VGSCs permitted calcium to permeate the channel as readily as sodium and suggested oxidative stress as a factor in the loss of channel selectivity (Fig. 4). Additionally, the passage of sodium through VGSCs enhanced the release of mitochondrial calcium and contributed to the elevated intracellular calcium levels (Tang et al. 2008a).

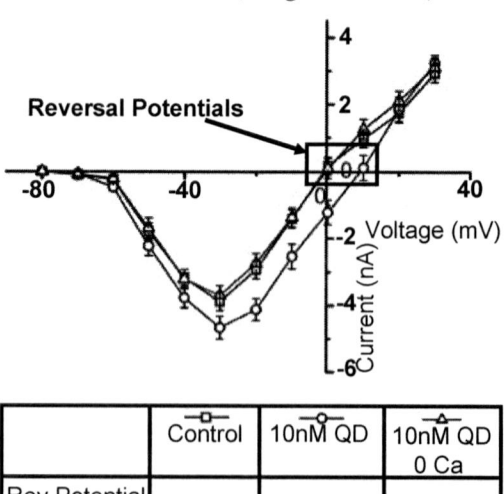

Fig. 4. Sodium current in rat hippocampal neurons as a function of voltage in the presence of quantum dots. Reversal potential is the potential at which pore permeation is reversed. The table below the graph indicates the reversal potentials for the three treatments control, QDs and QDs without calcium. The addition of QDs increases the reversal potential. The addition of QDs in the absence of calcium returned the reversal potential to control. Tang and collaborators (2008a) suggest that QDs enable the calcium to permeate the VGSC and this permeation is reversed when calcium is not present. With permission from (Tang et al. 2008a).

The elevation of intracellular calcium levels in neurons not only effects neuronal survivability, but also neurotransmission. In a follow-up study, the effects of CdSe QDs and streptavidin-CdSe/ZnS QDs demonstrated altered synaptic transmission and plasticity in the hippocampal dentate gyrus area of anesthetized rats. Paired-pulse facilitation was also suppressed under QD exposure. As intracellular calcium levels enable the release of neurotransmitter at the synapse, the elevation of intracellular calcium observed previously may account for this suppression (Tang et al. 2009). While this study concluded that QDs, regardless of modification, could impair synaptic transmission and plasticity; other studies have reported no significant alteration of ion channel function from exposure to QDs (Kirchner et al. 2005). It is possible that QDs may affect ion channels and processes dependent on ion channel flux through the degradation and release of M+s. This will be discussed further in the next section.

NANOMATERIALS AND ION CHANNELS—POSSIBLE INTERACTION MECHANISMS

Given the biological evidence in the literature reviewed above, along with knowledge of the fundamental behaviors of nanomaterials in biological environments, we propose four potential mechanisms of nanomaterial interaction with ion channels. As shown in Fig. 5, mechanisms are free M+ effects, adsorption of target ions, direct particle-protein interaction and oxidative stress. Each mechanism can potentially act independently or in concert with one or more other mechanisms to cause alterations of VGIC dynamics.

Free Metal Ion Effects

It is becoming increasingly apparent that the cellular responses to some nanomaterials are in fact attributable to M+s released from the nanomaterial rather than any direct biological interaction of the particle phase itself. For example, the antibacterial action and cytotoxicity of nano-silver is believed to be primarily the result of silver ion (Ag^+) release and the subsequent binding of Ag^+ to thiol targets (Lubick 2008; Liu et al. 2010). The response of cells to nickel containing compounds is often correlated with the concentration of nickel ions produced by dissolution processes (Liu et al. 2007). Most relevant to ion channels is the study that yttrium ions released from arc synthesized SWNTs are responsible for VGCC channel inhibition and not tubular graphene as previously thought (Jakubek et al. 2009) (Fig. 2). An emerging theme is the effect of ion release from nanomaterials on cellular function. This is particularly relevant as

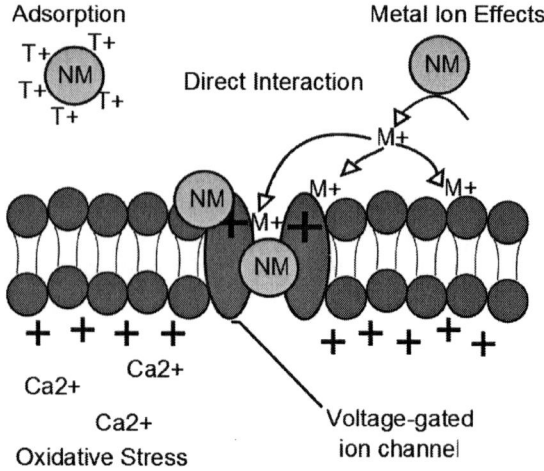

Fig. 5. **Summary of proposed mechanisms of nanomaterial interactions with voltage-gated ion channels.** The membrane above is depolarized and the channel is shown in the open conformation. Adsorption of target ions (T+), which are rendered biologically unavailable for permeation through the pore due to adsorption onto nanomaterials (NMs). Direct Interaction: NMs interact directly with the pore and prevent the permeation of T+ through the pore, or bind to the protein complex, alter the structure and disable channel function. Metal ion effects: NMs release metal ions (M+) that enter the pore to affect inhibition, bind to channel components outside the pore to alter channel dynamics or locally concentrate near the membrane by binding to membrane proteins to affect membrane potential and thereby movement of the voltage sensor. Oxidative stress: The increase of cytosolic calcium associated with oxidative stress affects channel dynamics (Unpublished).

M+s have been shown to block the pore, replace target ions and alter the channel dynamics of VGICs. This section discusses M+ release from nanomaterials and mechanisms of M+ effects on ion channels.

Evidence of Metal Ion Release from Nanoparticles

In general, a wide variety of commercially important nanomaterials may coexist with free ions or other soluble species in physiological solution. Many nanomaterials are oxides, and often undergo slow dissolution under physiological or environmental conditions; good examples are ZnO and NiO. Many nanomaterials contain zero-valent metals, which with the exception of gold, are capable of oxidative dissolution to produce ions. The release of M+s from nanometals, such as nanosilver, decreases with increasing pH, increases with increasing temperature, is dependent on oxidation and therefore the concentrations of dissolved oxygen in the environment and is decreased by the addition of humic or fulvic acids. (Liu and Hurt 2010). This suggests that under physiological conditions many nanometals will release ions with the potential to act as conventional M+

toxicants. Even carbon-based nanomaterials contain metal catalyst residues that may include Ni, Y, Co, Mo, or Fe. These catalyst particle residues have been shown to significantly release soluble metal forms (Kagan et al. 2006; Liu et al. 2007; Pulskamp et al. 2007) through oxidative attack on metal catalyst residues that are not fully encapsulated by graphenic carbon shells (Liu et al. 2007; Liu et al. 2008). The release of M+s from nanotubes has been shown to inhibit VGCC function (Jakubek et al. 2009).

In addition to carbon nanomaterials and nanometals, QDs have also been shown to release M+s. A bare core QD is composed solely of its primary semiconducting components making it susceptible to interactions with its environment, particularly water and oxygen. Oxidation of the QD will reduce fluorescence and cause degradation (Jasieniak and Mulvaney 2007). Passivation techniques involve the modification of the surface with mercaptopropionic acid, silanization and polymer coatings (Kirchner et al. 2005). The most widely used surface modification is CdSe with a ZnS shell. While shells reduce ion release, weathering under acidic (pH ≤ 4) or basic (pH ≥ 10) conditions destabilizes the shell resulting in cadmium/selenite ion release (≤ 1 min) (Mahendra et al. 2008) Cadmium release from QDs can also occur through intracellular degradation (Kirchner et al. 2005). Release, however, can be mediated by humic acids, bovine serum albumin (cellular media component), oxalate, nitrilotriacetic acid, EDTA, citrate and cysteine which are presumed to protect QD surface sites from etching (Mahendra et al. 2008).

Effect of Metal Ions on Voltage-Gated Ion Channels

In order to predict the interaction of M+s (and potentially nanomaterials) with ion channels, two key factors must be considered: the enthalpy of hydration and hard-soft characteristics of M+s. Through solvation, water molecules form a shell around the ion. The enthalpy of hydration is indicative of the attraction of the M+ for oxygen groups and is consequently inversely proportional to the water substitution rate. Hydration energies are highest (meaning the ion is more hydrated) for small ions with large ionic charge ($Mg^{2+}, Mn^{2+}, Ni^{2+}, Co^{2+}, Zn^{2+}$). This affects the metal interaction with ion channels as the solvated water molecules are replaced by dipolar groups of the channel's inner pore and therefore influences whether an ion will permeate or block the channel (Thévenod 2010).

Additionally, hard-soft ionic categorizations influence ionic complexation. A hard metal preferably retains its valence electrons and is not easily polarized. Traditionally, hard ions are from the alkaline earth metals, lanthanoids and aluminum. They are small in size, high in charge and prefer to complex with hard bases such as oxygen (water) or fluoride through electrostatic interaction. In contrast, soft ions are transition metals such as Au^+ Cu^+ Ag^+ or Hg^+. They are relatively large, easily polarized and prefer to complex with soft bases such as phosphorus, arsenic, nitrogen (amines, histidine) or sulfur (sulfhydryl, disulfide, thioether) through covalent interaction. The distinction between hard and soft is not well defined and some ions such as Zn^{2+} and Ni^{2+} fall in between. See Table 1 for a summary of ion radii, enthalpy of hydration and ligand preference (Elinder and Århem 2004; Thévenod 2010). These characteristics influence the interaction of M+s released from nanomaterials with ion channels.

Table 1. Summary of Ionic Radii, Hydration Energies, Ligand Preferences and Effects on VGCC.

	Radius (pm)	Δh_{hyd} (kJ/mol)	Ligand Preference	
Target Ions				
Na^+	98	−405	TSYDENQ	
K^+	133	−321	TSYDENQ	
Ca^{2+}	106	−1592	TSYDENQ	
Metal ions				VGCC IC50 mM
Ba^{2+}	143	−1304		1.5
Sr^{2+}	127	−1445		2.6
Cd^{2+}	103	−1806		0.001
Co^{2+}	82	−2054	KRHWCM	0.06
Zn^{2+}	83	−2044	KRHWCM	0.03
Ni^{2+}	78	−2106	KRHWCM	1.04
Cu^{2+}	72	−2100	KRHWCM	1.04
Y^{3+}	106	−3620		0.001

Target ions of VGICs: sodium, potassium and calcium provide a comparison for metal ions. Atomic radius is given in pm. Table highlights trends mentioned in the text. As charge is increased at constant radius, the enthalpy of hydration (ΔH_{hyd}) increases. For a given charge, as the radius decreases, ΔH_{hyd} increases. The higher the ΔH_{hyd}, the lower the rate at which dehydration occurs in the pore and is one factor contributing to ion permeation or pore blockage. Barium and strontium with lower ΔH_{hyd} are known to permeate VGCCs, where as cadmium and yttrium block VGCCs. Moreover cadmium and yttrium are the same size as calcium suggesting the possibility of biomimicry. Ligand preference, refers to known preferred complexation ligands, in which the letters denote standard amino acids. Table adopted from (Burgess 1978; Nachshen 1984; Elinder and Århem 2004).

Pore block. VGICs are designed for high throughput transport of target ions. Ion selectivity is the function of the channel's selectivity filter. Through similarities in size and charge, it is possible for M+s to enter the pore through a type of biomimicry. In this way, M+s mimic target ions and hijack the machinery of the pore. Pore block refers to the entry and binding of M+s to the pore resulting in decreased current. Current may decrease to such a degree that the channel is effectively blocked.

For example, the release of yttrium ions from water soluble SWNTs effectively blocks VGCCs (Fig. 2). Being the same size, yttrium replaces calcium in the pore through biomimicry. Current flow in the pore decreases due to increased adherence of highly valent yttrium ions to the four glutamate residues (EEEE) of the VGCC selectivity filter. The IC50 value for yttrium is 0.75µM (Jakubek et al. 2009). Similarly, the selectivity filter of VGSCs involves two regions: a ring of four amino acids (EEDD) in the outer vestibule of the pore, and an inner ring, also of four amino acids (DEKA). Amino acids (TVGYG) make up the selectivity filter of the VGKCs. Many ions released from common nanomaterials are effective voltage-gated channel blocking agents.

Altered Channel Dynamics. Metal ions can affect channel dynamics without ever entering the pore by altering the rates of activation or inactivation. There are three such mechanisms of M+ interaction with voltage-gated channels: charge screening, electrostatic modification of the voltage sensor and non-electrostatic binding effects (Ballatori 2002; Elinder and Århem 2004). Each of these methods may act alone or together to affect ion channel dynamics.

At rest, under physiological conditions, a negative membrane potential defines the extracellular space to be more positively charged than the intracellular space. When M+s are introduced to the system, the additional positive charge in the extracellular media may alter the membrane potential. The charge screening mechanism of M+ interaction with voltage-gated channels describes this phenomenon (Elinder and Århem 2004). Since high concentrations of M+s are necessary to have an effect through charge screening alone, it seems an unlikely mechanism for nanomaterial interactions.

In the second mechanism, electrostatic modification of the voltage sensor, the binding of released M+s to the membrane surface fixes charges, increases the ionic concentration near the surface and therefore reduces membrane potential. The reduction in membrane potential affects the action of the voltage sensor (Elinder and Århem 2004). The rate of channel opening subsequently decreases, and the rate of channel closure increases. In some cases, the action of the ion is dynamic and channel state specific. Zinc is believed to be attracted to the negatively charged component of the gating apparatus when the channel is at rest (Hille 2001). When the channel is activated, however, the positive charge of the voltage sensor prevents zinc binding. In this way, zinc decreases the rate of channel opening; however the rate of channel closure remains unaffected. Contrastingly, intracellular zinc does not affect opening but slows closing (Hille 2001; Elinder and Århem 2004).

Non-electrostatic binding effects is the third mechanism. Here, M+s bind close to the pore to affect gating in a non-electrostatic way. In general, the binding of the ion to a portion of the channel apparatus enables ion-mediated steric hindrance or mass effects that inhibit normal function of the channel (Elinder and Århem 2004). The effect of M+s on channel gating is dependent on channel type, M+ type and concentration. While the aforementioned summarizes general mechanisms, exploration into the effect of specific M+s in a given system must be addressed in order to fully account for the ionic effects on gating.

Adsorption of Target Ions

Adsorption refers to the binding of small molecules, proteins or ions to material surfaces including nanomaterial surfaces. This effect can have a two-fold mode of action to alter channel dynamics. In the first, target ions are chelated from the external solution and are no longer bioavailable for transport through the channels. In the second, M+s are adsorbed onto the material surface for later exposure.

Common chelators are BAPTA (calcium specific chelator) and EDTA (metal ion chelator). *In vitro*, NiO, ZnO, TiO_2, CeO_2 and Fe_2O_3 nanometal oxides significantly reduce calcium and phosphorus levels in the media. Sodium levels are unaffected (Horie et al. 2009). Carbon nanotubes can also be sorbents of target ions (Liu et al. 2008). As the concentration of calcium ions outside the cell is roughly 1.2 mM and the concentration inside the cell is 100 nM, it is unlikely that nanomaterials, unless highly concentrated, would adsorb enough calcium to have an appreciable effect on VGCC function.

The second adsorption mechanism pertains to the adsorption of M+s onto the nanomaterial surface for exposure later on. The adsorption capacity of M+s to carbon nanomaterials follows Pb^{2+} > Ni^{2+} > Zn^{2+} > Cu^{2+} > Cd^{2+}. Adsorption is dependent on temperature and the acidity of the solution. The addition of function groups to the nanomaterial surface increases M+ adsorption capacity (Rao et al. 2007; Jakubek et al. 2009). In a particular case, the introduction of additional salts in solution decreased the adsorption of Y^{3+} to functional groups on SWNTs suggesting the possibility that M+s compete for adsorption sites and can become bioavailable as cellular conditions change (Jakubek et al. 2009). As mentioned previously, certain M+s inhibit ion channel conductance and the conditions under which nanomaterials will adsorb and release M+s should be monitored.

Direct Particle-Protein Interactions

It has been suggested that some nanomaterials interact directly with ion channels. Many of the principles that apply to the interaction of M+s with ion channels and adsorption mechanisms can be combined and extended to understand the potential for nanomaterials to interact directly with ion channels. Such interaction could include a physical occlusion of the pore or electrostatic alterations. The channel pore is on the scale of nanometers. QDs have been shown to enter the cylindrical cavity of proteins on the same size scale (Ishii et al. 2003), and early work suggested that SWNTs physically occlude the VGKCs (Park et al. 2003). While the mode of action remains a mystery, more recent work has shown the non-ROS induced inhibition of VGKCs with carboxylated MWNTs of 40–50 nm outside diameter (Xu et al. 2009a).

Nanomaterials in media form aggregates (Gramowski et al. 2010) dependant on ionic strength, surface charge, and surface coating (Jiang et al. 2009). Given channel dimensions are fixed, agglomeration of nanomaterials increases the hydrodynamic radius of the material and consequently decreases the likelihood of physical interactions. Moreover, in media, nanomaterials are observed to adsorb proteins (Guo et al. 2008) and ions (Rao et al. 2007) increasing the effective size even more. This suggests that if direct interactions are to occur, they would not be mediated by physical occlusion, but rather through electrostatic modification through coupling to the cell membrane (Cellot et al. 2008) or nanomaterial mediated steric hindrance of the area near ion channels.

Oxidative Stress

The generation of ROS by nanomaterials and the material features contributing to ROS continues to be of debate. An increase in intracellular calcium, production of ROS and a decrease in mitochondrial membrane potential are characteristics of oxidative stress. In some cases cellular exposure to nanomaterials induces rises in intracellular calcium levels. Some studies have suggested ROS production upon exposure to nanomaterials results in altered channel function *in vitro* (Tang et al. 2008a), whereas others observed a disconnect between the two (Xu et al. 2009a; Gramowski et al. 2010). This raises the possibility that nanomaterial induced ROS production may contribute to alterations in ion channel function through downstream mechanisms, however more research is needed.

Applications to other Areas of Health and Disease

The effects of nanomaterials on ion channel dynamics are not only of interest for toxicology, but also pharmacology. With further study and understanding, the mechanisms underlying nanomaterial induced ion channel effects can be used to develop novel treatments for ion channel diseases and conditions. Recently, the remote control activation of ion channels to elicit a behavioral response in C-elegans is an exciting first step (Huang et al. 2010). Moreover, the field of nanofluidics is moving closer to reliable generation of nano-based synthetic ion channels which will have an impact not only biologically but also for the fields of filtration, and the development of biosensors (Harrell et al. 2004).

Key Facts about Ion Channels

- Ion channels are found in every cell of every biological organism.
- The primary function of VGICs is to selectively transport ions across the cellular membrane as a result of membrane depolarization events.
- Multiple binding sites within the selectivity filter of the channel enable the high throughput passage of target ions. In some cases, ion passage is achieved in less than 1 µs.
- Within the nervous system, different members of each ion channel family are localized to different regions, presumably, because their channel properties are suited for the physiological functions at those sites.
- Ion channels are the second largest target class of pharmaceuticals.

Summary Points

- Voltage-gated ion channels are ubiquitous among cellular life and are responsible for ion homeostasis in nervous tissue.
- While voltage-gated ion channels have different target ions, the channel structure and function is homologous suggesting materials may interact with each channel through similar mechanisms.
- Ion disregulation is responsible for, or exacerbated by, many disease conditions and many pharmaceutical drugs consequently target ion channels.
- Quantum dots, carbon nanomaterials and nanometal oxides have been shown to alter the channel dynamics of voltage-gated ion channels.
- The release of free metal ions, adsorption of ions or metals onto material surfaces, direct interaction of material with channels, and oxidative stress are possible mechanisms by which nanomaterials interact with ion channels.

Definitions

Action Potential: is an electrical message of constant amplitude and velocity propagated down an axon through the successive flow of ions through ion channels.

Adsorption: the binding of small molecules, proteins or ions to a material or nanomaterial surface.

Gating: refers to the opening and closing of ion channel pores.

IC50: is the concentration of ion or toxin necessary to inhibit 50% of available channels. It is subsequently used as a marker to evaluate effectiveness of channel inhibitors.

Ion Channel: pore forming membrane proteins that allow the transfer of ions as a probabilistic function of ionic concentration and electrochemical gradient. Ion channels are found throughout cellular life and enable cellular functions.

Abbreviations

ΔH_{hyd}	:	Enthalpy of Hydration
M+	:	Metal Ion
MWNT	:	Multi-Walled Carbon Nanotube
PEG	:	Polyethelene Glycol
QD	:	Quantum Dot
ROS	:	Reactive Oxygen Species
SWNT	:	Single-Walled Carbon Nanotube
VGCC	:	Voltage-Gated Calcium Ion Channel
VGIC	:	Voltage-Gated Ion Channel
VGKC	:	Voltage-Gated Potassium Ion Channels
VGSC	:	Voltage-Gated Sodium Ion Channels

References

Ballatori, N. 2002. Transport of Toxic Metals by Molecular Mimicry. Environ. Health Perspect. 110: 689–694.

Belyanskaya, L., S. Weigel, C. Hirsch, U. Tobler, H.F. Krug and P. Wick. 2009. Effects of carbon nanotubes on primary neurons and glial cells. Neurotoxicology 30: 702–711.

Börjesson, S.I. and F. Elinder. 2008. Structure, Function, and Modification of the Sensor in Voltage-Gated Ion Channels. Cell Biochem. and Biophys. 52: 149–174.

Burgess, J. 1978. Metal Ions in Solution. Ellis Horwood.

Catterall, W.A. 2000. Structure and regulation of voltage-gated calcium channels. Annual Review of Cell Develop. Bio. 16: 521–555.

Catterall, W.A. and A.P. Few. 2008. Calcium channel regulation and presynaptic plasticity. Neuron. 59: 882–901.

Cellot, G., E. Cilia, S. Cipollone, V. Rancic, A. Sucapane, S. Giordani, L. Gambazzi, H. Markram, M. Grandolfo, D. Scaini, F. Gelain, L. Casalis, M. Prato, M. Giugliano and L. Ballerini. 2008. Carbon Nanotubes might improve neuronal performance by favouring electrical shortcuts. Nat. Nanotechnol. 4: 126–133.

Elinder, F. and P. Århem. 2004. Metal ion effects on ion channel gating. Quarterly Reviews of Biophysics. 36: 373–427.

Elinder, F., J. Nilsson and P. Århem. 2007. On the opening of voltage-gated ion channels. Phys. and Behav. 92: 1–7.

Gramowski, A., J. Flossdorf, K. Bhattacharya, L. Jonas, M. Lantow, Q. Rahman, D. Schiffmann, D.G. Weiss and E. Dopp. 2010. Nanoparticles Induce Changes of the Electrical Activity of Neuronal Networks on Microelectrode Array Neurochips. Environ Health Perspect. 118: 1363–1369.

Guo, L., A. Von Dem Bussche, M. Beuchner, A. Yan, A.B. Kane and R.H. Hurt. 2008. Adsorption of essential micronutrients by carbon nanotubes and the implications for nanotoxicity testing. Small. 4: 721–727.

Harrell, C.C., P. Kohli, Z. Siwy and C.R. Martin. 2004. DNA-Nanotube Artificial Ion Channels. J. Am. Chem. Soc. 126: 15646–15647.

Hille, B. 2001. Ion Channels of Excitable Membranes. Sinauer Associates, Inc., Sunderland, MA USA.

Horie, M., K. Nishio, K. Fujita, S. Endoh, A. Miyauchi, Y. Saito, H. Iwahashi, K. Yamamoto, H. Murayama, H. Nakano, N. Nanashima, E. Niki and Y. Yoshida. 2009. Protein Adsorption of Ultrafine Metal Oxide and Its Influence on Cytotoxicity toward Cultured Cells. Chem. Research in Tox. 22: 543–553.

Huang, H., S. Delikanli, H. Zeng, D.M. Ferkey and A. Pralle. 2010. Remote control of ion channels and neurons through magnetic-field heating of nanoparticles. Nat. Nanotechnol. 5: 602–606.

Ishii, D., K. Kinbara, Y. Ishida, N. Ishii, M. Okochi, M. Yohda and T. Aida. 2003. Chaperonin-mediated stabilization and ATP-triggered release of semiconductor nanoparticles. Nature. 423: 628–632.

Jakubek, L.M., S. Marangoudakis, J. Raingo, X. Liu, D. Lipscombe and R.H. Hurt. 2009. The inhibition of neuronal calcium ion channels by trace levels of yttrium released from carbon nanotubes. Biomaterials 30: 6351–6357.

Jasieniak, J. and P. Mulvaney. 2007. From Cd-Rich to Se-Rich Γê Æ the Manipulation of CdSe Nanocrystal Surface Stoichiometry. J. Am. Chem Soc. 129: 2841–2848.

Jiang, J., G. Oberdörster and P. Biswas. 2009. Characterization of size, surface charge and agglomeration state of nanoparticle dispersions for toxicological studies. J. Nanoparticle Research 11: 77–89.

Kagan, V.E., Y.Y. Tyurina, V.A. Tyurin, N.V. Konduru, A.I. Potapovich, A.N. Osipov, E.R. Kisin, D. Schegler-Berry, R. Mercer, V. Castranova and A.A. Shvedova. 2006. Direct and indirect effects of single walled carbon nanotubes on RAW 264.7 macrophages: Role of iron. Tox. Letters 165: 88–100.

Kirchner, C., T. Liedl, S. Kudera, T. Pellegrino, A. Muñoz Javier, H.E. Gaub, S. Stölzle, N. Fertig and W.J. Parak. 2005. Cytotoxicity of Colloidal CdSe and CdSe/ZnS Nanoparticles. Nano Letters 5: 331–338.

Liu, J. and R.H. Hurt. 2010. Ion Release Kinetics and Particle Persistence in Aqueous Nano-Silver Colloids. Environ. Sci. & Tech. 44: 2169–2175.

Liu, J., D.A. Sonshine, S. Shervanti and R.H. Hurt. 2010. Controlled Release of Biologically Active Silver from Nanosilver Surfaces. ACS Nano. In Press:

Liu, X., L. Guo, D. Morris, A.B. Kane and R.H. Hurt. 2008. Targeted Removal of Bioavailable Metal as a Detoxification Strategy for Carbon Nanotubes. Carbon 43: 489–500.

Liu, X., V. Gurel, D. Morris, D.W. Murray, A. Zhitkovich, A.B. Kane and R.H. Hurt. 2007. Bioavailability of Nickel in Single-Wall Carbon Nanotubes. Adv Materials 2790–2796.

Liu, Z., G. Ren, T. Zhang and Z. Yang. 2009. Action potential changes associated with the inhibitory effects on voltage-gated sodium current of hippocampal CA1 neurons by silver nanoparticles. Tox. 264: 179–184.
Lubick, N. 2008. Nanosilver toxicity: ions, nanoparticlesεù+or both? Environ. Sci. & Tech. 42: 8617–8617.
Mahendra, S., H. Zhu, V.L. Colvin and P. Alvarez. 2008. Quantum Dot Weathering Results in Microbial Toxicity. Environ. Health Perspect. 42: 9424–9430.
Nachshen, D.A. 1984. Selectivity of the Ca Binding Site in Synaptosome Ca Channels. J. Gen. Phys. 83: 941–967.
Ni, Y., H. Hu, E.B. Malarkey, B. Zhao, V. Montana, R.C. Haddon and V. Parpura. 2005. Chemically Functionalized Water Soluble Single-Walled Carbon Nanotubes Modulate Neurite Outgrowth. J. Nanosci. and Nanotech. 5: 1707–1712.
Park, K.H., M. Chhowalla, Z. Iqbal and F. Sesti. 2003. Single-walled Carbon Nanotubes are a New Class of Ion Channel Blockers. J. Bio. Chem. 278: 50212–50216.
Pulskamp, K., S. Diabaté and H.F. Krug. 2007. Carbon Nanotubes show no sign of acute toxicity but induce intracellular reactive oxygen species in dependence on contaminants. Tox. Letters 168: 58–74.
Rao, G.P., C. Lu and F. Su. 2007. Sorption of divalent metal ions from aqueous solution by carbon nanotubes: A review. Separation and Purification Technology. 58: 224–231.
Tang, M., Z. Li, L. Chen, T. Xing, Y. Hu, B. Yang, D.-Y. Ruan, F. Sun and M. Wang. 2009. The effect of quantum dots on synaptic transmission and plasticity in the hippocampal dentate gyrus area of anesthetized rats. Biomaterials. 30: 4948–4955.
Tang, M., M. Wang, T. Xing, J. Zeng, H. Wang and D.-Y. Ruan. 2008a. Mechanisms of unmodified CdSe quantum dot-induced elevation of cytoplasmic calcium levels in primary cultures of rat hippocampal neurons. Biomaterials 29: 4383–4391.
Tang, M., T. Xing, J. Zeng, H. Wang, C. Li, S. Yin, D. Yan, H. Deng, J. Liu, M. Wang, J. Chen and D.-Y. Ruan. 2008b. Unmodified CdSe Quantum Dots Induce Elevation of Cytoplasmic Calcium Levels and Impairment of functional Properties of Sodium Channels in Rat Primary Cultured Hippocampal Neurons. Environ. Health Perspect. 116: 915–922.
Thévenod, F. 2010. Catch me if you can! Novel aspects of cadmium transport in mammalian cells. Biometals 23: 857–875.
Xu, H., J. Bai, J. Meng, W. Hao, H. Xu and J.-M. Cao. 2009a. Multi-walled carbon nanotubes suppress potassium channel activities in PC12 cells. Nanotechnol. 20: 285102–285110.
Xu, L.J., J.X. Zhao, T. Zhang, G.G. Ren and Z. Yang. 2009b. In vitro study on influence of nano particles of CuO on CA1 pyramidal neurons of rat hippocampus potassium currents. Environ. Toxicol. 24: 211–217.
Yang, Z., Z. Liu, R.P. Allaker, P. Reip, Z. Ahmad and G.G. Ren. 2010. A review of nanoparticle functionality and toxicity on the central nervous system. J of the Roy Soc, Interface 7: S411–S422.

Yu, F.H., V. Yarov-Yarovoy, G.A. Gutman and W.A. Catterall. 2005. Overview of Molecular Relationships in the Voltage-Gated Ion Channel Superfamily. Pharma Reviews 57: 387–395.

Zhao, J., L. Xu, T. Zhang, G. Ren and Z. Yang. 2009. Influences of nanoparticle zinc oxide on acutely isolated rat hippocampal CA3 pyramidal neurons. Neurotox. 30: 220–230.

8

Biomolecular Engineering for the Regulation of Alpha-synuclein Nanostructure—toward the Development of Alpha-synuclein Targeting Nanomedicine

Natsuki Kobayashi,[1,a] Jihoon Kim[1,b] and Koji Sode[1,c,]*

ABSTRACT

α-Synuclein is the causative protein of several neurodegenerative diseases, designated as synucleinopathies, such as Parkinson's disease (PD) and dementia with Lewy Body (DLB). α-Synuclein is one of the natively unfolded proteins (disordered proteins), which have little or no ordered structure under physiological *in vitro* conditions. However, this protein undergoes a conformational change to amyloid fibrils via the formation of oligomer and

---

[1]Department of Biotechnology, Graduate School of Engineering, Tokyo University of Agriculture & Technology, 2-24-16 Naka-cho, Koganei, Tokyo 184-8588 Japan.
[a]E-mail: nat77@med.showa-u.ac.jp
[b]E-mail: 50008831107@st.tuat.ac.jp
[c]E-mail: sode@cc.tuat.ac.jp
*Corresponding author

List of abbreviations after the text.

protofibril intermediates. Considering that α-synuclein is a natively unfolded protein, we are particularly interested in the regulatory factors maintaining the unfolded status as well as triggering the formation of supra-molecular nanostructures. This information will lead to novel strategies for designing molecules to prevent α-synuclein-induced neurodegeneration. Here we summarize our challenges in understanding the regulation of the nanostructure formation of α-synuclein via molecular engineering approaches at the primary and quaternary structural levels. At the primary structure level, we have carried out extensive mutational studies within a characteristic region of α-synuclein as well as in truncated α-synuclein molecules. At the quaternary structure level, we have investigated the protein/peptide interaction between the wild type and mutant/modified α-synucleins, as well as partial peptides, and *in silico*-designed peptides. We also discuss about DNA aptamers as a potential candidate in the regulation of α-synuclein nanostructure. These ligands, which regulate α-synuclein nanostructure, are expected to be further applied as nano-medicines.

INTRODUCTION

Human α-synuclein is the causative protein of several neurodegenerative diseases, designated as synucleinopathies, such as Parkinson's disease (PD) and dementia with Lewy Bodies (DLB) (Recchia et al. 2004). α-Synuclein is a natively unfolded protein of unknown function that is highly expressed in the neurons of the central nervous system. α-Synuclein's causative role in PD pathogenesis is strongly supported by the fact that it is the major fibrillar protein component of Lewy Bodies (LBs) in both sporadic and familial PD. It is also supported by the fact that three different α-synuclein missense mutations (A30P, A53T, and E46K) as well as the duplication and triplication of its locus cause autosomal-dominant PD (Polymeropoulos et al. 1997; Kruger et al. 1998; Singleton et al. 2003; Chartier-Harlin et al. 2004; Zarranz et al. 2004).

α-Synuclein is a small (14 kDa) protein, characterized by six imperfect repeats (consensus sequence KTKEGV) within the N-terminal part of the polypeptide (residues 1–95), and an acidic carboxyl-terminal region (residues 96–140). The central region of α-synuclein (residues 61–95) is known as the non-amyloid β component of Alzheimer's disease amyloid (NAC) and is characterized as a core region responsible for fibril formation (Fig. 1). In aqueous solution, α-synuclein is natively unfolded and hence

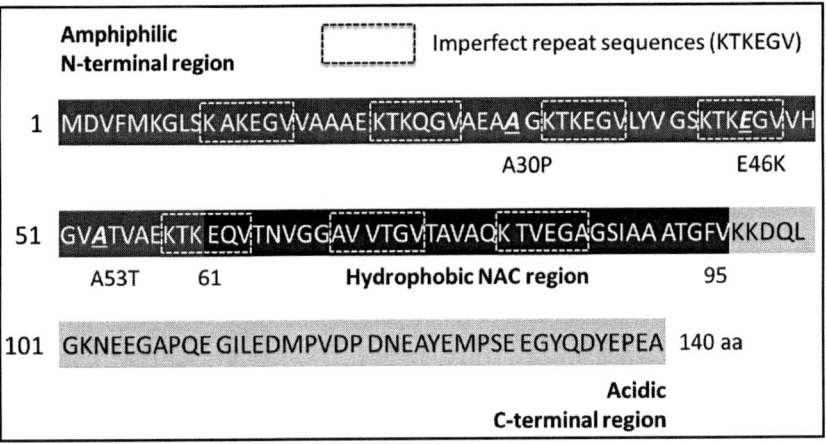

Fig. 1. The primary structure of α-synuclein. α-Synuclein can be divided into three characteristic regions: an amphiphilic N-terminal region, a hydrophobic NAC region, and an acidic C-terminal region. *This figure is unpublished.*

highly dynamic. On aggregation, α-synuclein undergoes conformational changes into amyloid fibrils by forming oligomeric and protofibril intermediates (Uversky and Fink 2002).

Earlier research achievements have reported factors regulating such nanostructure formation of α-synuclein. The *in vitro* analyses of the familial PD-related α-synuclein mutants A30P and A53T demonstrated that these amino acid substitutions resulted in α-synuclein molecules that are much more prone to form amyloid fibril (Conway et al. 1998). α-Synuclein amyloid formation was reported to be prevented by mixing with β-synuclein, the non-amyloidogenic homolog of α-synuclein. These pioneer achievements inspired us to carry out further molecular engineering approaches, at both primary and quaternary structural levels, as they may provide insight to novel strategies for designing molecules targeting α-synuclein and regulating its nanostructure formation, potentially leading to the future development of nanomedicines for PD and other synucleinopathies.

In this chapter, we summarize our challenges in understanding the regulation of α-synuclein nanostructure formation by molecular engineering approaches, at both primary and quaternary structure levels (Fig. 2). At the primary structure level, we have carried out extensive mutagenesis studies within a characteristic region of α-synuclein as well as truncated α-synuclein molecules. At the quaternary structure level, we have investigated the protein/peptide interaction between wild type and mutant/modified α-synucleins, as well as partial peptides, and *in silico*-designed peptides.

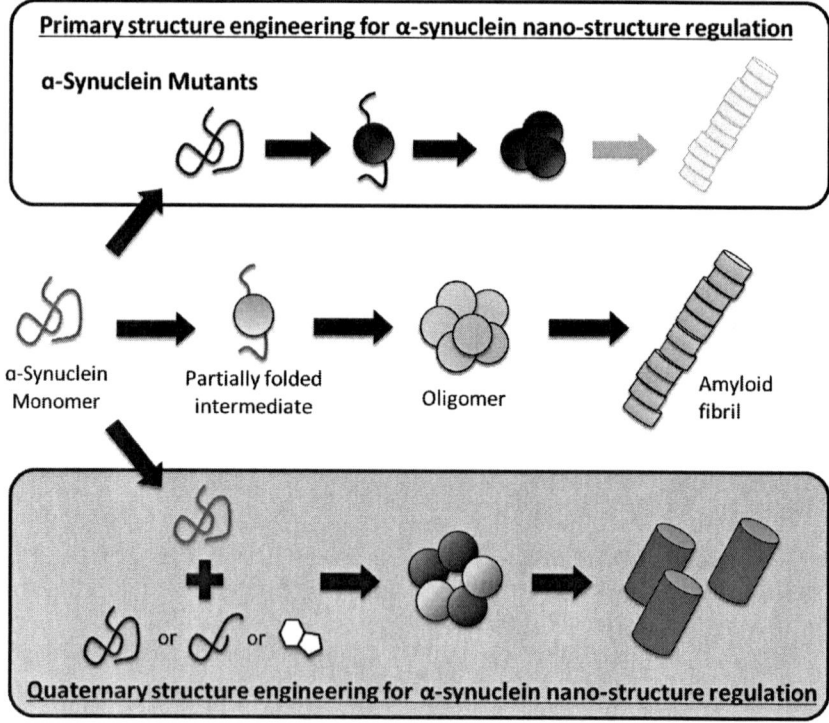

Fig. 2. The strategies of primary and quaternary structure engineering for α-synuclein nanostructure regulation. For primary structure engineering, α-synuclein mutants were constructed. For quaternary structure engineering, additives were utilized, such as α-synuclein mutants, peptides and low molecular weight compounds. *This figure is unpublished.*

PRIMARY STRUCTURE ENGINEERING FOR α-SYNUCLEIN NANOSTRUCTURE REGULATION

We first focused on the characteristic primary structure of α-synuclein. Regarding the N-terminal region, mutations on the imperfect repeat sequence were investigated. Among the residues in the central region, mutational analysis was carried out on the hydrophobic residues. Finally, investigation of the C-terminal region was carried out by constructing and characterizing C-terminal truncated α-synucleins. An overview of the constructed mutant and truncated α-synucleins are summarized in Fig. 3.

Mutational Analysis of the N-terminal KTKEGV Repeat Sequence

The N-terminal and NAC regions contain seven imperfect repeats (XKTKEGVXXXX) that include a highly conserved KTKEGV hexameric

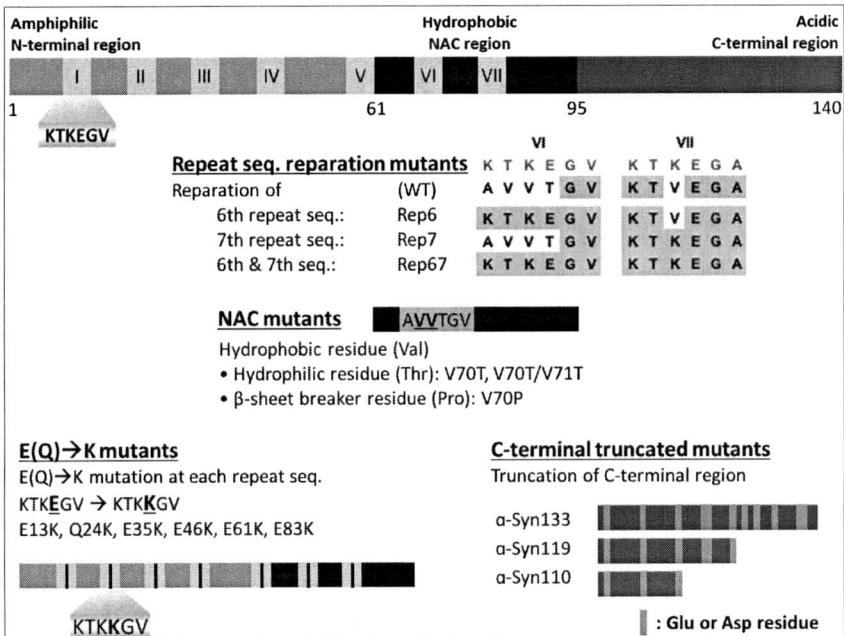

Fig. 3. The overview of constructed mutants and truncated α-synucleins. (i) Repeat sequence reparation mutants including E(Q)→K mutations on the imperfect repeat sequence and the repeat sequence reparative mutation to repair the fully conserved repeat sequence (KTKEGV) in either or both of 6th and 7th repeat sequences. (ii) NAC mutants were constructed by mutation on hydrophobic residues. (iii) C-terminal truncated mutants, α-Syn133, α-Syn119 and α-Syn110 were constructed by truncation at the carboxyl-terminal of the residues 133, 119 and 110 respectively. *This figure is unpublished.*

motif. However, the role of the KTKEGV motif in the α-synuclein structure has yet to be reported, particularly with respect to the protein folding process and maintaining the random unfolded structure. The characteristic hexameric motifs in the two repeats located within the NAC region are not conserved. In order to investigate the role of these two non-conserved hexameric motifs on the aggregation and fibril formation ability of α-synuclein, we carried out a site-directed mutagenesis study on this region (Sode et al. 2007).

We constructed three mutant α-synucleins in which either or both of the non-conserved hexameric motifs in the 6th and 7th repeats were substituted to the conserved KTKEGV sequence. The resulting mutants are Rep6 (Ala69Lys/Val70Thr/Val71Lys/Thr72Glu), Rep7 (Val82Lys), and Rep67 (Ala69Lys/Val70Thr/Val71Lys/Thr72Glu/Val82Lys). Comparison of the fibrillation patterns of wild type (WT) and mutant α-synucleins shows that Rep6 and Rep67 had much lower fibril formation rates than WT, with maximal fibril formation values less than 10% of the WT value

(Fig. 4). Our results indicated that the introduction of KTKEGV sequence at the NAC region resulted in drastic decrease in the fibril formation ability without losing its primary random structure and its water solubility. This fact indicated that KTKEGV is a sequence that may have a crucial role in preventing amyloid fibril formation, maintaining the random structure in a water soluble form.

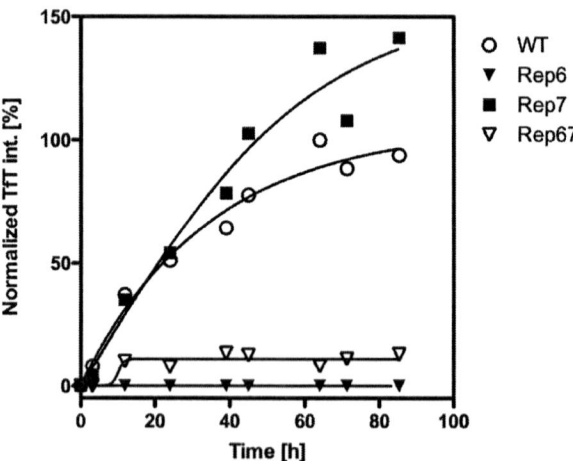

Fig. 4. Effect of amino acid substitution into imperfect repeat sequence on amyloid fibril formation of α-synuclein. Time course of fibril formation of α-synuclein and its mutants as determined by ThT fluorescence assay analysis: WT (*white circles*), Rep6 (*black inverted triangles*), Rep7 (*black squares*), and Rep67 (*white inverted triangles*). This figure is unpublished.

The E46K mutation, which is located in the fourth imperfect repeat sequence and is related to rare familial forms of early-onset PD, was reported to increase the ability of fibril formation and aggregation *in vitro* and in cultured cells (Fredenburg et al. 2007). Examination of PD-linked and engineered mutants suggests that the N-terminal region may be important in modulating aggregation and fibrillation. The presence of seven imperfect repeats in α-synuclein raises the question of whether mutations corresponding to E46K in the other imperfect KTKE(Q)GV repeats cause similar effects on aggregation and fibrillation, as well as their propensities to form α-helices. We therefore substituted the amino acid residue at the position corresponding to E46 in each of the seven imperfect repeats to a Lys residue (Harada et al. 2009).

Consistent with a previous report (Fredenburg et al. 2007), the E46K mutant has a shorter lag-time and higher maximal fluorescence intensity than WT. The Q24K mutation, located in the N-terminal region, greatly reduced the lag-time and had much lower maximal fluorescence intensity compared with that of WT. To compare the total aggregation of WT and mutant α-synucleins, a light scattering assay was performed. The ΔOD_{330} values of most of the E/K mutants were more than twice that of WT, except for Q24K, which was approximately 1.5 times that of WT. We also investigated the morphology of α-synuclein fibrils and aggregates by AFM imaging to understand better why the maximal fluorescence intensity decreased while the light scattering increased for the E(Q)/K mutants. The results of AFM imaging (Fig. 5) demonstrate that the effect of these mutations on the morphology varies according to their location. The fibrils of all mutants were different from those of WT, and mature twisted fibrils can be observed in the Q24K mutant. These data indicated that the aggregation ability of the Q24K mutant may be lower than that of the other E/K mutants, even though the Q24K mutant can form amyloid fibrils. This suggests that the N-terminal region may be important in modulating the morphology, as well as aggregation and fibrillation.

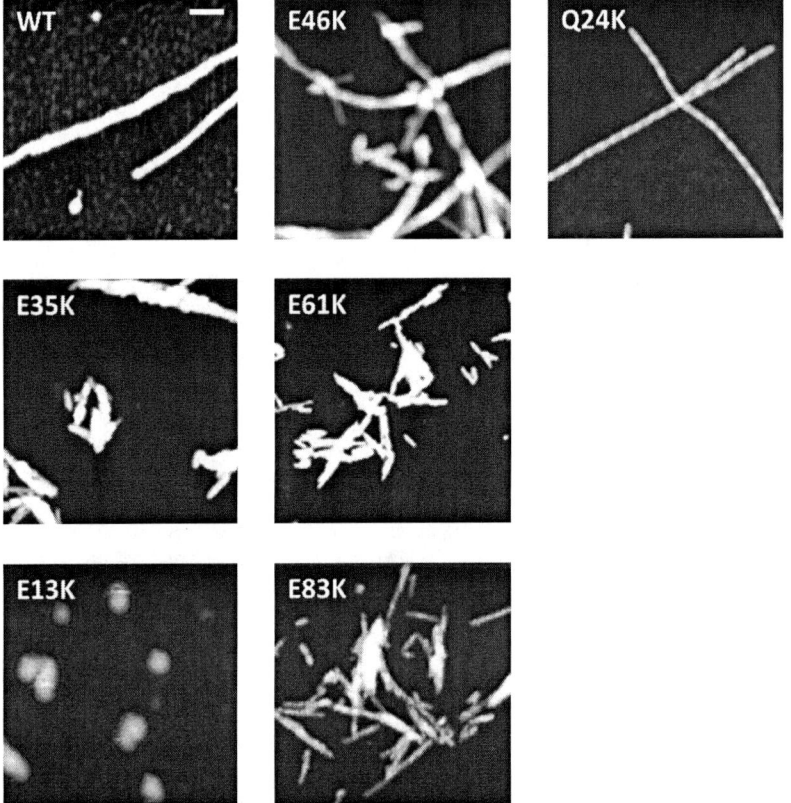

Fig. 5. AFM images of α-synuclein and its mutants. The samples for AFM imaging were prepared by incubation for 5 days. Scale bar = 250 nm. *This figure is unpublished.*

Amino Acid Substitutions in NAC Region

We focused on the NAC region, where hydrophobic amino acid residues are repeated. We assumed that the hydrophobic feature of this region and its potential β-strand formation may trigger the β-strand-dominant folding status of α-synucleins. Introduction of the β-sheet breaker Pro would be expected to decrease the β-sheet formation ability, while the introduction of Thr decreases the hydrophobicity of the region. Therefore, we constructed three different variants, Val70Pro, Val70Thr, and the Val70Thr/Val71Thr double mutant (Sode et al. 2005).

Fibril formation of Val70Thr, Val70Pro, and Val70Thr/Val71Thr was shown to be slower than that of WT (Fig. 6), especially Val70Thr/Val71Thr, which shows no obvious fluorescence enhancement even after incubation for 50 h. In contrast, WT shows fluorescence enhancement after incubation for 10 h. Finally, the maximal value of fluorescence intensity in Val70Thr/Val71Thr is less than one-tenth that of WT.

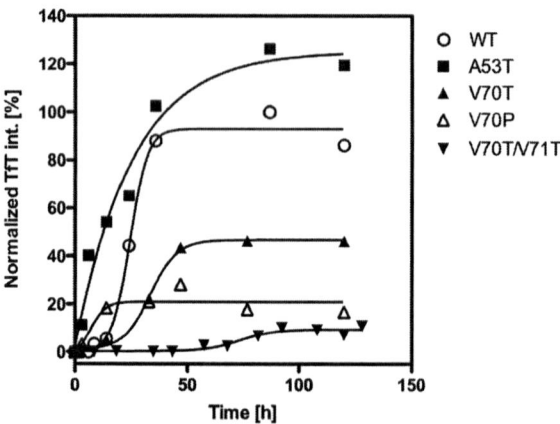

Fig. 6. Effect of amino acid substitution into NAC region on amyloid fibril formation of α-synuclein. Time course of fibril formation of α-synuclein and its mutants as determined by ThT fluorescence assay analysis: WT (*white circles*), Ala53Thr (*black squares*), Val70Thr (*black triangles*), Val70Pro (*white triangles*), and Val70Thr/Val71Thr (*black inverted triangles*). This figure is unpublished.

Although the amino acid substitution of Val70 to Pro or Thr resulted in reduction of fibril formation, both mutants showed similar or higher nonstructural aggregation compared with WT. These results indicate that introducing amino acid substitution in α-synuclein may not only influence the status of the folded structure and fibril formation but also the stability of the unfolded status. However, the Val70Thr/Val71Thr double mutant showed both a reduction in fibril formation and nonstructural aggregation. The simultaneous substitutions of two hydrophobic residues to hydrophilic residues may have prevented α-synuclein fibril formation by increasing the stability of the non-structural natively unfolded status.

Construction and Characterization of C-terminal Truncated α-synucleins

It is known from patients suffering from α-synucleinopathy that the C-terminal-truncated forms of α-synuclein are found in LBs (Li et al. 2005). The propensity of α-synuclein to aggregate into amyloid fibrils was demonstrated by *in vitro* studies to be accelerated by the truncation of its C-terminal region. Furthermore, truncated α-synuclein has been identified in normal brain. These observations indicate that the truncation of the C-terminal region of α-synuclein is related to the pathogenesis of PD.

We characterized the aggregation and fibrillation of truncated α-synuclein110, 119 and 133 (α-Syn110, 119 and 133), where the number corresponds to the position of the carboxyl terminus (Kim et al. 2010). The acidic C-terminal region of α-synuclein contains ten Glu and five Asp residues. Removing these negatively charged amino acids would be expected to decrease the repulsion between the C-termini, thus enhancing the interaction between hydrophobic regions contained in the NAC region, leading to faster aggregation and/or fibrillation. The truncated α-Syn110 showed a much faster rate of fibril formation than WT, while that of α-Syn119 was in between the two. Surprisingly, the fibril formation of α-Syn133 was slower than that of full-length α-synuclein. However, it was determined that the final total quantity of fibril and aggregates of α-Syn133 is at the same level as for WT. Considering that the fibril elongation process is influenced by the status of soluble oligomers, the decreased kinetics of α-Syn133 fibril formation might reflect the change of its oligomeric status.

Full-length and truncated α-synuclein samples were incubated four days with shaking at 37°C in the presence of PC12 cells. The cytotoxicity of these protein samples was then evaluated by adenylate kinase assay (AK assay). In contrast to the full-length α-synuclein, the C-terminal-truncated protein was highly toxic to PC12 cells (Fig. 7), in good agreement with cytotoxicity evaluation results of α-Syn119 based on CC8 assay.

These results indicate that all three characteristic regions of α-synuclein are responsible for maintaining the protein's natively unfolded status and also forming amyloid nanostructures.

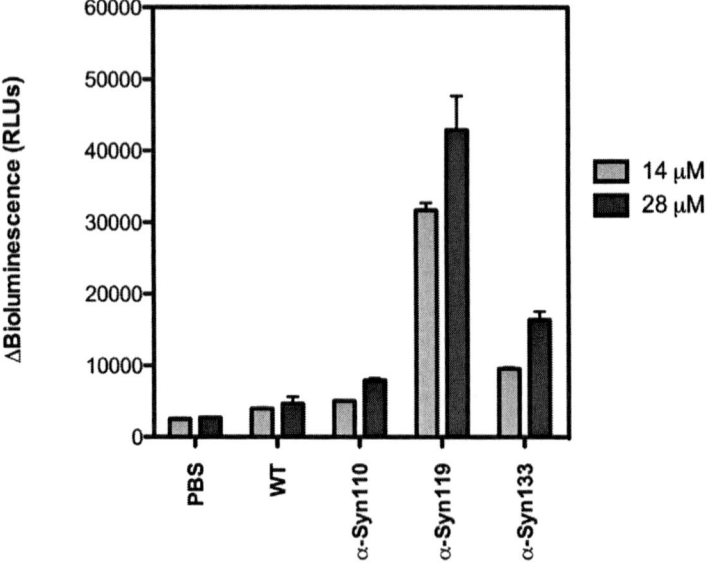

Fig. 7. Cytotoxicity evaluation of truncated α-synuclein. Full-length α-synuclein and truncated α-synuclein (final concentration 14 or 28 μM) were added to PC12 cells with shaking at 37°C. After 96 h of incubation, the release of adenylate kinase from the damaged cells was measured by the luminescence of luciferase. n = 3 and error bar = standard deviation. *This figure is unpublished.*

QUATERNARY STRUCTURE ENGINEERING FOR α-SYNUCLEIN NANOSTRUCTURE REGULATION

Three different quaternary structure engineering strategies were employed to regulate α-synuclein nanostructure. Whereas α-synucleins harboring mutations within the NAC region were less prone to form amyloid fibril, the C-terminal truncated α-synucleins exhibited higher propensity. The first approach was therefore to investigate the interaction of WT and mutant α-synucleins, and their effect on the formation of nanostructure. The second approach was to investigate the interaction of α-synuclein with peptides, whose sequences were either derived from partial α-synuclein sequences or designed *in silico*. The final approach was to investigate the effect of co-existence of WT α-synuclein with α-synuclein, or its peptide fragment, that is modified with an inhibitor of amyloid formation. The molecule chosen as the potential inhibitor of amyloid formation was pyrroloquinoline quinone.

The co-existence of wild Type α-synuclein with NAC Region-mutated α-synuclein or the C-terminal-truncated α-synuclein

As described above, the Val70Thr/Val71Thr double mutant of α-synuclein is less prone to form amyloid fibril. To investigate its dominant negative effect on fibril formation, the double mutant was co-incubated with either the WT α-synuclein or Ala53Thr, the mutant α-synuclein observed in the familial PD. The Val70Thr/Val71Thr double mutant exhibited an inhibitory effect on the fibril formation and aggregation of both WT and Ala53Thr, as well as in the formation of the protofibril intermediate (Fig. 8) (Sode et al. 2005). The formation of both protofibril and fibril are believed to result from the recognition and interaction of a particular structure within the partially folded α-synuclein. Although Val70Thr/Val71Thr α-synuclein may form a partially folded intermediate, the instability of this intermediate structure may prevent the formation of a stable oligomer, thereby preventing fibril formation. In a mixture of Val70Thr/Val71Thr and either WT or Ala53Thr, such an "unstable" intermediate structure of Val70Thr/Val71Thr might be recognized as similar to an intermediately folded structure of either WT or Ala53Thr and incorporated in the growing oligomer. However, due to the unstable nature of this intermediate structure formed by Val70Thr/Val71Thr, the constructed oligomer composed of both Val70Thr/Val71Thr and either WT or Ala53Thr would be easily disrupted, thus preventing fibril formation.

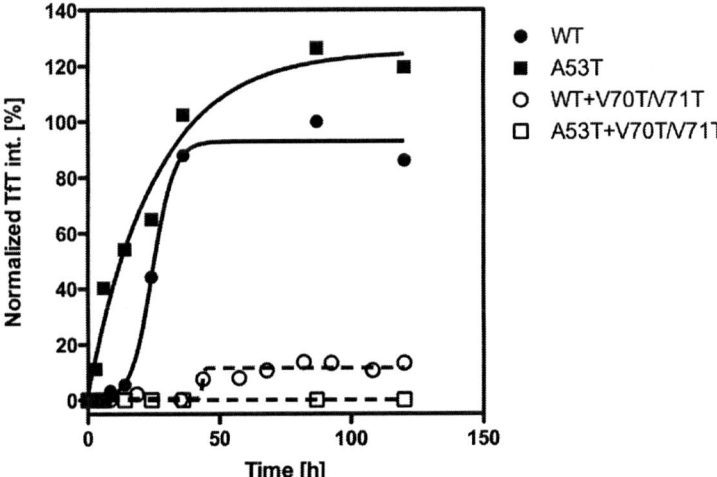

Fig. 8. Inhibition effect of Val70Thr/Val71Thr on WT and Ala53Thr fibril formation. Time course of fibril formation of α-synuclein and its mixture with mutants as determined by ThT fluorescence assay analysis: WT (*black circles*), Ala53Thr (*black squares*), the equimolar mixtures WT+Val70Thr/Val71/Thr (*white circles*), and Ala53Thr+Val70Thr/Val71Thr (*white squares*). *This figure is unpublished.*

The aggregation of full-length α-synuclein has been reported to be accelerated by the addition of C-terminal-truncated variants (Murray et al. 2003). This study suggested a C-terminal-truncated α-synuclein seeding mechanism, in which the truncated fragments nucleate full-length α-synuclein aggregation through the formation of hybrid protofibrils and then fibrils, thus shortening the lag-time of fibril formation. These truncated variants, α-Syn119 and α-Syn133, co-exist with full-length α-synuclein in both the normal and pathogenic brain. To investigate the effect of these truncated variants on the fibril formation of α-synuclein, mixtures were monitored by ThT fluorescence and light scattering analysis. When α-Syn110 and α-Syn119, but not α-Syn133, coexisted with WT at a ratio of 1:1, the fibrillation rate increased, the fibrillation lag-time was significantly reduced, and the total α-synuclein fibril formation was dramatically enhanced (Fig. 9). The co-existence of full-length and truncated α-Syn110 or α-Syn119 has a synergistic effect on aggregation and fibrillation.

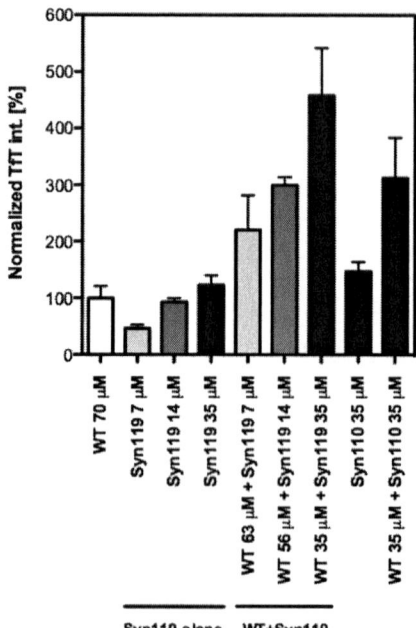

Fig. 9. Impact of C-terminal-truncated α-synuclein on full-length α-synuclein amyloid formation. The amount of amyloid after 80 hours of incubation of the sample with 70 μM full-length α-synuclein alone shown as the control (100%, *white bar*). 7, 14, and 35 μM of α-Syn119 and 35 μM of α-Syn110 (*black bar*) and the mixture protein (35 μM full-length α-synuclein+35 μM α-Syn119, 56 μM full-length α-synuclein+14 μM α-Syn119, 63 μM full-length α-synuclein+7 μM α-Syn119 and 35 μM full-length α-synuclein+35 μM α-Syn110, *gray bar*) was incubated in triplicate in PBS buffer with 0.02% NaN_3, pH 7.4. The fibril amount was determined by ThT analysis. n=3 and error bar = standard deviation. *This figure is unpublished.*

The Co-existence of α-synuclein Partial Peptide with α-synuclein

The NAC peptide and shorter peptide fragments of NAC has been reported to form fibrillar structures, however, the characteristics of these fibrils were different from those of full-length α-synuclein. Furthermore, such a peptide fragment not only forms fibril by itself but also promotes fibrillation of full-length α-synuclein (Giasson et al. 2001). We therefore expect to regulate nanostructures by using peptide fragments of the mutated NAC region. As mentioned above, the fibril formation propensity of α-synuclein was regulated by combining with Val70Thr/Val71Thr, the α-synuclein containing a double mutation within the NAC region. Therefore, the nanostructure formation of α-synuclein was investigated in the presence of a variety of NAC region-derived peptide sequences.

We investigated the effects of two peptide fragments of the NAC region on nanostructures of full-length α-synuclein amyloid fibril (Kobayashi et al. 2008). We used two peptide fragments; one corresponding to residues 66–82 of WT (WT_{66-82}) and the other has the corresponding amino acid sequence of the double mutant Val70Thr/Val71Thr (DM_{66-82}). The aggregates formed by co-incubation of full-length α-synuclein and either WT_{66-82} or DM_{66-82} was also analyzed by AFM (Fig. 10). We found that the peptide fragments affect the fibril morphology of full-length α-synuclein. Moreover, these effects depend on the peptide sequences.

From the above result and previously reported findings (Sode et al. 2005), we propose a mechanism of these effects. Full-length α-synuclein forms amyloid fibril via a "partially folded intermediate" and a "fibril nucleus". In the presence of peptides, these peptides interact with the fibril nucleus of full-length α-synuclein, resulting in the formation of hetero-oligomers. The resulting structural change of the fibril nucleus induces the morphology of the full-length α-synuclein amyloid fibril to also change.

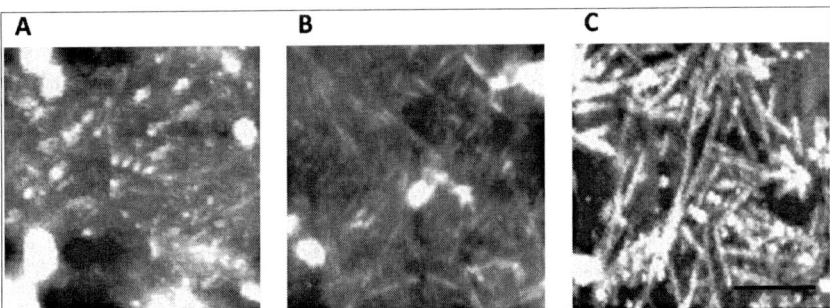

Fig. 10. Effects of peptides on full-length α-synuclein fibril morphology. The AFM images of the aggregates formed by full-length α-synuclein alone (A) and its mixture with WT_{66-82} (B) and DM_{66-82} (C), respectively. Scale bar=500 nm. *This figure is unpublished.*

A report by another group suggests that fibril seeds of various amyloid-forming proteins have a significant effect on the fibrillation of α-synuclein (Yagi et al. 2005). Similarly, the amyloid fibril morphology of prion protein is reported to be affected by the morphology and secondary structure of the template seeds. These reports support the idea that amyloid fibril morphology depends on the structure of fibril nucleus.

We then investigated the impact of the presence of *in silico* designed peptide sequences on the regulation of α-synuclein nanostructure (Abe et al. 2007). The computational screening would be useful to screen ligands against a defined target. Since N-mer peptides have a huge sequence variety (20^n), it is not sufficient to screen the huge sequence space randomly to obtain target peptides. We previously reported an evolutional screening method using genetic algorithms (GAs) that mimic Darwinian evolution (Ikebukuro et al. 2005; Noma and Ikebukuro 2006). GAs can greatly reduce the number of candidates to be evaluated and they have been effectively applied to screening binding poses for docking simulations and the optimization of lead compounds. In such screenings, the peptide sequence can be optimized by evaluating the biological or chemical activity. We previously proposed a peptide screening method using GAs combined with docking simulations and designated it as *in silico* panning (Yagi et al. 2007). We carried out *in silico* panning in order to screen for peptides that bind to the α-synuclein's amyloidogenic region and affect α-synuclein fibril formation (Abe et al. 2007). We selected an 11-mer peptide, α-synuclein 68–78 (GAVVTGVTAVA), as the target region and carried out a total of 6 rounds of screening. The screened peptides were ranked by their intermolecular energy and we evaluated the four top peptides (QSTQ, GSQQ, SQTQ, and AQTQ) and one inferior peptide (PTYF).

We co-incubated α-synuclein with 5-fold molar excess of the screened peptides to determine their effects on α-synuclein aggregation. The α-synuclein solutions containing any of the top-ranking peptides showed greater increases in the fluorescence intensity of ThT than that of the solution without peptide, indicating that the peptides promoted α-synuclein fibrillation. Especially, the α-synuclein solution with the top peptide, QSTQ, showed a fluorescence intensity that was more than 1.5 times stronger than that of the solution without peptide (Fig. 11).

The *in silico*-designed peptides may have induced conformational changes by binding to the hydrophobic region and accelerated the α-synuclein aggregation and subsequent fibrillation. Although the approaches based on both α-synuclein-derived partial peptides or *in silico*-designed peptides were not successful in inhibiting amyloid formation of α-synuclein, these approaches were revealed to be useful in constructing novel α-synuclein-based nanostructures. These strategies may be further applied for the design and preparation of nanostructure scaffolds.

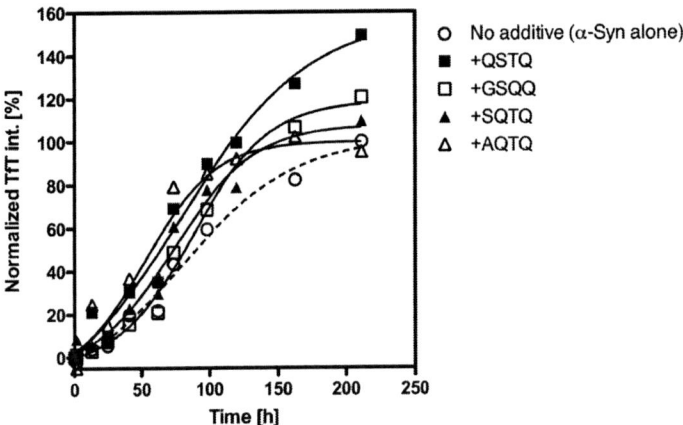

Fig. 11. Effect of *in silico* screened peptides on full-length α-synuclein fibrillation. Time course of fibril formation of α-synuclein and its mixture with peptides as determined by ThT fluorescence assay analysis: α-synuclein (140 μM) without peptide, (*white circles*); QSTQ, (*black squares*); GSQQ, (*white squares*); SQTQ (*Black triangles*); AQTQ, (*white triangles*). All peptide concentrations were 700 μM. *This figure is unpublished.*

The Effect of Co-existence of Wild Type α-synuclein with the Pyrroloquinoline Quinone-modified α-synuclein or its Partial Peptide Sequence

Considering that the formation of amyloid fibrils as well as their precursor oligomers is cytotoxic, agents that prevent the formation of oligomers and/or fibrils may lead to a novel therapeutic approach for PD and other α-synuclein-related diseases. Pyrroloquinoline quinone (PQQ) (Fig. 12) was reported to inhibit α-synuclein fibril formation (Kobayashi et al. 2006). PQQ prevents not only the amyloid fibril formation and aggregation of full-length α-synuclein but also of truncated α-synuclein, in a PQQ-concentration-dependent manner. Moreover, PQQ reduces the cytotoxicity of both full-length and truncated variants of α-synuclein. Since recent studies indicate that the oligomeric intermediate of fibrillation has a higher toxicity than the fibril, the presence of PQQ likely prevents the oligomer formation of both full-length and truncated α-synuclein.

Incubation of α-synuclein with PQQ for over 200 h resulted in the formation of PQQ-modified α-synuclein (α-Syn-PQQ), as confirmed by spectrophotometric analysis (Kobayashi et al. 2006). Considering that PQQ forms a conjugate with α-synuclein and that PQQ has reactive quinone groups like the oxidized form of baicalein, it is highly probable that PQQ binds with α-synuclein via a Schiff base. Interestingly, the incubation of α-Syn-PQQ did not cause an increase in ThT fluorescence intensity (Fig. 13), reflecting its inability to form amyloid fibrils and consistent with the

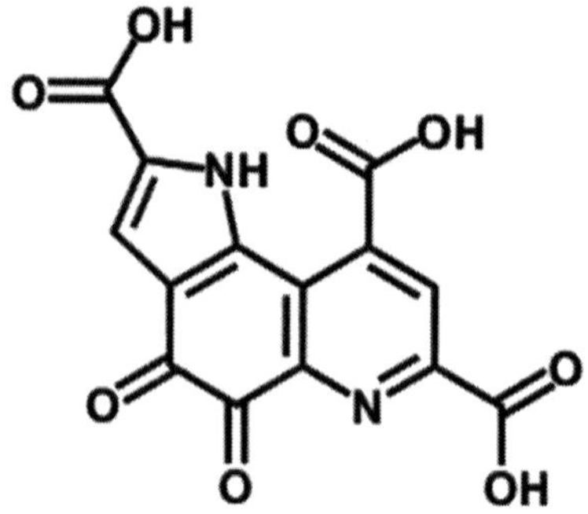

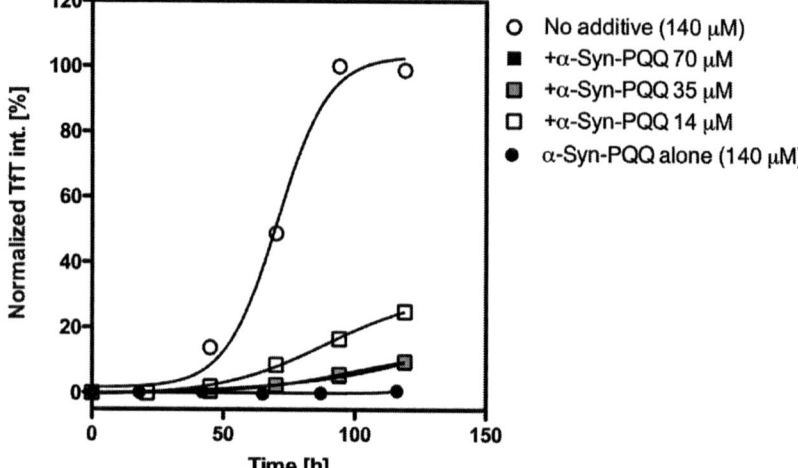

Fig. 12. Structure of pyrroloquinoline quinone (PQQ).

Fig. 13. Inhibition effect of PQQ-conjugated α-synuclein on WT fibril formation. The time course of fibril formation was monitored by ThT fluorescence intensity. The PQQ-modified α-synuclein (α-Syn-PQQ) was produced by co-incubating α-synuclein (140 μM) and PQQ (2 mM) at 37°C with shaking for over 100 h. The inhibition effect of the conjugate on the fibril formation of intact α-synuclein was investigated by mixing the α-Syn-PQQ (f.c. 70 μM, 35 μM or 14 μM) with α-synuclein (f.c. 140 μM): no additive (*white circles*), +70 μM α-Syn-PQQ (*black squares*), +35 μM α-Syn-PQQ (*gray squares*), +14 μM α-Syn-PQQ (*white squares*), and 140 μM α-Syn-PQQ alone (black *circles*). *This figure is unpublished.*

observed inhibitory effect of PQQ on amyloid formation. Furthermore, α-Syn-PQQ also prevented the amyloid fibril formation of unmodified α-synuclein in a dose-dependent manner.

The proteolytic products of α-Syn-PQQ also showed an inhibitory effect on the amyloid formation of full-length α-synuclein (Kobayashi et al. 2009). After Glu-C protease digestion of α-synuclein and α-Syn-PQQ, the proteolytic products were separated by RPC. One of the α-Syn-PQQ proteolytic products (SP-7) caused a marked decrease in ThT fluorescence when added to α-synuclein, whereas the corresponding proteolytic product from the unmodified α-synuclein did not (Fig. 14). Not all PQQ-modified peptides inhibit full-length α-synuclein fibrillation, indicating that this inhibitory effect is likely peptide sequence-specific. These results suggest that PQQ-modified α-synuclein partial peptides show a protein sequence-specific inhibitory effect, which would have significant advantages for developing future targeting nanomedicines.

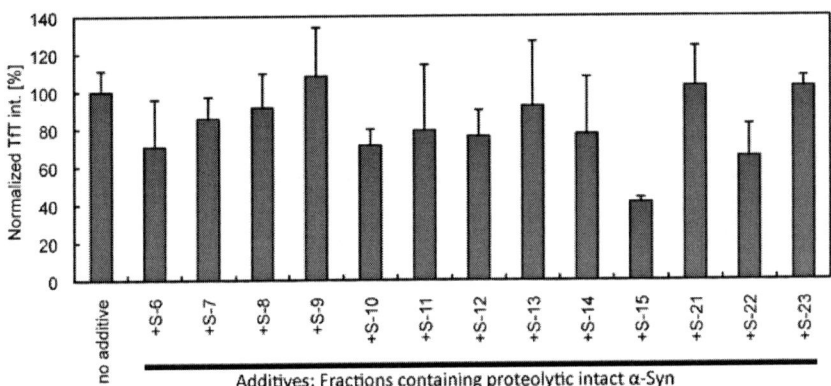

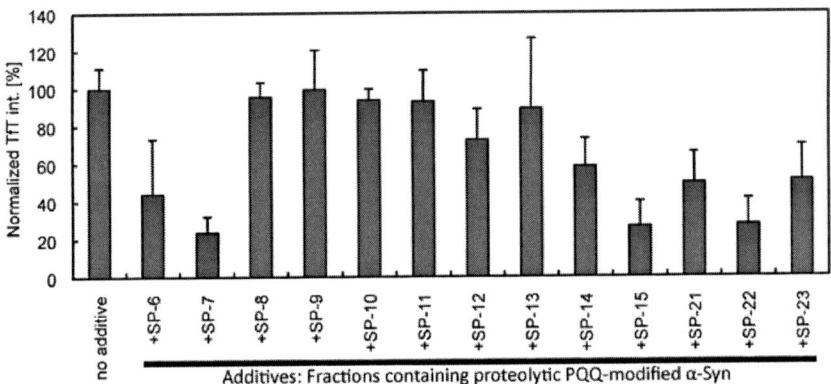

Fig. 14. The Inhibitory effects of the proteolytic products of α-synuclein and PQQ-modified α-synuclein on the amyloid fibril formation of full-length α-synuclein. Glu-C-digested α-synuclein and α-synuclein-PQQ were separated by means of RPC. After lyophilization, the effects of each fractionated Glu-C-digested α-synuclein (*top*) and α-synuclein-PQQ (*bottom*) on full-length α-synuclein amyloid formation were investigated. The vertical and horizontal axes indicate the relative final ThT intensity and the added fraction, respectively. The results are expressed as percentages of the value of α-synuclein alone (no additive), which is set at 100%. n=3 and error bar = standard deviation. *This figure is unpublished.*

Conclusion

In this chapter, our challenges in regulating the nanostructure formation of α-synuclein were summarized. Our challenges were categorized into two approaches, at the primary and quaternary structure levels.

By introducing amino acid substitutions or deletions into the sequence, the regulation of nanostructure formation is possible. The regulation of α-synuclein nanostructure at the quaternary structure engineering levels suggests the future potential of designing peptide-based nanostructure scaffold, and also α-synuclein-targeting nanomedicine. Especially, the investigation of WT α-synuclein with the PQQ-modified α-synuclein or its partial peptide sequence achieved the inhibition of amyloid formation, suggesting its potential as a protein sequence-specific targeting inhibitor.

The peptides and proteins are not the only potential candidates in the regulation of α-synuclein nanostructure. We are currently focusing and expecting the application of DNA aptamers. Aptamers are target-binding molecules that consist of DNA or RNA and can be selected from random sequence pools *in vitro* using a process referred to as "SELEX" (systematic evolution of ligands by exponential enrichment) (Ellington and Szostak 1990). Aptamers can be raised against almost any type of target, such as small molecules, chemical compounds, and proteins. They have the potential of undergoing structural change when they bind to the target. Aptamers have recently been obtained against amyloid-forming proteins such as amyloid β and prion (Ogasawara et al. 2007). We have now identified the first DNA aptamers against α-synuclein (Tsukakoshi et al. 2010). The identified aptamer, bound not only to the α-synuclein monomer but also to the α-synuclein oligomer. Because α-synuclein has a native, unfolded structure, the selected aptamer may recognize certain regions of the α-synuclein amino acid sequence.

We have been screening for aptamers against several targets and trying to apply those aptamers to the creation of novel biosensing systems. In fact, we have succeeded in obtaining new aptamers and developing original sensors in previous studies (Yoshida et al. 2006). These aptamers are expected to be applied not only to biosensors as molecular recognition elements for the detection of α-synuclein but future targeting nanomedicine.

Applications to Areas of Health and Disease

Several *in vitro* studies have suggested that α-synuclein's propensity to oligomerize and form fibrils may play a crucial role in its toxicity (Rajagopalan and Andersen 2001). Thus, the agents that prevent the formation of oligomers and/or fibrils might provide a novel therapeutic approach in PD and other α-synuclein-related LB diseases (Zhu et al.

2004). Considering that α-synuclein is a natively unfolded protein, we are particularly interested in the regulatory factors maintaining the unfolded status as well as triggering the formation of supra-molecular nanostructures. This information may lead to novel strategies for designing molecules to prevent α-synuclein-induced neurodegeneration.

Key Facts of Amyloid Fibril

- Amyloid fibril is a proteinaceous fibrillar aggregate with a diameter of ~10 nm, composed of a cross-β-sheet structure.
- Amyloid fibril formation is caused by protein misfolding and thought to be related to disease onset, such as Alzheimer's and Parkinson's diseases.
- α-Synuclein is the causative protein of several neurodegenerative diseases, such as Parkinson's disease.
- It has recently been shown that the oligomeric intermediate of fibril formation has high cytotoxicity.
- Several *in vitro* studies have suggested that α-synuclein's propensity to oligomerize and form fibrils may play a crucial role in its toxicity.
- Agents that prevent the formation of oligomers and/or fibrils might provide a novel therapeutic approach to treating PD and other α-synuclein-related neurodegenerative diseases.

Summary Points

- The primary structure of α-synuclein can be divided into three characteristic regions: an amphipathic repeat-containing N-terminal region, a central hydrophobic NAC region, and an acidic C-terminal region. All three regions of α-synuclein are responsible for maintaining the natively unfolded status and also in forming amyloid nanostructures.
- Primary structure engineering including introducing amino acid substitutions or deletions into α-synuclein enables us to regulate fibrillation and aggregation ability, amyloid nanostructure, and cytotoxicity of α-synuclein.
- Quaternary structure engineering of α-synuclein can regulate α-synuclein nanostructure. For quaternary structure engineering, we investigated the effects of the interaction of α-synuclein with (i) mutant α-synucleins, (ii) peptides derived from partial sequence of α-synuclein, or designed *in silico*, and (iii) modified α-synuclein or its partial peptide sequence.
- Nanostructure regulation of amyloid-forming proteins including α-synuclein would be applied for novel therapeutic strategy against conformational disease. Quaternary structure engineering ligands that

show the protein sequence-specific inhibitory effect will be applied for the future targeting nanomedicine.
- Not only peptides and proteins, but also DNA aptamers, target-binding DNA ligands, are focused on and expected as potential candidates in target-specific nanomedicine.

Definitions of Key Terms

ThT: is a dye utilized for characterizing the presence of amyloid fibrils and their rates of formation. ThT dye binds amyloid fibril specifically, and fluorescence emission at 482 nm is enhanced when excited at 450 nm.

OD: measurement reflects amount of insoluble aggregate (including amyloid fibril and amorphous aggregate).

Adenylate kinase (AK) assay: measures quantitatively the release of AK from the damaged cells to determine cytotoxicity of compounds against cells. The released AK converts ADP to ATP, which is measured by the bioluminescence assay, thus allowing the accurate and sensitive determination of cytotoxicity.

PQQ: is a redox cofactor in bacteria and is also found in plants and animals. PQQ has been drawing attention as a novel nutritional factor, due to its positive effect on cellular growth and its potential as a key pharmacological compound; the focus has been its characteristic as a radical scavenger.

Abbreviations

PD	:	Parkinson's disease
DLB	:	dementia with Lewy body
LB	:	Lewy body
NAC	:	non-amyloid β component of Alzheimer's disease amyloid
WT	:	wild type
OD	:	optical density
AFM	:	atomic force microscopy
AK assay	:	adenylate kinase assay
ThT	:	Thioflavin T
GAs	:	genetic algorithms
PQQ	:	pyrroloquinoline quinone
SELEX	:	systematic evolution of ligands by exponential enrichment

References

Abe, K., N. Kobayashi, K. Sode and K. Ikebukuro. 2007. Peptide ligand screening of alpha-synuclein aggregation modulators by *in silico* panning. BMC Bioinformatics 8: 451.

Chartier-Harlin, M.C., J. Kachergus, C. Roumier, V. Mouroux, X. Douay, S. Lincoln, C. Levecque, L. Larvor, J. Andrieux, M. Hulihan, N. Waucquier, L. Defebvre, P. Amouyel, M. Farrer and A. Destee. 2004. Alpha-synuclein locus duplication as a cause of familial Parkinson's disease. Lancet 364: 1167–1169.

Conway, K.A., J.D. Harper and P.T. Lansbury. 1998. Accelerated in vitro fibril formation by a mutant alpha-synuclein linked to early-onset Parkinson disease. Nat. Med. 4: 1318–1320.

Ellington, A.D. and J.W. Szostak. 1990. *In vitro* selection of RNA molecules that bind specific ligands. Nature 346: 818–822.

Fredenburg, R.A., C. Rospigliosi, R.K. Meray, J.C. Kessler, H.A. Lashuel, D. Eliezer and P.T. Lansbury Jr. 2007. The impact of the E46K mutation on the properties of alpha-synuclein in its monomeric and oligomeric states. Biochemistry 46: 7107–7118.

Giasson, B.I., I.V. Murray, J.Q. Trojanowski and V.M. Lee. 2001. A hydrophobic stretch of 12 amino acid residues in the middle of alpha-synuclein is essential for filament assembly. J. Biol. Chem. 276: 2380–2386.

Harada, R., N. Kobayashi, J. Kim, C. Nakamura, S.W. Han, K. Ikebukuro and K. Sode. 2009. The effect of amino acid substitution in the imperfect repeat sequences of alpha-synuclein on fibrillation. Biochim. Biophys. Acta 1792: 998–1003.

Ikebukuro, K., Y. Okumura, K. Sumikura and I. Karube. 2005. A novel method of screening thrombin-inhibiting DNA aptamers using an evolution-mimicking algorithm. Nucleic. Acids Res. 33: e108.

Kim, J., R. Harada, M. Kobayashi, N. Kobayashi and K. Sode. 2010. The inhibitory effect of pyrroloquinoline quinone on the amyloid formation and cytotoxicity of truncated alpha-synuclein. Mol. Neurodegener. 5: 20.

Kobayashi, M., J. Kim, N. Kobayashi, S. Han, C. Nakamura, K. Ikebukuro and K. Sode. 2006. Pyrroloquinoline quinone (PQQ) prevents fibril formation of alpha-synuclein. Biochem. Biophys. Res. Commun. 349: 1139–1144.

Kobayashi, N., S. Han, C. Nakamura and K. Sode. 2008. Nanostructure Fabrication Based on Engineered α-Synuclein. Nanobiotechnol. 4: 50–55.

Kobayashi, N., J. Kim, K. Ikebukuro and K. Sode. 2009. The Inhibition of Amyloid Fibrillation Using the Proteolytic Products of PQQ-Modified a-Synuclein. The Open Biotechnology Journal 3: 40–45.

Kruger, R., W. Kuhn, T. Muller, D. Woitalla, M. Graeber, S. Kosel, H. Przuntek, J.T. Epplen, L. Schols and O. Riess. 1998. Ala30Pro mutation in the gene encoding alpha-synuclein in Parkinson's disease. Nat. Genet. 18: 106–108.

Li, W., N. West, E. Colla, O. Pletnikova, J.C. Troncoso, L. Marsh, T.M. Dawson, P. Jakala, T. Hartmann, D.L. Price and M.K. Lee. 2005. Aggregation promoting C-terminal truncation of alpha-synuclein is a normal cellular process and is enhanced by the familial Parkinson's disease-linked mutations. Proc. Natl. Acad. Sci. USA 102: 2162–2167.

Murray, I.V., B.I. Giasson, S.M. Quinn, V. Koppaka, P.H. Axelsen, H. Ischiropoulos, J.Q. Trojanowski and V.M. Lee. 2003. Role of alpha-synuclein carboxy-terminus on fibril formation *in vitro*. Biochemistry 42: 8530–8540.

Noma, T. and K. Ikebukuro. 2006. Aptamer selection based on inhibitory activity using an evolution-mimicking algorithm. Biochem. Biophys. Res. Commun. 347: 226–231.

Ogasawara, D., H. Hasegawa, K. Kaneko, K. Sode and K. Ikebukuro. 2007. Screening of DNA aptamer against mouse prion protein by competitive selection. Prion. 1: 248–254.

Polymeropoulos, M.H., C. Lavedan, E. Leroy, S.E. Ide, A. Dehejia, A. Dutra, B. Pike, H. Root, J. Rubenstein, R. Boyer, E.S. Stenroos, S. Chandrasekharappa, A. Athanassiadou, T. Papapetropoulos, W.G. Johnson, A.M. Lazzarini, R.C. Duvoisin, G. Di Iorio, L.I. Golbe and R.L. Nussbaum. 1997. Mutation in the alpha-synuclein gene identified in families with Parkinson's disease. Science 276: 2045–2047.

Rajagopalan, S. and J.K. Andersen. 2001. Alpha synuclein aggregation: is it the toxic gain of function responsible for neurodegeneration in Parkinson's disease? Mech. Ageing Dev. 122: 1499–1510.

Recchia, A., P. Debetto, A. Negro, D. Guidolin, S.D. Skaper and P. Giusti. 2004. Alpha-synuclein and Parkinson's disease. FASEB J. 18: 617–626.

Singleton, A.B., M. Farrer, J. Johnson, A. Singleton, S. Hague, J. Kachergus, M. Hulihan, T. Peuralinna, A. Dutra, R. Nussbaum, S. Lincoln, A. Crawley, M. Hanson, D. Maraganore, C. Adler, M.R. Cookson, M. Muenter, M. Baptista, D. Miller, J. Blancato, J. Hardy and K. Gwinn-Hardy. 2003. alpha-Synuclein locus triplication causes Parkinson's disease. Science 302: 841.

Sode, K., S. Ochiai, N. Kobayashi and E. Usuzaka. 2007. Effect of reparation of repeat sequences in the human alpha-synuclein on fibrillation ability. Int. J. Biol. Sci. 3: 1–7.

Sode, K., E. Usuzaka, N. Kobayashi and S. Ochiai. 2005. Engineered alpha-synuclein prevents wild type and familial Parkin variant fibril formation. Biochem. Biophys. Res. Commun. 335: 432–436.

Tsukakoshi, K., R. Harada, K. Sode and K. Ikebukuro. 2010. Screening of DNA aptamer which binds to alpha-synuclein. Biotechnol. Lett. 32: 643–648.

Uversky, V.N. and A.L. Fink. 2002. Amino acid determinants of alpha-synuclein aggregation: putting together pieces of the puzzle. FEBS Lett. 522: 9–13.

Yagi, H., E. Kusaka, K. Hongo, T. Mizobata and Y. Kawata. 2005. Amyloid fibril formation of alpha-synuclein is accelerated by preformed amyloid seeds of other proteins: implications for the mechanism of transmissible conformational diseases. J. Biol. Chem. 280: 38609–38616.

Yagi, Y., K. Terada, T. Noma, K. Ikebukuro and K. Sode. 2007. *In silico* panning for a non-competitive peptide inhibitor. BMC Bioinformatics 8: 11.

Yoshida, W., K. Sode and K. Ikebukuro. 2006. Aptameric enzyme subunit for biosensing based on enzymatic activity measurement. Anal. Chem. 78: 3296–3303.

Zarranz, J.J., J. Alegre, J.C. Gomez-Esteban, E. Lezcano, R. Ros, I. Ampuero, L. Vidal, J. Hoenicka, O. Rodriguez, B. Atares, V. Llorens, E. Gomez Tortosa, T. del Ser, D.G. Munoz and J.G. de Yebenes. 2004. The new mutation, E46K, of alpha-synuclein causes Parkinson and Lewy body dementia. Ann. Neurol. 55: 164–173.

Zhu, M., S. Rajamani, J. Kaylor, S. Han, F. Zhou and A.L. Fink. 2004. The flavonoid baicalein inhibits fibrillation of alpha-synuclein and disaggregates existing fibrils. J. Biol. Chem. 279: 26846–26857.

9

Nanomaterials for Stem Cell Imaging in the Central Nervous System

Aniruddh Solanki,[1,a] Shreyas Shah,[1,b] Michael H. Koucky[2] and Ki-Bum Lee[3,*]

ABSTRACT

Stem cells hold great potential for the treatment of various injuries and degenerative diseases as they possess the unique ability to self-renew and differentiate into specific cell lineages. They hold great promise for treating diseases and injuries especially within the central nervous system. For instance, the role of stem cells in regenerative medicine has been suggested for degenerative diseases such as Parkinson's and Alzheimer's. Not only are they promising for degenerative diseases, but also for treating spinal cord injuries

---

[1]Department of Chemistry and Chemical Biology, Rutgers University 610 Taylor Road, Piscataway, NJ 08854, USA.
[a]E-mail: asolanki@rutgers.edu
[b]E-mail: spshah99@rutgers.edu
[2]Department of Biomedical Engineering, Rutgers University, 581 Taylor Road, Piscataway, NJ 08854, USA; E-mail: mhkoucky@eden.rutgers.edu
[3]Department of Chemistry and Chemical Biology, Rutgers Stem Cell Research Center, Rutgers University, 610 Taylor Road, Piscataway, NJ 08854, USA; E-mail: kblee@rutgers.edu
*Corresponding author

List of abbreviations after the text.

and traumatic brain injuries. Such injuries significantly impair the ability of the central nervous system to communicate, leading to a host of debilitating symptoms, eventually resulting in an ultimate loss of function. The discovery and applications of stem cells offers a flicker of hope to the patients tormented by these diseases and injuries. However, before the therapeutic applications of stem cells in regenerative medicine can be fully realized, several obstacles need to be overcome. One of the biggest obstacles is tracking and mapping the location of the transplanted stem cells within the central nervous system. The development of nanotechnology in recent years and its application to stem cell biology will play a significant role in addressing these challenges. This chapter details the unique properties and current applications of nanomaterials for labeling and tracking transplanted stem cells within the central nervous system using various imaging techniques.

INTRODUCTION

In order to make use of stem cells for nervous system repair, it is necessary to know the fate of stem cells and, therefore, the ability to track the stem cells after implantation *in vivo* is required. Nanoparticle-based imaging modalities are an extremely promising set of techniques which provide a unique set of attributes that may be useful for *in vivo* tracking of stem cells. The conventional method of tracking stem cells in the body is to label the stem cells *in vitro* with a fluorescent label such as BrdU or to genetically modify the cells to express GFP, introduce them into the body, and take tissue sections for immunohistochemistry. The main drawback of this procedure is that it is highly invasive, requiring the removal of significant amounts of tissue and, in most cases, sacrificing the study animal; thus making longitudinal studies more difficult.

Nanoparticles are interesting because they have a number of unique properties due to their size, which is between that of bulk materials and individual atoms or small molecules. While many of their properties may be described as intermediate between bulk materials and small molecules, other properties such as superparamagnetism and quantum confinement are unique to the nanoscale. In general, there are four groups of effects which give nanoparticles unique properties: effects due to small size *per se*, surface effects due to the enormous surface area-to-volume ratio of nanoparticles, quantum effects, and tunneling effects.

Nanoparticles have a number of properties which are quite useful for stem cell imaging. Their small size allows them to be easily taken up by the cell; furthermore, the particles can interact with individual biomolecules at the molecular level. Size can have a profound effect on the properties of nanoparticles and by controlling the conditions of nanoparticle synthesis, the size and shape can be controlled within a very tight distribution (Solanki et al. 2008). By making use of inorganic, organic, and surface chemistry, biomolecules such as antibodies (Wang et al. 2002) or peptides (Lewin et al. 2000) can be added to nanoparticles to include functionality, e.g., targeting or to improve cellular uptake.

Nanomaterials in Molecular Imaging

In the field of nanoparticle-based imaging, there are two classes which dominate: magnetic nanoparticles (MNPs) and quantum dots (QDs). Other materials have also been used for imaging, albeit to a lesser degree; these include gold nanoparticles, carbon nanotubes (CNTs), dendrimers, liposomes, and perfluorocarbons.

MNPs have been used as contrast agents in MRI for quite some time. The first class of MNPs used is known as superparamagnetic iron oxide nanoparticles (SPIONs). These are iron oxide particles small enough that each particle contains only one magnetic domain and, therefore, only one magnetic polarization vector. In the absence of an applied magnetic field, these polarization vectors align randomly, resulting in no net magnetic polarization; however, in the presence of an applied magnetic field, the polarization vectors align very strongly and create a large magnetic polarization. This phenomenon is known as superparamagnetism because, like in paramagnetism, the particles only exhibit a magnetic polarization in the presence on an applied magnetic field and, like in ferromagnetism, the particles exhibit a very large magnetization (Bean and Livingston 1959).

SPIONs are useful as contrast agents in imaging because when they are imaged by MRI they shorten the spin-spin relaxation time, the T2 signal. This means that the time for the spins of the protons in the water molecules to go out of phase with one another is shortened. The result of this is that any concentration of SPIONs in an MRI image will show up as a dark spot in a T2 weighted MR image (Thu et al. 2009).

There are several coatings which are commonly applied to MNPs. Gold coating makes the particles more chemically stable (Wang et al. 2006) and dextran coating reduces agglomeration (Lawaczeck et al. 2004). Furthermore, particles may be functionalized with antibodies to target specific tissues/cell types, TAT-peptides to increase cellular uptake, etc.

Quantum dots (QDs) are another class of nanoparticles useful for imaging. QDs are fluorescent nanoparticles composed of inorganic semiconductors, usually cadmium-selenide (CdSe) or cadmium-telluride (CdTe), although other formulations are possible. Compared to conventional fluorescent dyes, QDs have a narrow emission spectrum, a broad absorbance spectrum, a high quantum yield, high resistance to photobleaching, and long fluorescence lifetime (Solanki et al. 2008). The broad absorbance spectrum and narrow emission spectrum allow multiple structures to be simultaneously imaged and resolved. The high quantum yield allows a large fluorescent signal to be achieved from a low intensity laser, this property combined with the long fluorescent lifetime and resistance to photobleaching allow longitudinal studies to be performed with live-cell imaging.

QDs have such impressive fluorescence properties due to a phenomenon known as quantum confinement. When the radius of a nanoparticle is below a critical value, known as the Bohr radius, the band-gap of the bulk material is lowered into the visible or even infrared spectrum. Due to this effect, emission spectrum can be adjusted by changing both the chemical composition of the particle and the particle size. Because of the toxicity of cadmium, it is common to coat cadmium chalcogenides (CdSe or CdTe) with a zinc sulfide (ZnS) coating which improves chemical stability and allows functionalization by disulfide chemistry.

STEM CELL IMAGING AND TRACKING

Stem cells have been extensively studied due to their potential use in regenerative medicine (Guilak et al. 2009; Hansson et al. 2009). Stem cells are typically classified as embryonic stem cells, which are pluripotent and can differentiate into all cells types, or adult stem cells which are multipotent and show relatively limited potential for differentiation into organ specific cells; for instance, neural stem cells (NSCs) within the brain or spinal cord can differentiate into neuronal cells or glial cells (Gage 2000; Fuchs et al. 2004). Furthermore, pluripotency has been induced in somatic cells by forced expression of specific transcription factors, creating what are known as induced pluripotent stem cells (iPSCs) (Takahashi and Yamanaka 2006; Takahashi et al. 2007). The iPSCs can then be differentiated into specific cells types by providing them with appropriate signaling molecules and microenvironments (Dimos et al. 2008).

Transplanting progenitor and stem cells has been suggested as a promising therapeutic strategy not only for neurodegenerative diseases and devastating injuries to the nervous system, but also as a promising treatment modality for brain tumors (Sandu and Schaller 2010). One of the most important requirements for using stem cells in regenerative medicine and brain tumor therapy is the availability of techniques that would enable long-term, non-invasive detection of stem/progenitor cells and the monitoring of their migration, proliferation, and differentiation within the nervous system. Current strategies for monitoring the stem cells within the nervous system include various modes of imaging such as MRI, optical imaging and PET imaging. These techniques are extremely useful as they help trace the migration and location of the labeled stem cells implanted within the brain or spinal cord (Bulte et al. 2002) (Fig. 1).

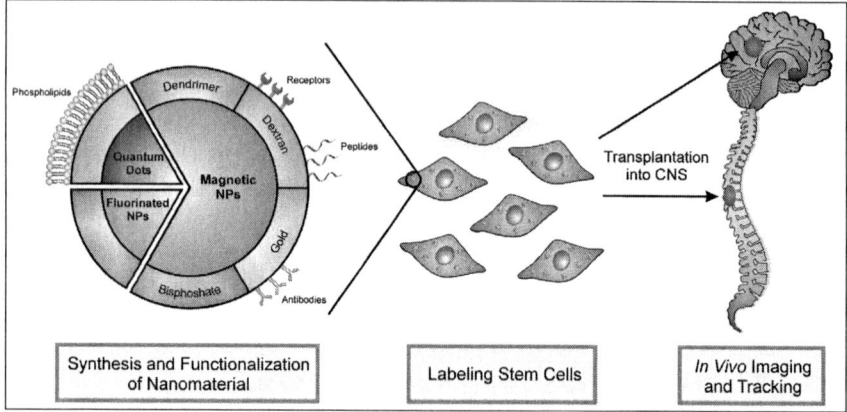

Fig. 1. Labeling of stem cells with functionalized nanomaterials for *in vivo* imaging and tracking. Stem cells can be labeled with a variety of nanomaterials. The surfaces of these nanomaterials can be modified in order to enhance their uptake and target specific cells. Once the stem cells are labeled they can be transplanted into the brain or spinal cord and their location can be mapped using various imaging techniques. *Data unpublished.*

NANOMATERIALS FOR STEM CELL IMAGING

Inorganic nanomaterials, especially MNPs and QDs are one of the most promising nanomaterials for stem cell labeling and imaging as they can be synthesized in large quantities from various materials with relative ease. Their properties can be tuned by changing their composition and size. They generally comprise of different metal, metal oxides, and semiconducting materials and their dimensions can be tuned from one to a few hundred nanometers with monodisperse size distribution. Table 1 summarizes the different types of nanomaterials used as imaging agents to track stem cells in the CNS.

Magnetic Nanoparticles

Magnetic iron oxide nanoparticles and their composites are emerging as novel contrast agents for MRI and are more sensitive than conventional gadolinium-based contrast agents (Lawaczeck et al. 2004). MRI, as an imaging technique, offers several advantages in preclinical and clinical settings. As compared to alternative techniques such as PET imaging, MRI offers high resolution, higher speed, relative accessibility and lower costs.

Table 1. Types of nanomaterials for tracking stem cells in the CNS.

Type of Nanomaterial	Stem Cell Types	CNS Location	Reference
Magnetic Nanoparticles			
Transferrin-tagged	OPC	Rat spinal cord	(Bulte et al. 1999)
HIV-Tat-derived	HPC, NPC	Mouse bone marrow	(Lewin et al. 2000)
Magnetodendrimers	MSC, NSC	Rat brain	(Bulte et al. 2001)
HEDP-coated	MSC, NSC	Rat brain	(Delcroix et al. 2009)
Gold-coated	NSC	Rat spinal cord	(Wang et al. 2006)
Endorem® (Dextran-coated)	ESC, MSC	Rat brain and spinal cord	(Jendelova et al. 2004)
Feridex® (Dextran-coated)	NSC	Human brain	(Zhu et al. 2006)
Quantum Dots	NSPC	Early mouse embryos	(Slotkin et al. 2007)
Fluorinated Nanoparticles	NSC	Rat brain	(Ruiz-Cabello et al. 2008)

Magnetic nanoparticles, quantum dots and fluorinated nanoparticles have been extensively used for labeling and tracking stem cells within the CNS. *Data unpublished.*
OPC: oligodendrocyte progenitor cells
HPC: hemotopoietic progenitor cells
MSC: mesenchymal stem cells
NSC: neural stem cells
NPC: neural progenitor cells
NSPC: neural stem and progenitor cells
ESC: embryonic stem cells
CNS: central nervous system
HIV-TAT: human immunodeficiency virus—transactivator of transcription
HEDP: hydroxyethylidene-1.1-bisphosphonic acid

The unique properties of magnetic nanoparticles enable precise control of size and composition, which offers great potential for MRI to track stem/progenitor cells with high specificity over longer time periods (Sykova and Jendelova 2006). SPIONs are well established contrast agents for labeling stem cells owing to their (i) biocompatibility, monocrystallinity, and small size; (2) the well understood and extensively studied magnetic properties; and (3) their proven ability to form covalent bonds with biomolecules such as proteins and antibodies.

One of the early studies involving neurotransplantation of oligodendrocyte progenitor cells (OPCs), utilized SPIONs for tracking the migration and myelination of OPCs (Bulte et al. 1999). The SPIONs were first coated with dextran to render them water soluble and then tagged with transferrin (Tfr) receptors, which led to higher uptake of the SPIONs by the OPCs. The magnetically labeled OPCs were transplanted into the spinal cord of myelin deficient rats to track their migration, and the extent of myelination within the spinal cord was monitored. Similarly, in another experiment, iron oxide nanoparticles, crosslinked using aminated dextran (CLIO), were tagged with an HIV-TAT peptide derivative for their efficient uptake by NPCs (Lewin et al. 2000). The HIV-TAT, a well-established membrane translocation signal, was covalently attached to the aminated dextran coating on the CLIO particles. To modify the NPs with a fluorescent label, FITC-derivatized HIV-TAT peptides were used. In addition to the fluorescent label, the SPIONs were modified with a chelator, DTPA, allowing the nanoparticles to be bound to the ^{111}In isotope for concomitant nuclear labeling.

Nanoparticles used for labeling stem cells are taken up by the cells mainly through endocytosis. Unfortunately, dextran coated SPIONs have been found to be unfavorable for inducing endocytosis; therefore relatively high amounts of dextran coated SPIONs are required for labeling stem cells. Furthermore, the dissolution of iron oxide within cells may increase the formation of reactive oxygen species and hydroxyl free radicals, leading to death of the labeled stem cells. To minimize such effects, iron oxide nanoparticles are often protected or coated with different materials. For instance, Bulte et al. developed magnetodendrimers (MD-100) as novel, label-free magnetic probes, specifically designed for higher uptake by any mammalian cell, including stem cells (Bulte et al. 2001). These dendrimer composites of iron oxide nanoparticles represent a versatile class of contrast agents for tracking stem cells using MRI. The MD-100, having an oligocrystalline structure of 7–8 nm, showed considerably improved magnetic properties over previously established contrast agents.

In case of other iron probes, toxic iron species may be released during the biodegradation process; however, in the case of MD-100, they accumulated within the endosomes, where the magnetic probe was shielded from the nucleus and cytoplasm until it degraded and metabolized into the normal iron pool. Human NSCs and MSCs were labeled with MD-100 *ex-vivo* and were found to retain their normal capacity for differentiation (Fig. 2C). This strategy was particularly advantageous as the NSCs and MSCs were labeled intracellularly, which prevented the possibility of nanoparticles being detached from cell membranes and non-specifically attaching to other cells *in vivo* (Fig. 2A, B). The labeled stem cells were then transplanted within the demyelinated rat brain. The labeled NSCs or NSC-derived OPCs were readily detected *in vivo* for six weeks post transplantation (Fig. 2D). The MD-100 has opened up the possibility of labeling a variety of mammalian cells for tracking them using MRI.

Core-shell structures are gaining tremendous popularity in nanomedicine due to their multifunctionality and ease of surface modification. Gold (Au) has been extensively used to coat nanoparticles, not only to render the particles inert, but also to provide a handle on the functionalization of nanoparticles due to the well-established thiol-gold chemistry. Magnetic iron oxide nanoparticles (MIONs) were coated with gold to form a typical core-shell structure (Wang et al. 2006). These Au-MIONs, of 20 nm hydrodynamic size, were used to label NSCs obtained

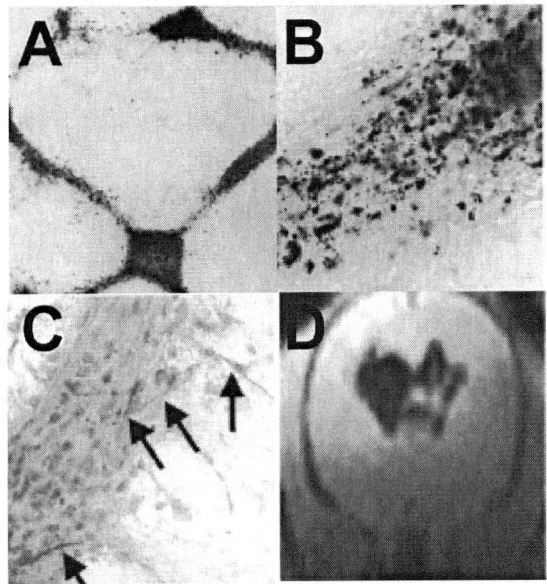

Fig. 2. Magnetodendrimer (MD-100)-labeled neural stem cells. (A) Prussian Blue stained MD-100-labeled NSCs after 24 hours. (B) MD-100-labeled cells at higher magnification. (C) Anti-MAP2 (neuronal marker) immunostaining of MD-100-labeled NSCs confirms that labeling NSCs does not affect the neuronal differentiation. (D) *In vivo* MR image of MD-100-labeled NSCs 42 days after intraventricular transplantation. *Reprinted by permission from Macmillan Publishers Ltd: Nature Biotechnology, (Bulte et al. 2001), Copyright 2001.*
MD-100: magnetodendrimers
NSC: neural stem cells
MAP2: microtubule-associated protein 2
MR: magnetic resonance

from the rat spinal cord. When the labeled NSCs were transplanted within the spinal cord, the Au-MION system offered a strong and durable signal such that, as little as 20 cells could be detected with relative ease under optimal conditions using MRI. In addition, the Au-MION labeling helped distinguish the labeled NSCs from the haemoglobin breakdown products *in vivo*. False positives obtained due to Prussian Blue staining of haemoglobin products in histology could be avoided by using a selective stain for gold.

In another study, Delcroix et al., coated SPIONs with HEDP to improve the uptake by MSCs and NSCs (Delcroix et al. 2009). It was observed that 90% of the MSCs were labeled with the HEDP-coated SPIONs. TEM and SEM results confirmed that the particles were taken up by the MSCs through endocytosis. The HEDP-coated SPIONs were found to be nontoxic and did not require any transfecting agents for labeling the stem cells. The HEDP coating further offers the possibility to functionalize SPIONS, increasing the range of its potential applications.

Quantum Dots

Quantum dots have opened the door to a range of applications in biological sciences including live monitoring of physiological events within cells by labeling specific cellular structures with QDs of different colors, tracing stem cell lineages, tracking cells *in vivo* and understanding the etiology and diagnosis of diseases. Their unique photophysical properties, which can be readily tuned by controlling their composition and size, coupled with their diverse biological applications make QDs attractive nanoprobes for labeling stem cells. Furthermore, dynamic changes occurring in the membranes of stem cells can be investigated at the molecular level by functionalizing the QDs with specific molecules that selectively bind to receptors on the membranes of stem cells. This helps in dynamically tracking the motions of such receptors, which gives remarkable insight into membrane receptor mediated stem cell behaviors.

QDs can be used efficiently to label NSCs and NPCs and study their migration and differentiation during mammalian development *in vivo*. However, labeling stem cells with QDs is challenging and not many techniques that efficiently label stem cells with QDs *in vivo* exist. To this end, Slotkin et al. developed a novel *in utero* electroporation and ultrasound guided *in vivo* delivery technique to directly and efficiently label neural stem cells within the developing mammalian CNS (Slotkin et al. 2007). The QDs used had a CdSe core and a ZnS shell, covered with an additional phospholipid core, and had diameters ranging from 10–20 nm. They were made water soluble by functionalizing them with -COOH as the terminal group. NSCs were directly labeled using ultrasound biomicroscopy (UBM)-guided injections of QDs into the parenchyma of the ventral telencephalon within the embryonic mouse brain (Fig. 3). The QD-loaded NSCs were found at significant distances away from the site of injection after 5 days and had differentiated into neurons (βIII tubulin+), oligodendrocytes (NG2+), and astrocytes (GFAP+). Thus, it was concluded that neither the QDs nor the labeling techniques *in vivo* had any effect on the migration or differentiation behavior of the labeled stem cells. The group then modified the *in vivo* electroporation techniques to label precursor cells (NPCs) within the CNS, surrounding the lateral ventricles. At first, it was found that less than optimal number of cells within the VZ and SVZ regions were loaded with QDs. This was attributed to the fact that the QDs were neutral in charge. By increasing the negative charge on phospholipid coats of the QDs using additional –COOH groups, a substantial increase in labeling efficiency was observed. Owing to the high photostability and quantum efficiency of the QDs, high resolution 3D images of the VZ and the SVZ NPCs were obtained using multiphoton imaging.

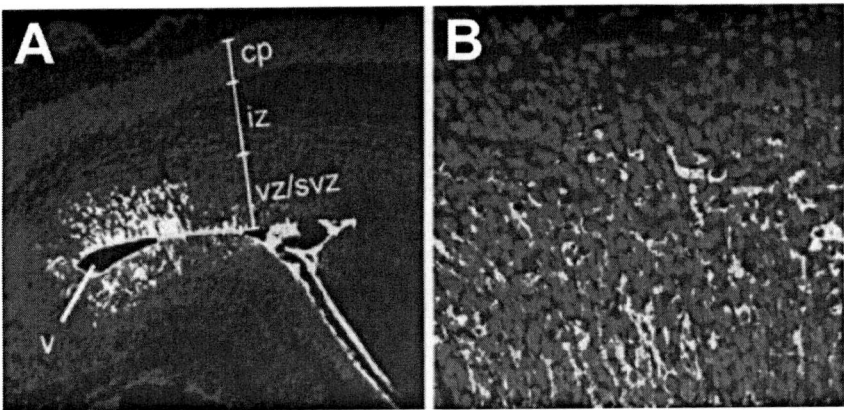

Fig. 3. Quantum dots for labeling stem cells in the brain. (A) Direct *in utero* for highly efficient labeling of NPCs in the ventricular zone and subventricular zone of the E16.5 mouse embryo with QDs (bright) using electroporation. (B) High magnification confocal optical image showing efficient QD (bright) labeling of NPCs surrounding the lateral ventricle. *Reprinted by permission from Wiley: Developmental Dynamics, (Slotkin et al. 2007), Copyright 2007.*
NPCs: neural precursor cells
QDs: quantum dots

Despite having substantial advantages over conventional labeling and tracking agents, toxicity remains one of the primary drawbacks of using QDs as labeling agents for stem and progenitor cells as these cells are extremely sensitive. QDs contain heavy metals such as cadmium and selenium, and the cytotoxicity is observed owing to the generation of Cd^{2+} and Se^{2+} ions within the stem cells. The cytotoxic effects of the QDs may not be a concern at lower concentrations as the ZnS shell considerably reduces the toxicity by blocking the oxidation of CdSe. However, these effects could be detrimental for embryo development when used in higher amounts.

Fluorinated Nanoparticles

Fluorinated nanoparticles have been slowly emerging as novel probes for MRI. Fluorine labeling is a relatively new technique to label stem cells and track ^{19}F *in vivo* using MRI. One of the most significant advantages of using ^{19}F for labeling stem cells is that ^{19}F is naturally abundant and offers negligible ^{19}F background signal, thus enabling "hot spot" MR imaging, analogous to its applications in nuclear medicine (Ruiz-Cabello et al. 2008). Ruiz-Cabello et al. synthesized and used cationic and anionic perfluor-15-crown-5-ether (PFCE) nanoparticles for labeling NSCs (Fig. 4A). The platform serves as an ideal ^{19}F tracer for cellular imaging as the PFCE NPs have a large number of chemically equal fluorine atoms which results in a ^{19}F spectrum consisting of single narrow resonance. It was observed that

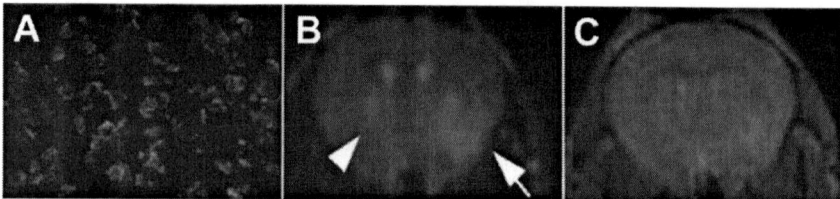

Fig. 4. Labeling and transplantation of neural stem cells with fluorinated nanoparticles. (A) NSCs labeled with PFCE-NPs (bright regions within the cells). (B, C) PFCE-labeled NSCs transplanted into a mouse brain after 1 hour (B) and 7 days (C). 4×10^4 NSCs were injected in the left hemisphere (arrowhead in B), and 3×10^5 NSCs were injected in the right hemisphere (arrow in B). The ^{19}F signal was superimposed on the MR images. *Reprinted by permission from Wiley: Magnetic Resonance in Medicine, (Ruiz-Cabello et al. 2008), Copyright 2008.*
NSCs: neural stem cells
NPs: nanoparticles
PFCE: perfluor-15-crown-5-ether
MR: magnetic resonance
DAPI: 4′,6-diamidino-2-phenylindole

the cationic PFCE NPs showed higher signal detection at concentrations as low as 0.6 mM. The labeling of NSCs was carried out without the use of transfection agents and the viability and proliferation of the PFCE-labeled NSCs was the same as that seen with unlabeled NSCs, proving that the PFCE NPs were non-cytotoxic. It was further observed that the NSCs retained the PFCE NPs for at least two weeks after intrastriatal transplantation, which provided the "hot spot" MR imaging of their cellular distribution *in vivo* (Fig. 4B,C). The sensitivity of detecting PFCE-labeled NSCs was found to be significantly lower compared to that obtained with SPOINs. Nevertheless, the technique may be particularly useful for tracking NSCs *in vivo* in cases of traumatic injury and hemorrhage, where using SPION-labeled NSCs is difficult, owing to the false positives obtained by the iron within the hemoglobin.

TRACKING STEM CELLS IN THE NERVOUS SYSTEM

Various types of nanoparticles have been used to label stem cells for tracking *in vivo*. These stem cells are typically transplanted within the brain or spinal cord as treatments for neurodegenerative diseases, ischemia and traumatic injuries, and as therapeutics for brain tumors. Tracking the fate of stem cells *in vivo* within the different regions of the nervous system, under different conditions, requires longer imaging times, which may be achieved with relative ease using nanomaterials.

Stem Cell Tracking in the Brain

Stem cells have a tremendous potential to migrate to pathological regions within the CNS. Adult stem cells such as NSCs and MSCs have been investigated extensively in terms of their migration and therapeutic potential within the CNS (Sykova and Jendelova 2007). Although MSCs are not found in the CNS, it has been observed that MSCs show the tendency to respond to microenvironmental signals and transdifferentiate into neuronal-like cells when transplanted into rat brains (Jendelova et al. 2004). These stem cells are typically labeled with nanomaterials for MR imaging. For example, it has been observed that NSCs, when implanted in the diseased or injured brain, can travel great distances and integrate into the local neural environment (Yip et al. 2003). Similarly, MSCs labeled with magnetic nanoparticles, when transplanted within a rat experimental model of stroke, migrated from the site of injection into the lesion. The MSCs could be tracked for three weeks after transplantation. MRI showed a decrease in the concentration of magnetic nanoparticles at the injection site and an increase within the lesion after the second and third weeks following injection (Sykova and Jendelova 2007).

Tracking stem cell migration in the brain is not only limited to adult stem cells like NSCs or MSCs. Mouse ESCs labeled with commercially available SPION solution, Endorem, were used to investigate their potential to migrate to the lesion in the brain (Jendelova et al. 2004). A photochemical lesion as a model of thrombotic stroke was used in the study. The mESCs were also labeled with eGFP prior to labeling them with SPIONs. The labeled eGFPmESCs were implanted into the cortex or intravenously injected into the femoral vein. Massive migration of the eGFPmESCs from the site of injection (or implantation) to the lesion was observed using MRI. The hypointensive (T2) signal from the cells persisted for more than 50 days. Histology results further showed that a large number of Prussian Blue-positive and eGFP-labeled cells had entered the lesion, confirming that the hypointensive signal observed with MRI was due to the SPION-labeled eGFP ESCs. There was no difference between the numbers of stem cells infiltrating the lesion in the brain when using implantation or intravenous injection, thus the latter route was preferred due to the implications of future clinical studies. For preventing the formation of tumors, the mESCs were cultured in neuronal differentiation medium. It was observed that 10% of the experiments performed led to formation of tumor-like masses when the cells were implanted; while no formation of tumors in the lesion was observed when the cells were injected intravenously (Jendelova et al. 2004).

In a rare study, Zhu et al. used human NSCs, from patients suffering from brain trauma, and labeled them with SPIONs using a non-liposomal transfecting agent (Zhu et al. 2006). The SPION-labeled NSCs were cultured and implanted around the region of brain damage within the patient. Distinct hypointensive signals were found at the site of injection on the first day and the signals eventually faded. A week after implantation the signal was found in the periphery of the damaged tissue, suggesting the migration of NSCs from the site of implantation to the damaged tissue. The hypointense signal eventually diluted due to cell proliferation. One of the biggest challenges for using SPIONs for stem cell imaging is their engulfment by macrophages which may lead to false positives. However, through a series of careful experimentation, involving double-labeled NSCs, Zhu et al. excluded the possibility of the NSCs being engulfed by macrophages and concluded that the signals were indeed generated by the migrating human NSCs.

Stem Cell Tracking in Brain Tumor Therapy

Malignant brain tumors are one of the most difficult-to-treat tumors and despite all the advancements in therapy, mortality of these tumors remains high. Delivering therapeutics and confirming that they reach the tumor site is a significant challenge due to the presence of the blood brain barrier and the highly invasive nature of these tumor cells in healthy neural tissue. Additionally, the tumors leave the surrounding regions of the brain distorted and destroyed, which may leave the patient in a debilitated state even if the tumors are successfully removed by surgery. Because of this, NSCs and MSCs have been used as alternative therapeutic delivery vehicles to target tumor mass and promote tumor regression (Sandu and Schaller 2010). The NSCs have the added advantage of differentiating into glial and neuronal cells, thus replacing the diseased or damaged neural tissues. Significant success using NSCs has been achieved as they can travel great distances, in some cases to single tumor cells, and diffuse into the neuronal abnormalities (Yip et al. 2003; Thu et al. 2009). For instance, cytolytic viruses and genes coding for anti-tumor cytokines, various neurotrophic factors, and prodrug-converting enzymes have been engineered into NSCs for delivering them to tumors in the brain (Thu et al. 2009). Nevertheless, not much is known about the mechanism by which NSCs exert their effects *in vivo*. There is some concern that the NSCs themselves may lead to formation of additional tumors which may result from single NSCs that do not differentiate. Therefore, the ability to monitor the fate of NSCs and MSCs would be quite advantageous.

Molecular imaging techniques provide a sensitive means to track the stem cells *in vivo* for weeks and even months. This advancement has led to a deeper understanding of the behavior of engrafted (or injected) stem cells in terms of their localization, viability, migration and proliferation within the brain. A novel approach for live-imaging of stem cells within the brain involves the use of bioluminescence reporter genes under the control of engineered promoters. These can be introduced into the cells using vectors or nonvectors (Schaller et al. 2008). Such reporter systems are advantageous as they do not require excitement by an external source, as is required for fluorescence imaging. They only require the administration of a probe to induce the emission of light for optical imaging. However, such reporter systems may not be practical for use in clinical settings due to their poor tissue penetration into the tumor mass and low spatial resolution. To address these problems, nanoparticle-based tracking of stem cells, especially using MRI, has unique advantages. NSCs labeled with iron oxide particles have been found to retain their proliferative status, tumor tropism and maintain stem cell character, while allowing real-time tracking of their migration to the brain tumor sites (Thu et al. 2009). Similar results were also demonstrated by Zhang et al. in a rat model of gliosarcoma. One week after the tumor implantation into the rat brain, they used dye-coated SPIONs for labeling NPCs and MSCs and administered them subcutaneously into the cisterns of the rat brain and a tail vein injection, respectively. Using MRI, it was confirmed that the NSCs and the MSCs targeted tumor aggregates and successfully infiltrated the tumor tissue. This was further confirmed by Prussian Blue staining and fluorescence imaging (Zhang et al. 2004). Such studies not only demonstrate the usefulness of MRI, but also show the potential of nanomaterials as labels for tracking stem cells for brain tumor therapy.

Stem Cell Tracking in the Spinal Cord

Spinal cord injuries (SCI) and neurodegenerative diseases like multiple sclerosis (MS) are particularly difficult to treat. For stem cells to be a viable option for treating patients suffering from such conditions, tracking stem cells and monitoring their fate within damaged regions of the spinal cord is perhaps as important as in the brain. OPCs typically differentiate into oligodendrocytes, which are responsible for the formation of myelin, thus remyelinating the axons within the spinal cord. In the study by Bulte et al. (Bulte et al. 1999), OPCs were highly loaded with dextran coated iron oxide NPs, tagged with transferrin receptors (Tfr) or OX-26 which is an internalizing monoclonal antibody that binds noncompetitively with transferrin (Bulte et al. 1999), thereby increasing the receptor mediated uptake into OPCs. These OPCs were transplanted within the spinal cord of myelin-deficient rats and tracked using MRI. The labeling of OPCs with ligand (Tfr or OX-26)-tagged SPIONs completely retained their migratory behavior and myelinating capacity *in vivo* and did not behave differently as compared to unlabeled cells. The labeled OPCs migrated extensively (up to 8.4 mm) within the dorsal column of the rat spinal cord (Fig. 5A). Furthermore, an excellent correlation between the MR contrast and histopathological staining for iron and newly formed myelin was observed. MSCs, like NSCs, are also known to migrate to damaged tissues and secrete bioactive molecules which help in the repair and regenerative process. SPION-labeled MSCs were injected intravenously, in rats having injured spinal cords, through the femoral artery. The spinal injury was induced using a balloon compression method. MR images as well as Prussian Blue staining confirmed the presence of a large number of MSCs in the lesion (Jendelova et al. 2004) (Fig. 5B). Tracking stem cells within the injured spinal cord is a critical requirement for monitoring regenerative therapy, and the use of nanomaterials as stem cell labeling agents enables stem cell tracking for longer periods of time as compared to currently available contrast agents.

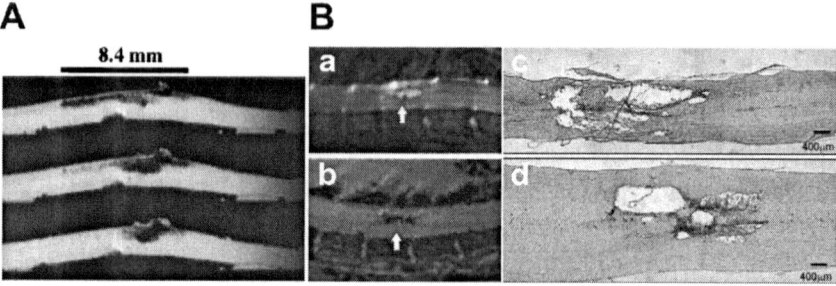

Fig. 5. Tracking stem cells in the spinal cord. (A) MR images of the spinal cord (three consecutive slices) show the cellular migration of the SPION labeled OPCs, over a distance of 8.4 mm, 10 days after transplantation. *Reprinted by permission from the National Academy of Sciences, USA: PNAS, (Bulte et al. 1999), Copyright 1999.* (B) SPION-labeled MSCs intravenously injected into a rat spinal cord. (a) The lesion in the spinal cord appears as a hyperintensive area (bright spot, arrow), (b) SPION-labeled MSCs are visible as a hypointensive spot (dark spot), (c) Prussian Blue staining of a lesion in the spinal cord of a control animal (without MSCs), (d) Prussian Blue staining of a lesion in the spinal cord of a rat injected with SPION-labeled MSCs. The lesion is populated with Prussian Blue-positive MSCs. *Reprinted by permission from Macmillan Publishers Ltd: Cell Death and Differentiation, (Sykova and Jendelova 2007), Copyright 2007.*
MR: magnetic resonance
SPION: superparamagnetic iron oxide nanoparticles
OPCs: oligodendrocyte progenitor cells
MSCs: mesenchymal stem cells

Applications to Areas of Health and Disease

While the most prominent use of nanomaterials for studying stem cells in the CNS has been for imaging and tracking purposes, the versatility and fine control over the intrinsic properties of nanomaterials affords several potentially useful designs and applications of nanomaterials in stem cells beyond the current use. With the advances in nanotechnology and molecular biology, multifunctional nanomaterials have also gained considerable attention as both diagnostic and therapeutic agents. Multifunctional nanomaterials are tailored to display more than one feature or function simultaneously, which provides an excellent tool for biological applications. Multifunctional nanoparticles can be designed to serve as both imaging agents and bimolecular carriers. Taking advantage of the surface chemistry and intrinsic magnetic properties, the synthesis and application of multifunctional magnetic nanoparticles has been demonstrated by numerous groups.

A number of studies have focused on the targeted delivery of chemotherapeutics, in which magnetic nanoparticle-based drug delivery allows for larger doses to be administered to the target site while allowing for imaging capabilities. Other studies have taken advantage of the intrinsic magnetic properties of these nanoparticles to induce hyperthermia, wherein exposure to an alternating magnetic field causes local heating (Thiesen and Jordan 2008). Hyperthermia has been shown to be advantageous for killing tumor cells or making the cells prone to further chemotherapeutic treatment. Besides magnetic nanoparticles, multifunctional quantum dots have also been demonstrated for effective therapy. Jung et al. illustrated the use of these nanoparticles for targeting and inhibiting brain tumor cells, which allowed for the simultaneous imaging and delivery of inhibitory biomolecules (small interfering RNA) (Jung et al. 2010). Such applications highlight the fact that besides imaging and tracking, nanomaterials also perhaps have the potential to manipulate stem cells in the nervous system by regulating both, the proliferation and differentiation pathways.

Key Facts

- The origin of quantum dots goes years back to work published by scientist Louis Brus in 1983, at Bell Labs in New Jersey. However, it was many years before another scientist, Mark Reed at Yale University, described these semiconductor crystals as "quantum dots", which when excited by light, emit colored light whose color depends upon the dots' size and material.

- To date, there are up to 40 FDA approved nanomedicine products in the market and even more in clinical trials. The SPIONs were first approved by the FDA in 1996 as tissue-specific MRI contrast agents and since have been used in various clinical applications.
- MRI is a medical imaging technique, discovered independently in 1946 by scientists Felix Bloch and Edward Purcell. They were later awarded the Nobel Prize in 1952. However, until 1970, MRI was only being used for physical and chemical analysis. It was in 1970, that Raymond Damadian showed that the nuclear magnetic relaxation times were different for normal tissues and tumors, thus encouraging scientists to use MRI to study diseases.
- Of the many malignant brain tumors, glioblastomamultiforme, is the most malignant, invasive and difficult-to-treat tumor. These tumors have a rapid growth rate, being capable of doubling in size within 10 to 20 days. Successful treatment of GBM is rare, with a mean survival of only 10 to 12 months.

Key Terms

- Nanomaterials have unique optical, electrical, magnetic, and chemical properties due to their nanoscale size.
- SPIONs are superparamagnetic iron oxide nanoparticles, used as contrast agents to track labeled cells *in vivo* using MRI. They are typically modified to interact with stem cells without affecting their biological activities.
- Quantum dots (QDs) are fluorescent nanoparticles having a narrow emission spectrum, broad absorbance spectrum, a high quantum yield and are resistant to photobleaching allowing a longer fluorescence lifetime.
- ESCs or iPSCs are pluripotent stem cells which can proliferate for unlimited generations and can differentiate into any cell type in the body.
- Adult stem cells such as NSCs and MSCs are multipotent can proliferate for a limited generations and can differentiate into organ specific cells.
- Cell labeling and transplantation is the process of labeling stem cells *in vitro* and transplanting or injecting them within the required organ.

Summary Points

- Several factors must be considered for applying nanotechnology to stem cell biology: Nanomaterials should be designed to interact with stem cells without affecting their biological activities; they should be appropriately modified to make them biocompatible and minimize

their toxicity; they must maintain their unique physical and chemical properties after surface modifications.
- Magnetic nanoparticles, the size of which can be precisely tuned, offer great opportunities of tracking stem cells *in vivo* using MRI by generating ultrasensitive images. SPIONs have also been used as multifunctional agents for tracking cells, as well as delivering biomolecules, and as agents for hyperthermia.
- Quantum dots, due to their unique photophysical properties, offer great opportunities for labeling stem cells *in vitro* and *in vivo*. In addition, multiplex imaging for labeling a variety of cells is possible using QDs.
- Using nanomaterials to track stem cells within the CNS in real-time is vital for developing therapeutic strategies for brain and spinal cord injuries.
- Stem cells labeled with nanomaterials can also be used as delivery agents to lesions within the brain and spinal cord.

Conclusions

Stem cell therapy is an attractive and rapidly growing field due to its potential in regenerative medicine, especially for applications involving the CNS. A crucial criterion for utilizing stem cell-based therapy for CNS research necessitates a long term, non-invasive means to monitor the survival and location of transplanted stem cells. A growing number of imaging strategies are beginning to utilize nanomaterials for tracking stem cells. A variety of nanoparticles with different compositions have been demonstrated for imaging and tracking stem cells (MSCs, NSCs, etc.) in both, the brain and the spinal cord. The majority of the nanoparticle-based imaging systems have focused on using magnetic nanoparticles and quantum dots due to the ease in synthesis, facile surface modification and excellent imaging properties using conventional imaging techniques (MRI, PET, optical imaging). Overall, nanomaterial-based approaches have revolutionized stem cell therapy in the CNS and will continue to have a major role in regenerative medicine.

Abbreviations

BrdU	:	bromodeoxyuridine
GFP	:	green fluorescent protein
eGFP	:	enhanced green fluorescent protein
MNPs	:	magnetic nanoparticles
QDs	:	quantum dots
CNTs	:	carbon nanotubes
MRI	:	magnetic resonance imaging

SPIONs	:	superparamagnetic iron oxide nanoparticles
TAT	:	transactivator of transcription
CdSe	:	cadmium-selenide
CdTe	:	cadmium-telluride
ZnS	:	zinc sulfide
NSCs	:	neural stem cells
iPSCs	:	induced pluripotent stem cells
PET	:	positron emission tomography
OPCs	:	oligodendrocyte progenitor cells
Tfr	:	transferrin
CLIO	:	crosslinked iron oxide
NPCs	:	neural progenitor cells
HIV	:	human immunodeficiency virus
DTPA	:	diethylenetriaminepenta-acetic acid
NPs	:	nanoparticles
FITC	:	fluorescein isothiocyanate
MD-100	:	magnetodendrimers
MSCs	:	mesenchymal stem cells
Au	:	Gold
MIONs	:	magnetic iron oxide nanoparticles
HEDP	:	hydroxyethylidene-1.1-bisphosphonic acid
TEM	:	transmission electron Microscopy
SEM	:	scanning electron microscopy
UBM	:	ultrasound biomicroscopy
NG2	:	nerve/glia antigen 2
GFAP	:	glial fibrillary acidic protein
NPCs	:	neural precursor cells
VZ	:	ventricular zone
SVZ	:	subventricular zone
PFCE	:	perfluor-15-crown-5-ether
ESCs	:	embryonic stem cells
mESCs	:	mouse embryonic stem cells
SCI	:	spinal cord injuries
MS	:	multiple sclerosis
OX-26	:	transferrin receptor antibody
HPC	:	hemotopoietic progenitor cells
NSPC	:	neural stem and progenitor cells
MAP2	:	microtubule-associated protein 2
DAPI	:	4',6-diamidino-2-phenylindole

References

Bean, C. and J. Livingston. 1959. Superparamagnetism. J. Appl. Phys. 30: S120–S129.
Bulte, J.W.M., T. Douglas, et al. 2001. Magnetodendrimers allow endosomal magnetic labeling and *in vivo* tracking of stem cells. Nat. Biotechnol. 19: 1141–1147.
Bulte, J.W.M., I.D. Duncan, et al. 2002. *In vivo* magnetic resonance tracking of magnetically labeled cells after transplantation. J. Cerebr. Blood F. Met. 22: 899–907.
Bulte, J.W. M., S.C. Zhang, et al. 1999. Neurotransplantation of magnetically labeled oligodendrocyte progenitors: Magnetic resonance tracking of cell migration and myelination. P. Natl. Acad. Sci. USA 96: 15256–15261.
Delcroix, G.J.R., M. Jacquart, et al. 2009. Mesenchymal and neural stem cells labeled with HEDP-coated SPIO nanoparticles: *In vitro* characterization and migration potential in rat brain. Brain Res. 1255: 18–31.
Dimos, J.T., K.T. Rodolfa, et al. 2008. Induced pluripotent stem cells generated from patients with ALS can be differentiated into motor neurons. Science 321: 1218–1221.
Fuchs, E., T. Tumbar, et al. 2004. Socializing with the neighbors: Stem cells and their niche. Cell 116: 769–778.
Gage, F.H. 2000. Mammalian neural stem cells. Science 287: 1433–1438.
Guilak, F., D.M. Cohen, et al. 2009.Control of Stem Cell Fate by Physical Interactions with the Extracellular Matrix. Cell Stem Cell 5: 17–26.
Hansson, E.M., M.E. Lindsay, et al. 2009. Regeneration Next: Toward Heart Stem Cell Therapeutics. Cell Stem Cell 5: 364–377.
Jendelova, P., V. Herynek, et al. 2004. Magnetic resonance tracking of transplanted bone marrow and embryonic stem cells labeled by iron oxide nanoparticles in rat brain and spinal cord. J. Neurosci. Res. 76: 232–243.
Jung, J.J., A. Solanki, et al. 2010. Selective Inhibition of Human Brain Tumor Cells through Multifunctional Quantum-Dot-Based siRNA Delivery. Angew. Chem. Int. Edit. 49: 103–107.
Lawaczeck, R., M. Menzel, et al. 2004. Superparamagnetic iron oxide particles: contrast media for magnetic resonance imaging. Appl. Organomet. Chem. 18: 506–513.
Lewin, M., N. Carlesso, et al. 2000. Tat peptide-derivatized magnetic nanoparticles allow *in vivo* tracking and recovery of progenitor cells. Nat. Biotechnol. 18: 410–414.
Ruiz-Cabello, J., P. Walczak, et al. 2008. *In Vivo* "Hot Spot" MR Imaging of Neural Stem Cells Using Fluorinated Nanoparticles. Magnet. Reson. Med. 60: 1506–1511.
Sandu, N. and B. Schaller. 2010. Stem Cell Transplantation in Brain Tumors: A New Field for Molecular Imaging? Mol. Med. 16: 433–437.
Schaller, B.J., J.F. Cornelius, et al. 2008. Molecular Imaging of Brain Tumors Personal Experience and Review of the Literature. Curr. Mol. Med. 8: 711–726.

Slotkin, J.R., L. Chakrabarti, et al. 2007. *In vivo* quantum dot labeling of mammalian stem and progenitor cells. Dev. Dynam. 236: 3393–3401.
Solanki, A., J.D. Kim, et al. 2008. Nanotechnology for regenerative medicine: nanomaterials for stem cell imaging. Nanomedicine-UK 3: 567–578.
Sykova, E. and P. Jendelova. 2006. Magnetic resonance tracking of transplanted stem cells in rat brain and spinal cord. Neurodegener. Dis. 3: 62–67.
Sykova, E. and P. Jendelova. 2007. Migration, fate and *in vivo* imaging of adult stem cells in the CNS. Cell Death Differ. 14: 1336–1342.
Takahashi, K., K. Tanabe, et al. 2007. Induction of pluripotent stem cells from adult human fibroblasts by defined factors. Cell 131: 861–872.
Takahashi, K. and S. Yamanaka. 2006. Induction of pluripotent stem cells from mouse embryonic and adult fibroblast cultures by defined factors. Cell 126: 663–676.
Thiesen, B. and A. Jordan. 2008. Clinical applications of magnetic nanoparticles for hyperthermia. Int. J. Hyperther. 24: 467–474.
Thu, M.S., J. Najbauer, et al. 2009. Iron Labeling and Pre-Clinical MRI Visualization of Therapeutic Human Neural Stem Cells in a Murine Glioma Model. Plos One 4:e7218.
Wang, F.H., I.H. Lee, et al. 2006. Magnetic resonance tracking of nanoparticle labelled neural stem cells in a rat's spinal cord. Nanotechnology 17: 1911–1915.
Wang, S.P., N. Mamedova, et al. 2002. Antigen/antibody immunocomplex from CdTe nanoparticle bioconjugates. Nano Lett. 2: 817–822.
Yip, S., K.S. Aboody, et al. 2003. Neural stem cell biology may be well suited for improving brain tumor therapies. Cancer J. 9: 189–204.
Zhang, Z.G., Q. Jiang, et al. 2004. *In vivo* magnetic resonance imaging tracks adult neural progenitor cell targeting of brain tumor. Neuroimage 23: 281–287.
Zhu, J.H., L.F. Zhou, et al. 2006. Tracking neural stem cells in patients with brain trauma. New Engl. J. Med. 355: 2376–2378.

10

Carbon Nanotubes and Neuronal Performance

Alessandra Fabbro,[1,a] Francesca Maria Toma,[2,a] Giada Cellot,[1,b] Maurizio Prato[2,b] and Laura Ballerini[1,c,]*

ABSTRACT

The application of nanotechnology to contemporary neuroscience promotes innovative solutions that may be useful for brain and spinal cord repair strategies after damage. Carbon nanotubes (CNTs) are a novel form of carbon made of rolled layers of graphite. Since their discovery CNTs appeared immediately interesting materials, due to their remarkable properties. In fact, depending on their structure, CNTs are both immensely strong and mechanically flexible. In addition, CNTs exhibit useful properties allowing nanotubes to conduit electrical current between electrochemical

---

[1]Life Science Department, University of Trieste, via Giorgeri 1, I-34127, Trieste, Italy.
[a]E-mail: afabbro@units.it
[b]E-mail: gcellot@units.it
[c]E-mail: lballerini@units.it
[2]Department of Chemical and Pharmaceutical Sciences, University of Trieste, Piazzale Europa 1, I-34127, Trieste, Italy.
[a]E-mail: fmtoma@units.it
[b]E-mail: prato@units.it
*Corresponding author

List of abbreviations after the text.

interfaces. CNTs are scaffolds composed of small tubes with diameters similar to those of neural processes, thus particularly attractive for neuroscience research applications. It is not clear whether CNTs, simply due to their intrinsic structure, could directly improve the rewiring of disconnected neuronal networks. In this chapter, we discuss the use of CNTs as scaffolds able to impact and sustain neuronal network formation and activity *in vitro*. The chemical properties of CNTs can be systematically varied by attaching different functional groups. We will discuss how such functionalized CNTs may control the outgrowth and branching patterns of neuronal processes. We will also discuss the direct and specific interactions between synthetic CNTs and biological cell membranes and their role in improving neuronal performance. Investigating novel nanomaterials represents a great promise, in the attempt to develop biologically compatible scaffolds for controlling tissue repair and tissue engineering. In these processes, a critical step to reproducibly fabricate biosolid nanodevices, such as neurological implants, is to further characterize the efficacy of purified CNTs to modulate neuronal adhesion and performance.

INTRODUCTION

CNTs have been at the forefront of nanotechnology due to their unique electrical, mechanical and thermal features, which allow the development of a variety of miniaturized devices with remarkable properties (Krishnan et al. 1998). More recently, CNTs have attracted tremendous attention for the development of nano-bio hybrid systems able to govern cell-specific behaviours in cultured neuronal networks (Mattson et al. 2000; Hu et al. 2004; Lovat et al. 2005; Ni et al. 2005; Galvan-Garcia et al. 2007; Cellot et al. 2009). Among the emerging applications of nanotechnology to neuroscience, of particular interest is the development of artificial nanomaterials, such as CNTs and nanofibres, as next generation-scaffolds for nerve tissue engineering (Gilmore et al. 2008), or the use of CNTs for long-term implants and neural interfaces (Keefer et al. 2008; Kotov et al. 2009). Alternatively, CNT soluble preparations have been proposed to exploit the fabrication of molecular sensing, diagnostics, or drug delivery devices (Pantarotto et al. 2004).

CNTs are cylindrical nanostructures characterised by a high aspect ratio, due to a radius of a few nanometers made of carbon in sp_2 hybridization state, and a variable length, usually below 1 μm. The application of CNTs in basic neuroscience research has been oriented towards the use of two different types of CNTs: (i) single-walled CNTs (SWCNTs), composed of a single graphene sheet rolled-up and closed at each end by a hemispherical fullerene cap; and (ii) multi-walled CNTs (MWCNTs), composed of numerous concentric graphite cylinders. Several features of CNT structure make them intriguing candidates for nervous system applications, e.g., their shape and size, that mimic the morphology of small neuronal processes (Gilmore et al. 2008) or their high electrical conductivity and surface area which may increase charge injection capacity of CNT-based microelectrodes (Kotov et al. 2009).

The continuously growing body of research exploring CNTs potentials for the development/improvement of neuronal interfaces and electrodes (Nguyen-Vu et al. 2006; Keefer et al. 2008; Kotov et al. 2009; Lu et al. 2010), implies that CNT-based devices have to fulfil the requirements of safety, biocompatibility, mechanical and electrochemical stability in a biological environment. What are the direct effects of CNTs on the neuronal (and glial) cells that they will intimately contact? Do CNTs modulate the interface between cells and materials? Do CNT interactions instruct cell-specific behaviours,ultimately affecting neuron performance?

The leading scope of this chapter is to highlight the impact of CNTs on the morphology (e.g., adhesion, growing and neurite/dendrites extension) and function (electrical regenerative properties and synaptic activity) of cultured neurons and networks.

THE RELEVANCE OF PROCESSING CNTs: POST-SYNTHETIC MODIFICATIONS

CNTs can be obtained by means of different production procedures, which in turn may affect CNT impact on biological systems in general and on neurons, in particular. To render CNT water-soluble, as preferably needed in biological applications, several post-synthetic approaches are available (Bardi et al. 2009) which again may indirectly mediate the biological effects of CNTs, together with their variable dimensions, lengths, walls number, purity degrees and metal content. Different techniques are available for CNTs synthesis (arc discharge, laser ablation and chemical vapour deposition; Jorio et al. 2008) leading to SWCNTs or to MWCNTs production. After synthesis, CNTs need additional processing to be purified from metallic residues and amorphous carbon, e.g., purification procedures by strong acids are commonly developed to oxidize the metallic residues present, even in high percentage, in SWCNTs (Liu et al. 1998). To overcome CNT insolubility, chemical functionalization is necessary. In general, we can distinguish between covalent and non-covalent functionalizations. Covalent functionalizations consist in the direct attachment of pendant groups on CNT surface (Tasis et al. 2006), also by introducing defects sites (like carboxylic residues) with piranha solution or mixture of nitric and sulfuric acids (Singh et al. 2009). Alternatively, non-covalent functionalizations deal with the formation of weak bonds between CNTs and other (macro)molecules, establishing hydrophobic, Van der Waals or π-π interactions (e.g., DNA or polymers with extended pi-conjugated systems; Hirsch 2002; Tasis et al. 2006; Singh et al. 2009). One of the more popular reactions to obtain biocompatible carbon nanotubes is the 1,3-dipolar cycloaddition that employs an aldheyde (or a ketone) and an α-aminoacid to induce the formation of a substituted pyrrolidinic ring on the CNT surface (Georgakilas et al. 2002; Scheme 1).

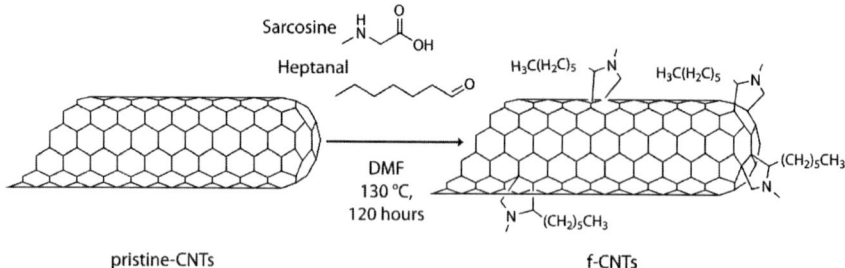

Scheme 1. **Scheme of the 1,3-dipolar cycloaddition reaction.** The 1,3-dipolar cycloaddition reaction on CNTs employs an aldheyde (or a ketone) and an α-aminoacid to induce the formation of a substituted pyrrolidinic ring on the CNT surface. The reaction is used to increase CNT solubility, allowing purification and improving the homogeneity of CNT suspensions/depositions (F.M. Toma., unpublished).

Both functionalized CNTs (f-CNTs) and as-prepared CNTs have been used as platforms to support and interface neuronal growth and activity, and as *soluble* agents able to cross biological membranes by passive diffusion (nano-needle system) or by active uptake (Ménard-Moyon et al. 2010).

An accurate chemical-physical characterization prior to CNTs use in neurobiological applications is highly relevant. Unfortunately, the traditional tools used in such characterizations, as nuclear magnetic resonance (NMR) or mass spectrometry, are not applicable to f-CNTs. In fact, ^{1}H NMR analyses of f-CNTs lead to ambiguous interpretations, even though new methodologies (i.e., diffusion NMR; Marega et al. 2009) may overcome these problems in the next future. So far, to fully characterize f-CNTs, alternative tools have to be employed, like thermogravimetric analysis (TGA), Raman and UV-Visible Spectroscopy, and not least Transmission and Scanning Electron Microscopy (TEM and SEM) together with Atomic Force Microscopy (AFM).

CNT PLATFORMS BIOCOMPATIBILITY AND IMPACT ON NEURONAL PERFORMANCE

Morphology and Growth

The assessment of CNT scaffolds ability to promote neuronal attachment and neurite extension and/or branching is particularly relevant in the perspective of employing these scaffolds to support neuronal network formation and to interface neuronal function. This issue has been addressed by many researchers, employing both SWCNTs and MWCNTs, as-prepared or functionalized (or co-deposited) with (conductive) polymers (Ni et al. 2005; Malarkey et al. 2009; Lu et al. 2010).

The simplest route to test CNT substrates for neuronal growth is to deposit a solution onto the traditional cell culturing supports, the glass coverslips. The first report on CNT biocompatibility dates back to more than 10 years ago, when the pioneering work of Mattson and colleagues (2000) showed that dissociated hippocampal neurons were able to attach and grow on MWCNT-layered substrates. These substrates were prepared by coating the glass with polyethylenimine (PEI), and then by depositing CNTs on it. Interestingly, already in this innovative work the idea that CNT functionalization may strongly affect CNT-mediated effects (later confirmed by many other works; see, e.g., Hu et al. 2004; Ni et al. 2005) emerged. In particular, the authors modified (non-covalently) f-MWCNTs with 4-hydroxynonenal and discerned their behaviour from that of non-functionalized CNTs. They found that neurons grown on f-CNTs constructed more elaborated neuritic trees, with an increased number of processes, longer neurites and higher branching, when compared with neurons grown on unmodified CNTs. Unfortunately, the authors did not provide a full chemical and physical characterization of the CNTs employed (Mattson et al. 2000).

Several other studies confirmed that CNTs allow neuronal growth. It is well established that positively-charged substrates (e.g., polylysine or polyornithine) promote neuronal growth and neurite extension, in this view Hu et al. (2004) tested at pH 7.35 negatively charged (-COOH), neutral (poly-m-aminobenzene sulfonic acid: PABS, zwitterionic) and positively charged (ethylenediamine: EN) f-MWCNTs. Neurons grown on the different substrates displayed no difference in the number of neurites per neuron, but the processes' average length was higher in neurons grown on positively-charged MWCNT-EN and the amount of neurites' branching progressively increased from negatively (MWCNT-COOH), to neutral (MWCNT-PABS), to positively charged (MWCNT-EN) CNTs (Fig. 1A).

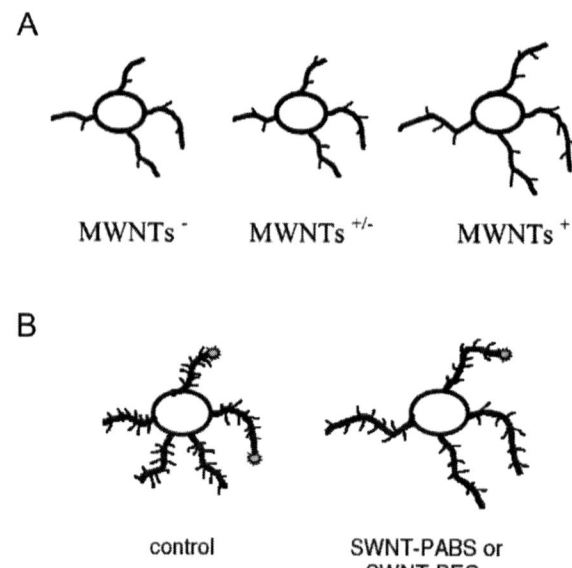

Fig. 1. **Effects of CNT on growth cones emission and neurite outgrowth and branching.** (a) Outgrowth and branching of neuronal processes can be controlled by functionalizing multi-walled carbon nanotubes (MWNT) substrates with differently charged chemical groups (-: negative; +/-: neutral; +: positive; modified with permission from Hu et al. 2004; the authors are gratefully acknowledged). (b) The treatment of neuronal cultures with differently functionalized (PABS or PEG) water-soluble single-walled carbon nanotubes (SWNT) can increase neurite length while decreasing growth cones number (modified with permission from Ni et al. 2005; the authors are gratefully acknowledged).

A more complex procedure for CNT substrate production was developed by a different laboratory (Lovat et al. 2005; Mazzatenta et al. 2007; Cellot et al. 2009): in these works CNTs were previously functionalized by 1,3-dipolar cycloaddition reaction and then a dimethylformammide (DMF) solution of f-CNTs was deposited onto the glass substrates and left evaporating at 80°C. CNT films were subsequently de-functionalized by exposure to high temperatures in inert atmosphere (Scheme 2). Using this procedure, Lovat et al. (2005) reported that the neuronal density of cultured hippocampal neurons seeded on de-functionalized CNTs is similar to that of neurons seeded on pure glass, and that the number of neurites per neuron was similar in both culture conditions.

The issue of CNTs electrical charge and conductivity as modulatory factors in mediating CNT effects on neuronal outgrowth was further explored by Malarkey and colleagues (2009). These authors sprayed a polyethylenglycol (PEG) f-CNTs solution onto glass coverslips producing films with increasing conductivities (from 0.3 to 42 S/cm), but similar roughness. Dissociated hippocampal neurons were cultured on these

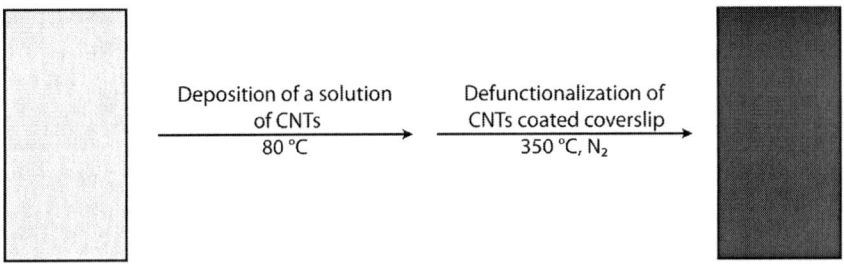

Scheme 2. **CNT deposition on glass substrates.** For deposition on glass coverslips, a dispersion of carbon nanotubes (CNTs; functionalized by the 1,3-dipolar cycloaddition reaction) in dimethylformammide is deposited on the glass support. After slow evaporation of the dimethylformammide solvent (80°C), the substrates are treated at 350°C under nitrogen atmosphere (N_2) for 20 min, to remove the functionalization, obtaining pure CNT-covered glass substrates (F.M. Toma, unpublished).

different substrates. Interestingly, the number of neurites was not affected by varying CNT conductivities, but the neurites' lengths and branching were significantly higher in neurons grown on PEG-CNTs (with the smallest conductivity, 0.3 S/cm) than those grown on PEI substrates (controls) or on PEG-CNTs with larger conductivity. These authors suggested that the selective effect related to the different conductivity of CNT substrates might be a possible explanation for the variable results obtained so far by testing CNT impact on neuronal growth (Mattson et al. 2000; Hu et al. 2004; Lovat et al. 2005).

The cited works (Mattson et al. 2000; Hu et al. 2004; Lovat et al. 2005; Malarkey et al. 2009) reported the effects of different f-CNTs on neuronal morphology, however an interesting question is that related to the effects of purified, spatially oriented, non-functionalized CNTs scaffolds on neuronal growth and viability. The recent work from Galvan-Garcia et al. (2007) showed that: i) highly oriented CNT sheets or yarns were fully biocompatible, allowing the attachment and survival of a large variety of cell phenotypes in culture; ii) the number and length of neuronal processes detected in cerebellum and cerebral cortex neurons grown on CNTs were comparable to those grown on polyornithine-covered glass; iii) remarkably, the area of growth cones at the tips of growing neuronal processes is almost twice on CNTs compared to controls. In the same study, the authors reported that primary sensory neurons (from dorsal root ganglions) are able to efficiently attach and extend their neurites along CNT yarns, closely following their surface topology. These findings encourage the idea that CNT-based substrates are suitable materials to be used in designing implantable devices aimed at favouring a spatially (and hopefully also functionally) controlled neuro-regeneration.

In this panorama, the work of Gheith et al. (2005) raised particular interest: they used layer-by-layer (LBL) technique to produce CNTs films to be used as biocompatible growing substrate. These authors reported that neuroblastoma/glioma hybrid cells were able to grow, survive and differentiate onto SWCNT/polymer multilayers. Afterwards, the same group (Gheith et al. 2006) studied the possibility to electrically stimulate neuronal cells laid on a 30 layers-LBL film. In this work, the deposition was made on glass substrates using positively charged CNTs and negatively charged polymer, exploiting different characteristics: i) this approach generated a hierarchical structure from the nano- to the micro-scale necessary for the interaction with living cells; ii) the polymers coupling with CNTs generated highly conductive films with high structural flexibility. After characterizing the normal behaviour of cultures by electrophysiological techniques, Gheith and co-authors extrinsically stimulated neurons by direct applications of voltage steps to the CNT films. More recently, Jan and Kotov (2007) proposed the use of LBL to induce differentiation of neural stem cells. In this work, SWCNTs were coated with poly (sodium, 4-styrene-sulfonate) and the LBL was assembled using positively charged PEI; the authors found that neural stem cells differentiated similarly on a 6 layers-LBL and on polyornithine.

A further development has recently emerged, that is the possibility to fuse extracellular matrix (ECM) molecules to synergistically integrate the effects of CNTs and ECM molecule on neuronal performance. In particular, two recent works explored this issue employing either MWCNTs or SWCNTs. In 2010, Cho and Borgens blended MWCNTs with type IV collagen and used this blend as substrate for PC12 cells culturing: the composite was obtained by purifying commercial MWCNTs, and mixing with a collagen IV solution at different CNTs concentrations (between 0.1 and 90%). Although composite films were not as favourable substrates for PC12 cells growth as collagen alone, 0.1–5% CNT blends were similar to control in supporting and promoting neurite elongation. The work also reported pioneering findings when blending CNTs with collagen IV, additionally providing an electrical stimulation of the preparation. The possibility of delivering electrical stimulation to neurons through CNT layers was already investigated by several groups (Liopo et al. 2006; Mazzatenta et al. 2007; Wang et al. 2006), who demonstrated that CNTs offer a suitable and efficient interface for the direct electrical stimulation of neuronal cells seeded on CNTs themselves; the work of Cho and Borgens (2010) expanded this knowledge, exploring the effects of neuronal stimulation via a CNT/ECM molecule blend. In this work, even if the presence of CNT decreased neuron elongation abilities with respect to pure collagen, this negative effect was overcome by the electrical stimulation via the substrate. At the same time, Tosun and McFetridge (2010) used a blend of SWCNTs plus type I collagen as growing substrate for PC12 cells, again demonstrating the full biocompatibility of this preparation. These recent works are the first evidences of the potential employment of CNT/ECM molecules blends in cell physiology and will probably be (and hopefully) expanded to neuronal physiology, investigating neuronal and synaptic properties.

CNT PLATFORMS BIOCOMPATIBILITY AND IMPACT ON NEURONAL PERFORMANCE: FUNCTIONAL PROPERTIES

To comprehend the mechanisms which mediate CNTs' effects on neuronal morphology is of paramount importance in designing neural devices/prosthesis: the implement of a controlled neurite growth and/or branching is expected to affect not only the neuronal re-growth ability, but also the neuronal signalling ability. In fact, since dendrites and axons are the structures that provide neurons with their computational complexity, CNTs-induced modulations may, in principle, alter signal integration at multiple levels. Non-functionalized CNTs may affect neuronal signalling regardless of morphological changes, by means of CNT/cell membrane "direct" interactions.

In 2005, Lovat and collaborators assessed for the first time how neurons reconstruct a functional network when integrated to non-functionalized CNTs. These authors observed that the growth of cultured hippocampal circuits on a conductive MWCNTs meshwork is always accompanied by a significant enhancement in the frequency of spontaneous synaptic activity, measured as post synaptic currents, when compared to control neurons (glass coverslips; Fig. 2). This boost in network spontaneous activity was accompanied by an increased frequency of action potential firing. MWCNTs and SWCNTs were equally able to improve synaptic activity in cultured networks (Mazzatenta et al. 2007). Immunocitochemistry experiments, that allow the visualization of cultured neurons (Lovat et al. 2005; Mazzatenta et al. 2007), ruled out the hypothesis that the increased network activity be related to an increased number of neurons adhering on CNT platforms. Labelling the cultures with antibodies targeted to microtubule-associated protein 2 (MAP-2) allowed a direct quantification of hippocampal cells. Neuronal density did not differ between the two groups (control and CNTs), furthermore, the general neuronal morphology, monitored as the cell body diameter and the number of neurites departing from the soma, were found similar in the two growing conditions (Lovat et al. 2005; Mazzatenta et al. 2007). Additionally, the membrane passive

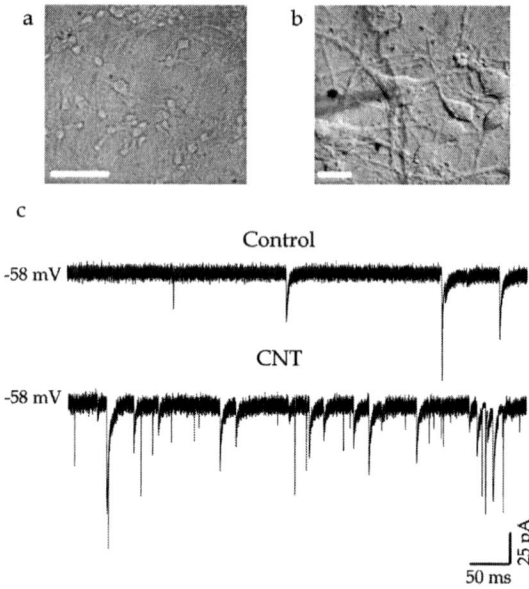

Fig. 2. **CNTs boost neuronal network activity.** (a) Bright field image of a dissociated hippocampal culture grown on CNT (carbon nanotubes) substrate, showing a healthy morphology (calibration bar: 100 µm). (b) Neuronal electrical activity is monitored by means of patch clamp technique. Note the glass pipette employed to record the activity from a visualized neuron (calibration bar: 10 µm). (c) Exemplificative voltage-clamp recordings of spontaneous activity. Note the typical strong increase in spontaneous postsynaptic currents in a CNT neuron compared to a control one (G. Cellot and L. Ballerini, unpublished).

properties of recorded neurons (capacitance, input resistance and resting membrane potential), generally accepted as useful indicators of the cellular dimensions and health, were fully comparable between control and CNTs neurons (Lovat et al. 2005; Mazzatenta et al. 2007; Cellot et al. 2009). These results thus indicated that neuronal density and morphology were not implied in the increased spontaneous activity detected in the presence of pure non-functionalized CNT scaffolds and that different phenomena, other from network sizes, seemed to take place in the interplay between neurons and CNTs.

In this context, SEM and TEM images demonstrated the occurrence of intimate and tight contacts between neuronal membranes and CNTs layers (Mazzatenta et al. 2007; Cellot et al. 2009, Fig. 3) and suggested a direct (even electrical) coupling of CNTs to neuronal processes, with potential effects on neuronal performance that relapsed at the network level. Sorkin and colleagues (2009) further supported this hypothesis, showing that process entanglement is crucial to neuronal anchoring to rough surfaces, thus contributing to neuronal physical interactions with CNTs.

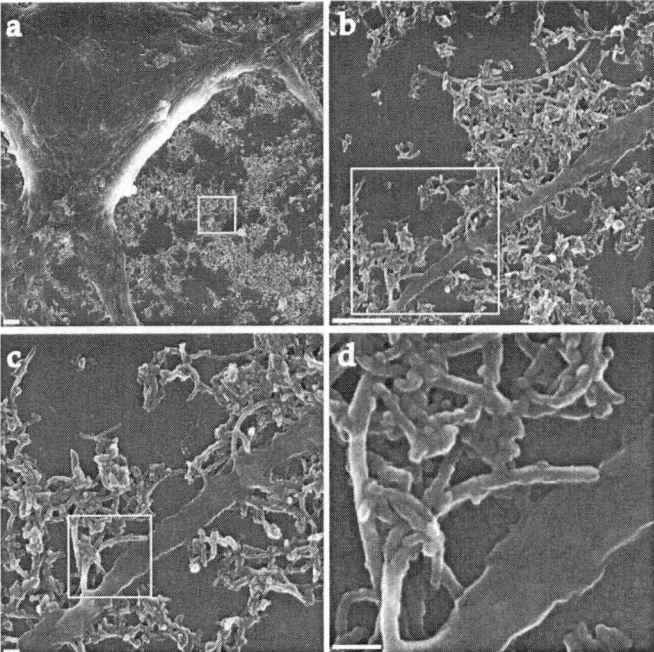

Fig. 3. **CNTs form hybrid structures with neuronal membranes.** SEM images showing a typical hippocampal neuron cultured on a multi-walled carbon nanotubes (MWCNTs)-covered substrate: each micrograph is the magnification of the square indicated in the previous one. In the higher enlargements, the numerous, tight points of contact between CNT and neuronal membranes are evident. Calibration bars: 1μm (a, b), 100 nm (c, d). (G. Cellot and F.M. Toma, unpublished).

The emergence of an electrical coupling between CNTs and neuronal membranes was explored in the work by Cellot and co-authors (2009). These authors used single-cell electrophysiology techniques and found that CNTs re-engineered the integrative electrical abilities of hippocampal neurons *in vitro*. In this study neurons, on CNT substrates (SWCNTs or MWCNTs) and in control (glass coverslips), were forced to fire short trains of action potentials at variable frequencies (20–100 Hz) to maximize the interactions between regenerative properties in proximal and distal areas of a single neuron (Larkum et al. 1999). The authors showed that CNT scaffolds favoured the generation of back-propagating action potentials, a neuronal regenerative property involved in the regulation of local synaptic feedback and in the release of messengers (Zilberter et al. 2005; Kuczewski et al. 2008). Ultimately, the CNT-induced modification of dendritic electrogenic properties resulted in an increased single-cell excitability, that, at least in part, may explain the boosting of spontaneous activity. Although speculative, the correlation between the effects detected at single-cell level and those displayed at the network level, were further supported through a combination of electrophysiological techniques and theoretical modelling (Cellot et al. 2009). The impact of the improved dendritic regenerative properties by CNTs and the emerging network behaviour was predicted by network modelling, where a circuit composed by randomly connected individual neurons was enriched with backpropagating action potentials. The resulting electrical activity generated by the modelled circuit showed the occurrence of prolonged duration in bursting of synaptic events. This prediction was confirmed in voltage-clamp recordings from neuronal networks developed on CNT substrates.

The work of Cellot et al. (2009) further addressed which CNTs' characterizing properties (i.e., their nanostructure or their electrical conductivity) are needed to successfully manipulate neuronal regenerative ability. Neurons were cultured on materials different from CNTs, but presenting either comparable nano-roughness or comparable conductivity: the self assembling peptide RADA 16 (mimicking CNT-like three-dimensional nanoscaffolds in the absence of electrical conductivity) or Indium Tin oxide (ITO, reproducing CNT-like conductivity in the absence of three-dimensional nanostructure). These substrates were unable to replicate CNTs effects, therefore the authors suggested that CNTs mediate their effects *via* the co-existence of conductivity and nanostructure.

So far, the molecular mechanisms underlying the generation of back-propagating action potentials in neurons grown on CNTs have not been totally understood. However, considered the good conductive properties of CNTs and their tight contacts with neuronal membranes, it is tempting to speculate that these hybrid structures facilitated the generation of back-propagating action potentials *via* a direct electrical shortcut between adjacent compartments on dendrites, as supported by additional theoretical modelling (Cellot et al. 2009).

NANO-MODIFICATION OF ELECTRODE SURFACES BY CNT LAYERS: IMPACT ON NEURONAL PERFORMANCE

The extraordinary strength, electrical conductivity and surface area of CNTs make them excellent candidates for interfacing with neural systems for the development of biocompatible and robust neuroprosthetic devices. The high CNT surface area can drastically increase charge injection capacity and decrease the interfacial impedance with neurons, all requirements for further miniaturization of electrodes (Kotov et al. 2009). The current state of the development of CNT/neural interface is far from the complete realization or even understanding of promising properties of CNTs in respect to neuro-electrode engineering. A significant advance was made by Keefer et al. (2008), who showed that CNT coatings decreased electrode impedance and increased charge transfer. There are a number of excellent chapters in this issue devoted to the nano-electrode developments, in this paragraph we briefly focused on nano-modified CNTs electrode surfaces and neuronal performance.

Recently, Lu and colleagues (2010) assessed the biocompatibility of polypyrrole (PPy)/SWCNT films, testing their suitability for nanosurface modifications of electrodes by electrodepositing them on Pt-electrode. After showing that PC12 cells extended longer neurites and complex neurite networks when grown on a PPy/SWCNT-covered ITO surface (compared to ITO alone), the authors studied the consequences of the implantation of PPy/SWCNT-covered Pt devices directly in the rat brain and compared it with the non-covered Pt implants. The authors demonstrated that, around the site of PPy/SWCNT implants, there was less gliosis and increased neuronal survival when compared to Pt alone. Unfortunately, the authors did not dissect the effects of PPy alone from those of SWCNTs alone in terms of gliosis and neuronal survival; nevertheless, this study paved the way to the development of a critical and systematic study of the effects of CNT-based materials directly on neuronal tissue *in vivo*, in the perspective of the employment of CNT-based materials to implement neuronal interfaces.

Several recent reports (Khraiche et al. 2009; Shoval et al. 2009) focussed on the use of CNTs as nano-structured electrodes. Khraiche and colleagues (2009) coated microelectrode arrays (MEAs) by depositing several drops of SWCNTs solution and observed that rat hippocampal neuronal networks developed on CNTs-MEAs showed an earlier onset of electrical activity in comparison with those developed on control electrodes. The authors suggested that the increase in surface roughness due to CNTs immobilized on the microelectrodes leads to an increased activation of adhesion molecules such as integrins, which might in turn promote a faster neuronal differentiation.

Shoval and co-authors (2009) grew CNTs directly on MEA electrodes to record the activity of whole-mount neonatal mouse retinas. After minutes from the placement of the retinas on the electrodes, the authors could monitor the typical retinal spontaneous bursting and propagating waves which displayed a higher signal-to-noise ratio in comparison to that detected from commercially available electrodes. Interestingly, the recorded signals underwent over a period of minutes to hours to a gradual increase in spike amplitude. The authors indicated this phenomenon as the result of a dynamic interaction between CNTs and neurons, determining an improvement in cell-electrode coupling. Nevertheless, it should not be excluded that CNTs can intrinsically modify neuronal network activity, thanks to their properties (see above).

SOLUBLE CNTs AND NEURONAL PERFORMANCE

In this paragraph we briefly address recent preclinical studies investigating the biological impact of f-CNTs when internalized by neurons in the perspective of employing such materials for molecular sensing, diagnostics or drug delivery devices.

The differential effects of various CNT functionalizations on neuronal morphological properties were explored employing *soluble* f-CNTs. Indeed, several studies reported that the treatment of neurons with soluble f-CNTs can affect neuronal performance. Ni and colleagues (2005) treated dissociated hippocampal cultures with water-soluble SWCNTs functionalized with PABS or PEG and observed that this treatment did not affect neuronal viability, promoted a decrease in the number of neurites per neuron, an enhancement of neurite length and a parallel reduction in the growth cones number. Using calcium imaging techniques, the authors showed that SWCNT-PEG were able to inhibit the cell depolarization-dependent calcium influx and this blockade might ultimately lead to the detected morphological alterations. The real mechanisms underlying the observed effects of soluble SWCNTs are still unresolved, although the hypothesis of the endocytosis block by functionalized SWCNTs has been formulated as a possible explanation (Malarkey et al. 2008). These authors used SWCNT-PEG to monitor plasma membrane/vesicular recycling by exposing neuronal cells to a fluorescent dye. The authors discovered that when neurons were incubated with SWCNT-PEG, they displayed an inhibited endocytotic process. The authors hypothesized that such inhibition might be due to the insertion of f-CNTs into the transiently opened vesicles fused with the membrane at the level of pre-synaptic terminals, and that the membrane resulting from blocked endocytotic vesicles would go to feed the neuritis extension.

Another aspect of neuronal performance which is altered by the treatment with soluble CNTs is ion channel activity. Xu and co-authors (2009) found that the incubation of PC12 cells cultures with soluble carboxy-terminated MWCNTs determines the time-dependent and most probably irreversible suppression of the current densities of three different types of voltage activated potassium channels. Although such results are in agreement with other reports, where the block of ion channels by soluble CNTs was demonstrated in diverse cell types (Park et al. 2003), other studies showed that the treatment with soluble CNTs seemed unable to alter neuronal performance. For instance, dissociated hippocampal cultures treated with MWCNTs functionalized with two different cell adhesion peptide sequences (GRGDSPC or IKVAVC, by the introduction of carboxylic groups together with the 1,3-dipolar cycloaddition reaction) showed physiological parameters in their neuronal membrane passive properties, spontaneous neuronal activity and current densities (Gaillard et al. 2009). Similarly, membrane passive properties, ionic conductance and action potential generation of neurons from chicken spinal cord cultures were unaffected by the incubation with SWCNTs with different agglomeration degrees, although the same CNTs preparation modified both basic cellular properties, such as the capacitance and the resting potential, and inward current conductance in dorsal root ganglion cultures (Belyanskaya et al. 2009).

After numerous morphological works reporting the CNTs effects on neurite elongation, the molecular mechanisms and the intracellular pathways involved in such CNT-mediated effects are emerging. Recently, Matsumoto and colleagues (2010) showed that water-soluble MWCNT (with solubility increased with free amino groups) were able to strongly potentiate the effect of NGF, added to the culture together with CNTs, in promoting neurite outgrowth in both dorsal root ganglion neurons and PC12 cells. Remarkably, this work showed for the first time that the CNT effects involve the ERK kinase intracellular signalling pathway. The investigations of the molecular mechanisms underlying the various effects of CNTs on neuronal morphology, physiology and function are still in their infancy and further work is needed to fully understand the intracellular mechanisms and complex pathways involved.

The above reported works showed conflicting results concerning the impact of soluble CNTs on neuronal performance. This could be accounted for by differences in the type of tested cells (PC12, hippocampal, spinal cord, dorsal root ganglion cells); however, as widely discussed elsewhere (Cellot et al. 2010), the lack of a detailed chemical and physical characterization of the CNTs employed (type, production process, functionalization, purity degree, etc.) and of relevant controls can prevent definite and clear interpretations of the induced biological responses.

Applications to other Areas of Health and Disease

The present chapter reports the fundamental, pioneering evidences, as well as the more recent advances, in the study of the interactions between CNTs and neurons. CNTs are biocompatible, they are able to boost spontaneous activity of neuronal networks *in vitro* and, in particular conditions (different functionalizations, solid substrates or soluble form; see above) they are suggested to be an efficient tool to selectively control neurite outgrowth and branching. These features, together with the unique mechanical, physical and chemical properties of CNTs, raise an intriguing question: can we use CNT-based devices to control neuronal behaviour *in vivo*, in particular in CNS disorders? The promising employment of CNT

Table 1. Summary of the main CNT preparation methods/protocols employed in the study of CNT effects on neuronal performance.

CNT Functionalization	CNT Type	Cells/Substrate	Effect	Reference
	MWCNT + 4-hydroxynonenal	Hippocampal neurons seeded onto CNT and PEI-coated glass cover slips	↑Elaborated neuritic trees ↑Number of processes ↑Neurites lenght ↑Branches occurrence	Mattson et al. 2000
	MWCNT-COO⁻ MWCNT-CO-PABS MWCNT-CO-EN⁺	Hippocampal neurons seeded onto CNT-coated glass coverslips	↑Processess lenght (MWCNT-EN) ↑Branching MWCNT-EN> MWCNT-PABS> MWCNT-COO⁻	Hu et al. 2004
	Defunctionalized SWCNTs/ MWCNTs	Hippocampal neurons seeded onto CNT-coated glass coverslips	↑Spontaneous synaptic activity ↑Generation of back-propagating AP probability	Lovat et al. 2005 Mazzatenta et al. 2007 Cellot et al. 2009
	SWCNT-CO-PEG	Hippocampal neurons seeded onto CNT-coated glass coverslips	↑Branches and neurites lenght depending on conductivity	Malarkey et al. 2009
	MWCNTs	DRG placed onto MWCNT sheet	↑ Area of the growth cones	Galvan-Garcia et al. 2007

	Material	Setup	Main findings	Reference
	SWCNTs + poly(N-cetyl-4-vinylpyridinium bromide-co-N-ethyl-4-vinylpyridinium bromide-co-4-vinylpyridine	NG108-15 neuroblastoma/glioma hybrid cells seeded onto CNT-LBL-coated glass	Cells are able to grow, survive and differentiate; possibility of electrical stimulation via CNT-LBL	Gheith et al. 2005 and 2006
	SWCNTs + PSS	Neuronal stem cells seeded onto CNT-LBL-coated glass	Biocompatibility, neurite outgrowth and expression of neuronal markers similar to commonly used substrates	Jan and Kotov et al. 2009
	MWCNTs + Collagen IV SWCNTs + Collagen I	PC12 cell culture seeded on Petri dishes	No evidence of morphological difference with respect to the control	Cho and Borgens 2010 Tosun and McFetridge 2010
CNT-covered Electrodes				
	SWCNT-COOH + Polypyrrole	CNT-coated platinum electrode implanted in rat brain	↑ Neuronal survival ↔ Gliosis	Lu et al. 2010
	SWCNTs	Rat hippocampus slices placed onto MEA	↑ Neuronal differentiation	Khraiche et al. 2009
	MWCNTs grown by CVD	Isolated mouse retina placed onto MEA	↑ Signal-to-noise ratio ↑ Spike amplitude ↑ Cell-electrode coupling	Shoval et al. 2009

Table 1. contd....

Table 1. contd....

CNT Functionalization	Soluble CNTs			Reference
	CNT Type	Cells/Substrate	Effect	
	SWCNT-CO-PABS SWCNT-CO-PEG	Hippocampal neuronal cultures grown onto PEI-coated coverslips	↓ Neurites per neuron ↑ Neurite lenght ↑ Growth cones	Ni et al. 2005
	SWCNT-CO-PEG	Hyppocampal neurons grown on glass coverslips	Inhibition of endocytic process	Malarkey et al. 2008
	MWCNTs-COOH	Undifferentiated PC12 cells grown on poly-lysine coated flasks	↓ Current densities of K^+ channels	Xu et al. 2009
	MWCNT-COOH + GRGDSP peptide MWCNT-COOH + IKVAV peptide	Hippocampal neurons grown on glass coverslips	No alterations in neuronal performance	Gaillard et al. 2009
	MWCNT-CONH(CH$_2$)$_2$NH$_2$ MWCNT-CONH(CH$_2$)$_8$NH$_2$	DRG neurons and PC12 cells grown on laminin-coated 96-well culture plates	↑ Effect of NGF	Matsumoto et al. 2010

AP, action potential; CVD, chemical vapour deposition; DRG, dorsal root ganglia; EN, ethylenediamine; LBL, layer-by-layer; MEA, microelectrode array; MWCNT, multi-walled carbon nanotubes; NGF, nerve growth factor; PC12 cells, pheochromocytoma 12 cells; PEG, polyethylene glycol; PEI, polyethylenimine; PABS, poly-m-aminobenzene sulfonic acid; PSS, poly(sodium, 4-styrene-sulfonate); SWCNT, single-walled carbon nanotubes.

Table 2. Key Facts of Neurons.

1. Neurons are electrically excitable cells that transmit and process information in the nervous system.
2. Neurons express a wide set of ion channels and receptors, determining their electrophysiological features and properties and playing critical roles in health and disease.
3. Neurons generate action potentials (short and stereotyped waves of electrical excitation, rapidly propagating along the neuron).
4. When an action potential arrives to specialized contact zones between neurons, termed synapses, the neuron releases its neurotransmitter, which binds to dedicated receptors on the downstream neuron. The neurotransmitter can be excitatory or inhibitory.
5. Neurons can be maintained in culture, where they form functional neural networks, and used as *in vitro* models for the study of neuronal functions.

This table lists the key facts of neurons, including their basic features like the formation of synapses and action potentials generation.

as suitable material for implementing neuronal interfaces and electrodes, delivering electrical stimuli or recording electrical signals, is beginning to be tested *in vivo* (Nguyen-Vu et al. 2006; Keefer et al. 2008; Lu et al. 2010) and the application of CNT-based materials could potentially carry great benefit particularly in the field of neural prosthesis for neuroregeneration in case of nervous system damage. In addition, CNTs in their soluble form have the striking potential to be used as efficient and reliable tools for the production of diagnostics, molecular sensing, or drug delivery devices (Pantarotto et al. 2004), thus widening the potential use of CNT to continuously growing areas of molecular biology, clinical neuroscience and bioengineering.

Definitions

Cultured neuronal networks: Population of neurons (from CNS explant) functionally interconnected via synapses between neuronal axons and dendrites and able to generate spontaneous, synaptic network activity, maintained *in vitro*.

Functionalization: Addition of functional chemical groups onto the surface of a molecule. Functionalization can be used, e.g., to improve molecule solubilisation/dispersion, or to allow subsequent reactions via the added chemical group.

Multi-walled carbon nanotube (MWCNT): Carbon nanotube made up of numerous concentric graphite cylinders, with diameters up to 100 nm.

Neural interfaces: Biomedical devices implanted in the central nervous system developed to control motor disorder or to translate willful brain processes into specific actions by the control of external devices.

Neurite: Projection emerging from the neuronal cell body. The term usually refers to immature/developing projections (axon and dendrites hardly distinguishable).

Single-walled carbon nanotube (SWCNT): Carbon nanotube made up of a single rolled-up graphite sheet, with open end or closed by fullerene-like caps, typically with a 0.8–2 nm diameter.

Summary Points

- Carbon nanotubes (CNTs) are made up by rolled-up graphene sheets. Their peculiar electrical, thermal and conductivity properties, their shape, roughness and size similar to neuronal processes and the characteristic fractal-like nanostructure of random CNTs dispersions, make CNTs among the most promising nanomaterials for the development of neuronal/nanomaterial hybrid networks and for biomedical applications.
- CNT-based nano-modified surfaces are fully biocompatible and support neuronal survival and growth *in vitro*.
- Variously functionalized CNT substrates support and promote neurite extension and/or branching (depending on the specific functionalization).
- CNT substrates boost spontaneous neuronal activity and increase single-cell excitability in cultured hippocampal neurons (an effect requiring both the conductive and nano-structural CNT properties), suggesting their capacity to re-engineer neuronal integrative abilities.
- CNTs can cover/modify electrodes to deliver electrical stimuli or to record neuronal electrical signals, improving neuronal-electrode interface contact and signal-to-noise ratio.

Abbreviations

AFM	:	Atomic Force Microscopy
BDNF	:	Brain-derived neurotrophic factor
CNT	:	Carbon nanotubes
DMF	:	Dimethylformammide
DRG	:	Dorsal root ganglia
EN	:	Ethylenediamine

ECM	:	Extracellular matrix
f-CNT	:	Functionalized carbon nanotubes
ITO	:	Indium tin oxide
LBL	:	Layer-by-layer
MAP-2	:	Microtubule-associated protein 2
MEA	:	Microelectrode array
MWCNT	:	Multi-walled carbon nanotubes
CNS	:	Central nervous system
NGF	:	Nerve growth factor
NMR	:	Nuclear magnetic resonance
PC12 cells	:	Pheochromocytoma 12 cells
Pt	:	Platinum
PEG	:	Polyethylene glycol
PEI	:	Polyethylenimine
PABS	:	Poly-m-aminobenzene sulfonic acid
PPy	:	Polypyrrole
SEM	:	Scanning electron microscopy
SWCNT	:	Single-walled carbon nanotubes
TGA	:	Thermogravimetric analysis
TEM	:	Transmission electron microscopy

Acknowledgments

Financial support from NEURONANO-NMP4-CT-2006-031847 and CARBONANOBRIDGE- ERC-2008-227135, to LB and MP, is gratefully acknowledged.

References

Bardi, G., P. Tognini, G. Ciofani, V. Raffa, M. Costa and T. Pizzorusso. 2009. Pluronic-coated carbon nanotubes do not induce degeneration of cortical neurons *in vivo* and *in vitro*. Nanomedicine 5: 96–104.

Belyanskaya, L., S. Weigel, C. Hirsch, U. Tobler, H.F. Krug and P. Wick. 2009. Effects of carbon nanotubes on primary neurons and glial cells. Neurotoxicology 30: 702–711.

Cellot, G., E. Cilia, S. Cipollone, V. Rancic, A. Sucapane, S. Giordani, L. Gambazzi, H. Markram, M. Grandolfo, D. Scaini, F. Gelain, L. Casalis, M. Prato, M. Giugliano and L. Ballerini. 2009. Carbon nanotubes might improve neuronal performance by favouring electrical shortcuts. Nat. Nanotechnol. 4:126–133.

Cellot, G., L. Ballerini, M. Prato and A. Bianco. 2010. Neurons Are Able to Internalize Soluble Carbon Nanotubes: New Opportunities or Old Risks? Small 6: 2630–2633.

Cho, Y., R.B. Borgens. 2010. The effect of an electrically conductive carbon nanotube/collagen composite on neurite outgrowth of PC12 cells. J. Biomed. Mater. Res. A. 95: 510–517.

Gaillard, G., G. Cellot, S. Li, F. M. Toma, H. Dumortier, G. Spalluto, B. Cacciari, M. Prato, L. Ballerini and A. Bianco. 2009. Carbon Nanotubes Carrying Cell-Adhesion Peptides do not Interfere with Neuronal Functionality. Adv. Mater. 21: 2903–2908.

Galvan-Garcia, P., E.W. Keefer, F. Yang, M. Zhang, S. Fang, A.A. Zakhidov, R.H. Baughman and M.I. Romero. 2007. Robust cell migration and neuronal growth on pristine carbon nanotube sheets and yarns. J. Biomater. Sci. Polym. Ed. 18: 1245–1261.

Georgakilas, V., K. Kordatos, M. Prato, D. M. Guldi, M. Holzinger and A. Hirsch. 2002. Organic Functionalization of Carbon Nanotubes. J. Am. Chem. Soc. 124: 760–761.

Gheith, M.K., V.A. Sinani, J.P. Wicksted, R.L. Matts and N.A. Kotov. 2005. Singlewalled carbon nanotube polyelectrolyte multilayers and freestanding films as a biocompatible platform for neuroprosthetic implants. Adv. Mater. 17: 2663–2667.

Gheith, M.K., T.C. Pappas, A.V. Liopo, V.A. Sinani, B.S. Shim, M. Motamedi, J.P. Wicksted and N.A. Kotov. 2006. Stimulation of Neural Cells by Lateral Currents in Conductive Layer-by-Layer Films of Single-Walled Carbon Nanotubes. Adv. Mater. 18: 2975–2979.

Gilmore, J.L., X. Yi, L. Quan and A.V. Kabanov. 2008. Novel nanomaterials for clinical neuroscience. J. Neuroimmune. Pharmacol. 3: 83–94.

Hirsch, A. 2002. Functionalization of single-walled carbon nanotubes. Angew. Chem. Int. Ed. 41: 1853–1859.

Hu, H., Y. Ni, V. Montana, R.C. Haddon and V. Parpura. 2004. Chemically functionalized carbon nanotubes as substrates for neuronal growth. Nano Lett. 4: 507–511.

Jan, E., N.A. Kotov. 2007. Successful Differentiation of Mouse Neural Stem Cells on Layer-by-Layer Assembled Single-Walled Carbon Nanotube Composite. Nano Lett. 7: 1123–1128.

Jorio, A., G. Dresselhaus and M.S. Dresselhaus. 2008. Carbon Nanotubes: Advanced Topics in the Synthesis, Structure, Properties and Applications. Springer, Heidelberg. Germany.

Keefer, E.W., B.R. Botterman, M.I. Romero, A.F. Rossi and G.W. Gross. 2008. Carbon nanotube coating improves neuronal recordings. Nat. Nanotechnol. 3: 434–439.

Khraiche, M., N. Jackson and J. Muthuswamy. 2009. Early onset of electrical activity in developing neurons cultured on carbon nanotube immobilized microelectrodes, 777–780. The Proceedings of the 31st Annual International Conference of the IEEE Eng. Med. Biol. Soc., Minneapolis, Minnesota, USA.

Kotov, N.A., J.O. Winter, I.P. Clements, E. Jan, B.P. Timko, S. Campidelli, S. Pathak, A. Mazzatenta, C.M. Lieber, M. Prato, R.V. Bellamkonda, G.A. Silva, N.W. Shi Kam, F. Patolsky and L. Ballerini. 2009. Nanomaterials for Neural Interfaces. Adv. Mater. 21: 3970–4004.

Krishnan, A., E. Dujardin, T.W. Ebbesen, P.N. Yianilos and M.M.J. Treacy. 1998. Young's modulus of single-walled nanotubes. Phys. Rev. B. 58: 14013–14019.

Kuczewski, N., C. Porcher, N. Ferrand, H. Fiorentino, C. Pellegrino, R. Kolarow, V. Lessmann, I. Medina and J.L. Gaiarsa. 2008. Backpropagating action potentials trigger dendritic release of BDNF during spontaneous network activity. J. Neurosci. 28:7013–7023.

Larkum, M.E., K.M. Kaiser and B. Sakmann. 1999. Calcium electrogenesis in distal apical dendrites of layer 5 pyramidal cells at a critical frequency of back-propagating action potentials. Proc. Nat. Acad. Sci. 96: 14600–14604.

Liopo, A.V., M.P. Stewart, J. Hudson, J.M. Tour and T.C. Pappas. 2006. Biocompatibility of native functionalized single-walled carbon nanotubes for neuronal interface. J. Nanosci. Nanotechnol. 6: 1365–1374.

Liu, J., A.J. Rinzler, H. Dai, J.H. Hafner, R.K. Bradley, P.J. Boul, A. Lu, T. Iverson, K. Shelimov, C.B. Huffman, F. Rodriguez-Macias, Y.S. Shon, T.R. Lee, D.T. Colbert and R.E. Smalley. 1998. Fullerene pipes. Science 280: 1253–1256.

Lovat, V., D. Pantarotto, L. Lagostena, B. Cacciari, M. Grandolfo, M. Righi, G. Spalluto, M. Prato and L. Ballerini. 2005. Carbon nanotube substrates boost neuronal electrical signaling. Nano Lett. 5: 1107–1110.

Lu, Y., T. Li, X. Zhao, M. Li, Y. Cao, H. Yang and Y.Y. Duan. 2010. Electrodeposited polypyrrole/carbon nanotubes composite films electrodes for neural interfaces. Biomaterials 31: 5169–5181.

Malarkey, E.B., R.C. Reyes, B. Zhao, R.C. Haddon and V. Parpura. 2008. Water soluble single-walled carbon nanotubes inhibit stimulated endocytosis in neurons. Nano Lett. 8: 3538–3542.

Malarkey, E.B., K.A. Fisher, E. Bekyarova, W. Liu, R.C. Haddon and V. Parpura. 2009. Conductive single-walled carbon nanotube substrates modulate neuronal growth. Nano Lett. 9: 264–268.

Marega, R., V. Aroulmoji, F. Dinon, L. Vaccari, S. Giordani, A. Bianco, E. Murano, M. Prato. 2009. Diffusion-Ordered NMR Spectroscopy in the Structural Characterization of Functionalized Carbon Nanotubes. J. Am. Chem. Soc. 131: 9086–9093.

Matsumoto, K., C. Sato, Y. Naka, R. Whitby and N. Shimizu. 2010. Stimulation of neuronal neurite outgrowth using functionalized carbon nanotubes. Nanotechnology 21: 115101.

Mattson, M.P., R.C. Haddon and A.M. Rao. 2000. Molecular functionalization of carbon nanotubes and use as substrates for neuronal growth. J. Mol. Neurosci. 14: 175–182.

Mazzatenta, A., M. Giugliano, S. Campidelli, L. Gambazzi, L. Businaro, H. Markram, M. Prato and L. Ballerini. 2007. Interfacing neurons with carbon nanotubes: electrical signal transfer and synaptic stimulation in cultured brain circuits. J. Neurosci. 27: 6931–6936.

Ménard-Moyon, C., E. Venturelli, C. Fabbro, C. Samorì, T. Da Ros, K. Kostarelos, M. Prato and A. Bianco. 2010. The alluring potential of functionalized carbon nanotubes in drug discovery. Expert Opin. Drug Discov. 5: 691–707.

Nguyen-Vu, T.D.B., H. Chen, A.M. Cassell, R. Andrews, M. Meyyappan and J. Li. 2006. Vertically aligned carbon nanofiber arrays: an advance toward electrical-neural interfaces. Small 2: 89–94.

Ni, Y., H. Hu, E.B. Malarkey, B. Zhao, V. Montana, R.C. Haddon and V. Parpura. 2005. Chemically functionalized water soluble single-walled carbon nanotubes modulate neurite outgrowth. J. Nanosci. Nanotechnol. 10: 707–712.

Pantarotto, D., R. Singh, D. McCarthy, M. Erhardt, J.P. Briand, M. Prato, K. Kostarelos and A. Bianco. 2004. Functionalized carbon nanotubes for plasmid DNA gene delivery. Angew. Chem. Int. Ed. Engl. 43: 5242–5246.

Park, K.H., M. Chhowalla, Z. Iqbal and F. Sesti. 2003. Single-walled carbon nanotubes are a new class of ion channel blockers. J. Biol. Chem. 278: 50212–50216.

Shoval, A., C. Adams, M. David-Pur, M. Shein, Y. Hanein and E. Sernagor. 2009. Carbon nanotube electrodes for effective interfacing with retinal tissue. Front. Neuroengineering 2, 4.

Singh, P., S. Campidelli, S. Giordani, D. Bonifazi, A. Bianco and M. Prato. 2009. Organic functionalisation and characterisation of single-walled carbon nanotubes. Chem. Soc. Rev. 38: 2214–2230.

Sorkin, R., A. Greenbaum, M. David-Pur, S. Anava, A. Ayali, E. Ben-Jacob and Y. Hanein. 2009. Process entanglement as a neuronal anchorage mechanism to rough surfaces. Nanotechnology 20: 015101.

Tasis, D., N. Tagmatarchis, A. Bianco and M. Prato. 2006. Chemistry of Carbon Nanotubes. Chem. Rev. 106: 1105–1136.

Tosun, Z., P.S. McFetridge. 2010. A composite SWNT-collagen matrix: characterization and preliminary assessment as a conductive peripheral nerve regeneration matrix. J. Neural. Eng. 7: 066002.

Wang, K., H.A. Fishman, H. Dai and J.S. Harris. 2006. Neural stimulation with a carbon nanotube microelectrode array. Nano Lett. 6: 2043–2048.

Xu, H., J. Bai, J. Meng, W. Hao, H. Xu and J.M. Cao. 2009. Multi-walled carbon nanotubes suppress potassium channel activities in PC12 cells. Nanotechnology 20: 285102.

Zilberter, Y., T. Harkany and C.D. Holmgren. 2005. Dendritic release of retrograde messengers controls synaptic transmission in local neocortical networks. Neuroscientist 11: 334–344.

Section 2: Therapeutics

11

Nanotechnologies for Treatment of Stroke and Spinal Cord Injury

Šárka Kubinová[1,]* and Eva Syková[2]

ABSTRACT

Treatment of stroke and spinal cord injury, as well as other central nervous system disorders, is a major challenge for regenerative medicine. To cover some of the latest nanotechnology trends in the treatment of stroke and spinal cord injury, this chapter will mainly focus on the use of nanotechnology for cell tracking and tissue engineering. Cell transplantation is emerging as a promising therapeutic option that may replace lost cells or support endogenous regeneration. To track transplanted cells in a living organism, superparamagnetic iron oxide nanoparticles can be used for cell labeling and their non-invasive MRI monitoring *in vivo*. To improve the delivery of drugs or transplanted cells, targeted magnetic cell delivery may, due to a localized magnetic field, promote the accumulation of magnetically labeled cells

---
[1] Institute of Experimental Medicine, Academy of Sciences of the Czech Republic, Videnska 1084, 14220 Prague 4, Czech Republic; E-mail: sarka.k@biomed.cas.cz
[2] Institute of Experimental Medicine, Academy of Sciences of the Czech Republic, Videnska 1084, 14220 Prague 4, Czech Republic; Department of Neuroscience and Center for Cell Therapy and Tissue Repair, 2nd Medical Faculty, Charles University, 15000 Prague 5, Czech Republic; E-mail: sykova@biomed.cas.cz
*Corresponding author

List of abbreviations after the text.

or magnetic nanoparticles carrying drugs in the targeted area. Alternatively, systemic ultra-small superparamagnetic iron oxide nanoparticles administration utilizes their preferential uptake by monocytes/macrophages for magnetic resonance visualization of the inflammatory response after stroke or other central nervous system disorders. Spinal cord injury often results in scar barriers and cavity formation that hinder axons from regenerating across the lesion. One aspect of nanotechnology is the development of biologically compatible nanofiber scaffolds that mimic the structure of the extracellular matrix and can serve as a permissive bridge for axonal regeneration, as well as for drug or cell delivery.

INTRODUCTION

The regenerative capacity of the adult human central nervous system (CNS) is very limited. In contrast to the peripheral nerves, the inhibitory environment of the CNS hinders axons from regenerating, which is the main reason for permanent paralysis and loss of sensation following injury. Despite intensive research, there is currently no effective therapy for the repair and regeneration of injured CNS tissue. Nanotechnologies that comprise designing materials and devices at the cellular, molecular and atomic levels can provide more precise, targeted and effective approaches in CNS treatment. Increasing attention has focused on the development of novel therapeutic methods and materials that aim to regenerate damaged CNS tissue by using recent advances in drug and gene delivery, cell therapy and tissue engineering.

Superparamagnetic iron oxide nanoparticles (SPION) represent a significant application of nanotechnology in CNS diagnosis as well as treatment. SPION were primarily developed as contrast agents, but their ability to label cells that can be non-invasively monitored in a living organism by MRI has been widely used for the *in vivo* tracking of transplanted cells (Sykova and Jendelova 2007). In addition, targeted magnetic nanoparticles or magnetically labeled cells may facilitate the accumulation of drugs or cells in the desired target area with the aid of a local magnetic field. Another approach using SPION in cell labeling is the systemic administration of ultra-small SPION (USPION), which utilizes their preferential uptake by monocytes/macrophages for MR visualization of the inflammatory response after stroke or other CNS disorders (Hoehn et al. 2007).

Nanofibers have been developed as suitable materials for various tissue engineering applications. The structure of the nanofibrous network mimics the fibrous architecture of the extracellular matrix and their high surface area to volume ratio has been shown to support the adhesion, proliferation and differentiation of various cells. Recent trends in techniques for producing nanofibers for tissue engineering include the development of biomimetic scaffolds that not only provide structural support for living cells, but can also serve as a delivery system for drugs, growth factors or cytokines that may further promote cell function and tissue regeneration. Currently, electrospun and self-assembled nanofibers are being extensively studied as potential scaffolds for neural tissue engineering (Kubinova and Sykova 2010). An important role in nanomaterial research for neural tissue engineering applications is also played by carbon nanotubes and carbon nanofibers, especially because of their good mechanical properties and electrical conductivity.

MAGNETIC NANOPARTICLES FOR CELL TRACKING

Magnetic particles are attractive candidates for various biomedical applications, such as MRI contrast agents, cell separation, cell labeling, drug delivery, magnetic nanoparticle-mediated gene transfer (magnetofection), or the induction of hyperthermia in experimental cancer therapy (Gupta et al. 2007).

To address cell tracking, SPION are emerging as a suitable probe for cell labeling and their non-invasive MRI monitoring in the living organism; they are biocompatible, easily detectable, able to incorporate into the cell cytoplasm or bind to the cell surface without affecting cell viability, can be degraded by cell metabolism, and, moreover, their surface can be modified by various functional groups or ligands for specific cell labeling (Veiseh et al. 2010). SPION usually consist of a crystalline iron oxide core coated with

Table 1. Key features of stroke.

Stroke is the most common cause of disability, the second leading cause of death in Europe, and the third leading cause of death in the U.S. Approximately 1.1 million strokes occur in Europe and 600,000 in the U.S. each year, of which 25% are fatal.
Stroke is caused by the interruption of the blood supply into the brain and can lead to permanent neurological deficits, complications or even death.
There are two types of stroke: ischemic stroke caused by the occlusion of blood vessels and hemorrhagic stroke caused by the rupture of blood vessels in the brain.
The treatment of acute ischemic stroke involves thrombolytic therapy; other possibilities are neuroprotectives (antioxidants) and secondary prevention using anticoagulants and antithrombotics as well as the early initiation of rehabilitation.
Cell transplantation has been shown to be an effective strategy in animal models of stroke. Several clinical trials are currently underway to evaluate the safety and efficacy of cell therapy in stroke patients (Locatelli et al. 2009).

dextran or another polymer shell (Fig. 1A). As a negative contrast agent, SPION produce a strong signal loss in T2*-weighted MRI relaxation, and can thus be detected in the tissue as a hypointense (dark) MR signal (Fig. 2 B, D). The SPION coating helps to prevent their aggregation, induces the efficient internalization of the contrast agent into the cell and minimizes any deleterious effects on cellular function. According to their size after coating, SPIO particles can be distinguished as standard SPION with a diameter between 50–150 nm, USPION with a diameter 10–50 nm, or micron size paramagnetic iron oxide (MPIO) particles. Systemically administered SPIO particles are mainly taken up from the blood by phagocytosis by the reticulo-endothelial system, while the particle size as well as the surface properties play a great role in determining the effectiveness of phagocytosis.

Commercially available dextran-coated SPION, e.g., ferumoxide (Feridex®, Endorem®) and ferucarbotran (Resovist®), have been approved as intravenous contrast agents by the US FDA for imaging of liver tumors. Unlike SPION, which are rapidly eliminated from the blood by mononuclear cells, smaller USPION (e.g., Combidex®, Sinerem®) have a longer blood half-life that enables their widespread tissue distribution, utilized for the diagnosis of lymph node or bone marrow metastasis.

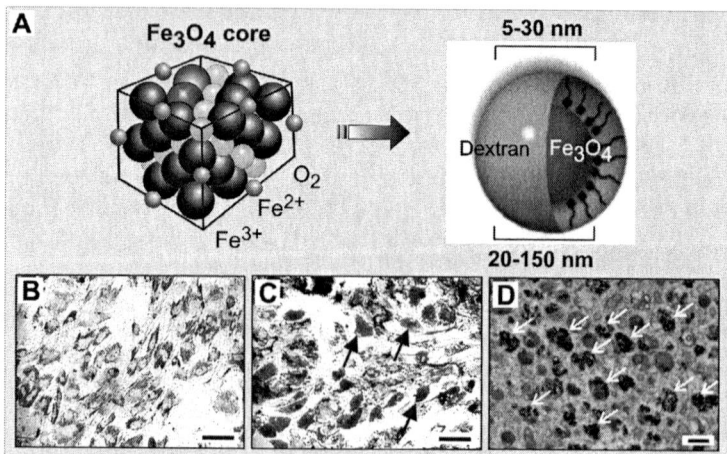

Fig. 1. **Superparamagnetic iron oxide nanoparticles for cell labeling.** (A) Scheme of an iron nanoparticle. The contrast agent Endorem® consists of a superparamagnetic Fe_3O_4 core coated by a dextran shell. (B) Mesenchymal stem cell (MSC) culture labeled with SPION and stained with Prussian blue. (C) MSC culture labeled with superparamagnetic iron oxide nanoparticles (SPION) differentiated into adipocytes (arrows). (D) Transmission electron micrograph of clusters of SPION in the cell cytoplasm (arrows). Scale bar: 100 µm (B, C) and 50 nm (D). Unpublished results.

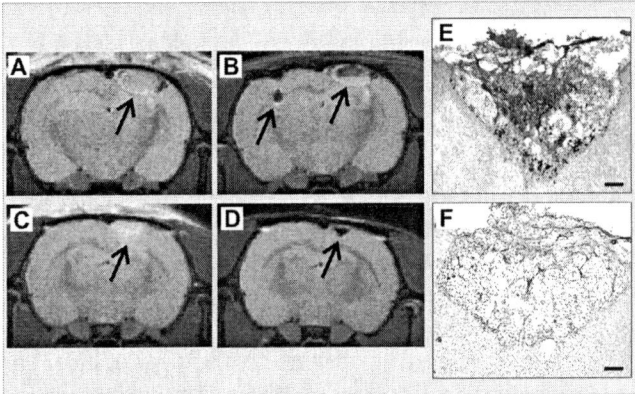

Fig. 2. Cell migration into a cortical ischemic lesion. (A-D) Magnetic resonance imaging (MRI) of a rat brain with a cortical photochemical lesion (arrow), which is an experimental model of stroke. (A, C) The control lesion is visible on MRI as a hyperintensive (light) area. (B) The photochemical lesion two weeks after the intracerebral transplantation of Endorem®-labeled rat mesenchymal stem cells (MSCs) into the contralateral hemisphere. The site of implantation is indicated by the arrow on the left side. The Endorem®-labeled MSCs migrated into the lesion and were detected as a hypointensive signal (arrow on the right side). (D) Endorem®-labeled MSCs detected in the lesion two weeks after intravenous transplantation. (E) MRI cell detection *in vivo* was confirmed by histological detection of iron using Prussian blue staining, which revealed a massive invasion of Endorem®-labeled MSCs into the lesion site. (F) Prussian blue staining of a control lesion without cell transplantation. Scale bar: 100 μm (E, F). Modified by Jendelova et al. (2004).

Depending on the strategy and purpose of the SPION application, there are two different labeling approaches that are used for non-invasive MRI cell tracking in the CNS:

1. *In situ* cell labeling by systemic SPION administration or the direct injection of SPIO particles into the target area or tissue.
2. *In vitro* SPION cell labeling to monitor transplanted cells in a living organism.

In situ Cell Labeling

During the acute stage of an ischemic stroke, brain inflammation is present, and it has been shown that an increase in systemic inflammatory parameters correlates with lesion volume and stroke severity. The preferential uptake of intravenously administered USPION by phagocytic cells, such as monocytes/macrophages, allows for MR visualization of the inflammatory response when the macrophages invade the inflamed area. This feature has been studied extensively in experimental models of stroke, when the accumulation of iron-positive macrophages has been detected by MRI in the ischemic border zone and correlated with immunohistochemical and/or iron-sensitive Prussian blue staining

(Hoehn et al. 2007). This concept also has been transferred into the clinic, where the potential usefulness of a systemic injection of USPION for MRI visualization of macrophages in an ischemic lesion was tested in humans in a single-center, clinical phase II study (Saleh et al. 2004). Monitoring the extent of ischemia after stroke may be important for estimating stroke severity and also for the development and monitoring of the effect of neuroprotective or anti-inflammatory agents.

In addition to stroke, the concept of using USPION for the *in vivo* monitoring of macrophage activity in pathological tissue provides an opportunity for the imaging of other CNS disorders, such as brain tumors, multiple sclerosis, atherosclerotic plaques, CNS trauma and epilepsy. Moreover, SPION conjugated to target-specific ligands, such as proteins or antibodies, have been used for selective *in vivo* tumor imaging as well as for drug and gene delivery (Veiseh et al. 2010).

In vivo Tracking of Transplanted Cells

The transplantation of various stem or progenitor cells is a promising therapeutic strategy for treating a variety of CNS disorders, including Alzheimer's and Parkinson's diseases, amyotrophic lateral sclerosis, sclerosis multiplex, stroke and CNS injury. Transplanted cells may not only replace lost cells, but also provide trophic factors and support endogenous regeneration. The cell types most frequently used for transplantation in brain or spinal cord injury (SCI) include embryonic stem cells (ESCs), adult stem/progenitor cells such as hematopoetic stem cells, neural stem/progenitor cells (NSPCs), mesenchymal stem cells (MSCs), umbilical cord blood cells, adipose tissue cells, and Schwann cells and olfactory ensheathing cells (OECs), which are used mainly for spinal cord repair (Xu et al. 2009).

In addition to cell type, the optimal route of cell delivery represents a crucial aspect of cell transplantation and can be dependent on a variety of factors, such as blood vessel permeability, edema and inflammation development, cell type, etc. In experimental animal models, transplanted cells are mostly delivered into the CNS via intraparenchymal, intracerebroventricular, intravenous (i.v.) or intraarterial (i.a.) delivery. An alternative approach for SCI could be cell transplantation through the cerebrospinal fluid via lumbar puncture. It was demonstrated that all delivery methods result in cell targeting into the lesion, however, with various effectiveness. The most effective delivery method, intraparenchymal, may be harmful and is limited to the area that can be reached by the injection. On the other hand, i.v. administration is non-invasive but has a lower targeted effectiveness, as most of the transplanted cells are captured from the blood stream by filtering organs such as the lung, liver or spleen. In contrast, i.a. delivery eliminates the initial cell trapping in systemic organs and offers direct targeting of damaged organs;

moreover, this type of application is clinically appropriate (Walczak et al. 2008). Currently, clinical trials using i.a. as well as i.v. administration of human MSCs in stroke patients are now underway (Lee et al. 2010).

As clinical translation of cell therapy requires identifying safe and efficient methods of cell delivery to the site of damage, it is therefore of key importance to track the delivery of the transplanted cells and to non-invasively monitor their homing, survival and proliferation within the targeted area. To screen cells *in vivo*, several techniques have been described for non-invasive cell tracking, such as quantum dots, fluorescence, bioluminescence, nuclear medicine techniques and MRI. Among them, MRI offers several advantages that enable its application in human medicine.

Dextran-coated SPION have been commonly used for labeling various cells. The commercial contrast agent Endorem® was demonstrated to be suitable for labeling human and rat MSCs (Fig. 1B-D), rat OECs, human chondrocytes and human ESCs (Sykova et al. 2005), as it can be incorporated into cells in culture by endocytosis without the use of transfection agents.

Another labeling approach is based on a combination of dextran-coated SPION (Feridex®, Sinerem®) and transfection agents (e.g., poly-L-lysine or Lipofectamine®) that increase the uptake of the nanoparticles into cells. Other labeling approaches involve magnetofection, electroporation, the specific targeting and endocytosis of nanoparticles through the use of transferrin receptors, or transduction agents such as HIV-derived TAT protein (Gupta et al. 2007). For clinical application, however, it would be advantageous to use a contrast agent that does not need any further additives to facilitate cellular uptake. To improve labeling efficiency, SPION coated with poly-L-lysine, D-mannose or poly(N,N-dimethylacrylamide) have been developed, which, when compared with Endorem®, exhibited better internalization into the cells and easier MRI detection with a lower concentration of iron within the cells (Horák et al. 2009). An innovative approach for the development of contrast agent labeling that goes beyond simply cell localization is the monitoring of functional cell status, in which the contrast agent becomes detectable by intrinsic activation mechanisms due to particular enzyme expression or via up-regulation by the activation of the corresponding promotor in transgenic cells (Hoehn et al. 2007).

To study cell therapy in animal model of stroke, Endorem®-labeled human MSCs, Endorem®/BrdU co-labeled rat MSCs, and Endorem®-labeled mouse GFP-ESCs were grafted into rats with a cortical photochemical lesion. The cells were administered either intracerebrally into the hemisphere contralateral to the lesion or i.v., and their migration into the lesion site was monitored by MRI using a 4.7 T Bruker spectrometer (Fig. 2). The lesion was visible on MRI images already 2 hrs after induction as a hyperintense signal and remained visible throughout the measurement period (Fig. 2A, C). Grafted labeled cells migrated to the lesion regardless of the route of application; their presence was detected in the lesion as a MRI hypointense signal (Fig. 2B, D) and confirmed histologically by Prussian blue staining for iron (Fig. 2 E), anti-BrdU staining for dividing cells and by GFP labeling. The hypointense MR signal remained detectable for more than 7 weeks (Jendelova et al. 2004; Sykova and Jendelova 2007).

Many other studies have demonstrated that transplanted SPION-labeled stem cells migrate into an ischemic brain area; MRI detection was confirmed by histological analysis with iron staining (Fig. 3). Zhang et al. (2003) showed the migration of iron-labeled neural progenitor cells (NPCs) transplanted into the cisterna magna toward the ischemic area in rats with middle cerebral artery occlusion (MCAO). Ferumoxide-labeled human neural stem cells have been i.v. injected 24 hr after MCAO induction in rats; the migration of engrafted cells was observed on MRI 3 days after injection, and 28 days post-injection, the ferumoxide-labeled neural stem cells differentiated into neurons and astrocytes (Song et al. 2009). In another study, the migration of human MSCs labeled with Feridex® into the MCAO lesion was visualized by MRI up to 10 weeks following transplantation (Kim et al. 2008). Further work used i.a. or i.v. delivery of SPION-labeled MSCs into the ischemic area in rats with a MCAO lesion, while MRI cell monitoring was correlated with transcranial laser Doppler flowmetry. In this work, however, the successful delivery of labeled MSCs could be visualized with MRI only after i.a. delivery, and cerebral engraftment was associated with impeded cerebral blood flow (Walczak et al. 2008).

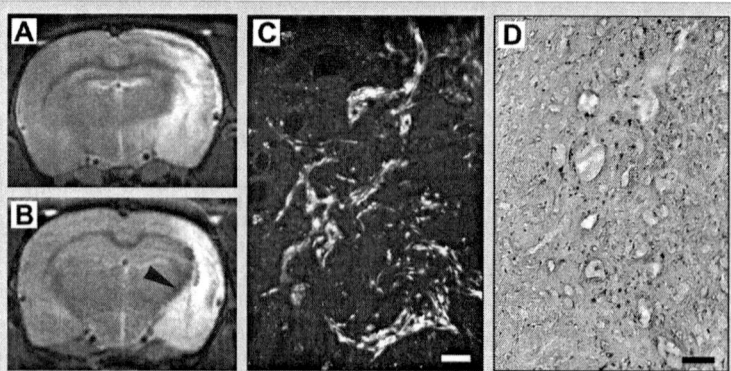

Fig. 3. Cell migration into a stroke lesion. Endorem®-labeled mesenchymal stem cells transplanted intravenously into rats with middle cerebral artery occlusion. (A) The ischemic lesion is visible on MRI as a hyperintensive area. (B) Transplanted cells detected 2 weeks after transplantation in the border of the lesion (arrow). (C) The presence of transplanted cells around the lesion was detected via green fluorescent protein. (D) Prussian blue staining confirmed the presence of iron-positive cells around the lesion. Scale bar: 50 μm (C, D). Unpublished results.

In the rat spinal cord, dextran-coated SPION-labeled Schwann cells and OECs have been shown to remyelinate axons after transplantation into demyelinated lesions (Dunning et al. 2004). In our study, the intravenous injection of Endorem®-labeled MSCs significantly improved the recovery of hindlimb motor function in rats with a spinal cord balloon induced compression lesion (Fig. 4A), a relevant model of SCI. The fate of i.v. transplanted MSCs was followed by *in vivo* (Fig. 4C) MRI; staining for Prussian blue (Fig. 4E) confirmed migration of many labeled cells in the lesion site. The lesion cavities were significantly smaller than in control animals (Fig. 4D, E), and there was a significantly better recovery of motor and sensory functions than in control animals (Hejcl et al. 2008).

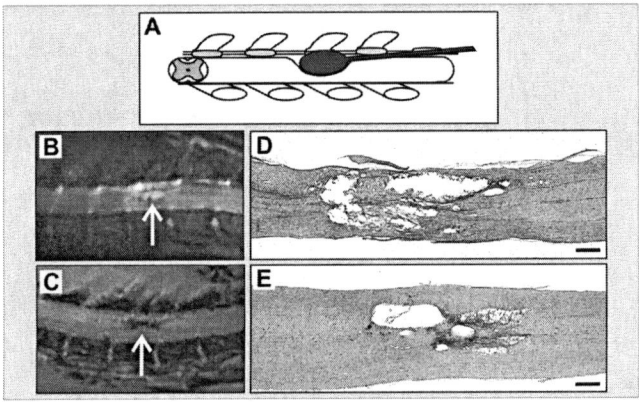

Fig. 4. Cell migration into a spinal cord lesion. (A) Balloon compression lesion, which is a model of spinal cord injury. A French Fogarty catheter is inserted into the dorsal epidural space through a hole in the Th10 vertebral arch. A spinal cord lesion is made by balloon inflation at the Th8-Th9 spinal level. (B) Longitudinal section of a rat spinal cord compression lesion on MRI, 5 weeks after compression. The lesion is seen as a hyperintensive area (arrow). (C) Longitudinal MRI of a spinal cord compression lesion populated with intravenously (i.v.) injected Endorem®-labeled mesenchymal stem cells (MSCs), 4 weeks after implantation. The lesion with Endorem®-labeled cells is visible as a dark hypointensive area (arrow). (D) Prussian blue staining of a spinal cord compression lesion (control animal). (E) Prussian blue staining of a spinal cord lesion with i.v. injected Endorem®-labeled MSCs. Scale bar: 500 μm (D, E). Modified by Jendelova et al. (2004).

Although the SPION labeling method is widely used in preclinical and also clinical trials of cell therapy, there are, nonetheless, possible confounding factors leading to the misinterpretation of MR signals and that thus require further analysis. In many cases, an intrinsic MRI signal can originate from iron dissociated from hemoglobin, hemorrhage, tumor, traumatic experimental procedures, or macrophages engulfing the nanoparticles or transplanted cells. On the other hand, the loss of MRI signal can be caused by the biodegradation of the nanoparticles or by their dilution due to cell proliferation.

TARGETED DELIVERY

Due to the magnetic properties of SPIO particles, a magnetic field can be used to direct the localization of magnetically labeled cells in a targeted area. For example, the cells can be taken up from the blood stream using a ferromagnetic stent targeted to a damaged artery, or cells administered into the cerebrospinal fluid can be directed into a SCI with the use of an external magnet placed or implanted above the injury site.

The blood brain barrier (BBB) is a major restriction for the diffusion of many potential therapeutic agents that thus cannot reach the CNS. Polymeric nanoparticles as drug carriers crossing the BBB are being extensively developed for the delivery of anti-tumor agents and also for drug or gene delivery in the treatment of other CNS disorders, such as Alzheimer's or Parkinson's diseases.

Interestingly, nanoparticles enabling the local sustained delivery of a drug have been shown to treat acute SCI. Standard treatment of an acute traumatic SCI includes the administration of a high dose of the anti-inflammatory drug methylprednisolone to reduce neurological deficits. To eliminate systemic administration that can lead to a variety of side effects and high risk of infections, poly(lactic-co-glycolic) acid (PLGA) nanoparticles loaded with methylprednisolone were embedded in agarose gel and placed into the injury site. This treatment reduced lesion volume significantly and improved behavioral outcomes when compared to systemic or local steroid therapy (Kim et al. 2009).

NANOFIBERS AS SCAFFOLDS FOR SPINAL CORD INJURY TREATMENT

Although the adult spinal cord is currently known to be capable of functional reorganization and axonal sprouting and, like other regions of the adult CNS, the adult spinal cord contains NSPCs, the inhibitory environment of the mature mammalian CNS, the glial scar barrier as well as cyst formation hinder the neurons from regenerating across the site of injury. Cell transplantation has emerged as a promising therapeutic approach for SCI. In an acute SCI, transplanted cells may replace dead cells and provide various bioactive factors that promote endogenous regeneration and prevent apoptosis and cavity formation. However, in the case of large or chronic spinal cord lesion, when the cystic cavity is developed, cell transplantation alone is not sufficient for tissue regeneration, and tissue repair requires "bridging" the lesion with a 3D permissive environment that would fill the tissue gap and concomitantly support axonal re-growth and the re-establishment of damaged connections. The scaffold design must then consider parameters such as biocompatibility, controlled porosity and permeability, suitable mechanical properties comparable with the neural tissue, biodegradability and, additionally, support for cell attachment, growth and differentiation.

As scaffolds in neural tissue engineering, synthetic as well as biologically derived hydrogels have been widely used, as they can form a soft and porous structure ideal for implantation into neuronal tissue of both the brain and spinal cord (Hejcl et al. 2010). In addition to widely used polymeric hydrogels, nanofiber materials have emerged as a promising advance in tissue regeneration. The structure of the nanofibrous network is similar to the native extracellullar matrix, and the high surface area to volume ratio has been shown to support the adhesion, proliferation and differentiation of various cells (Fig. 5A-C). Nanofibrous materials have been developed as scaffolds for various tissues, such as bone, cartilage, skeletal and smooth muscle, blood vessel, skin, cardiac and neural tissue, but also for enzyme immobilization and the delivery of drugs and bioactive molecules (Kubinova and Sykova 2010). The main processing techniques used to produce nanofibers for neural tissue engineering application are electrospinning and molecular self-assembly.

Electrospun Nanofibers

Electrospinning is a widely used technique for the production of nanofibers from various natural or synthetic polymers, but also from metal, ceramic and glass materials. In the electrospinning process, nanofibers, with a fiber diameter ranging from ten to hundreds of nanometers, are created as polymeric jets from the surface of a polymeric solution in a high intensity electrostatic field when the electric field overcomes the polymer surface tension. Adjusting the processing conditions and the properties of the polymer solution enables one to control the properties of the resultant nanofibrous meshes, such as fiber diameter, porosity, mechanical properties and surface topography. Regarding the nanofiber formation, the polymer jets can be ejected either at the tip of a capillary tube (needle or capillary spinners) or from liquid surfaces on a rotating spinning roller (needleless technology) (Kubinova and Sykova 2010).

Spatial orientation of nanofibrous scaffolds can be formed by modifying the nanofiber-collecting devices using, e.g., a high speed rotating cylinder collector. Aligned nanofibrous scaffolds that serve as contact guidance for directing cell growth (Fig. 5C) have been studied as promising scaffolds, especially for vascular or peripheral nerve regeneration.

A variety of electrospun nanofibers from synthetic or naturally occurring polymers, e.g., poly(L-lactic acid), poly(ε-caprolactone) (PCL), PLGA, collagen, gelatin, chitosan or their blends, have been explored as

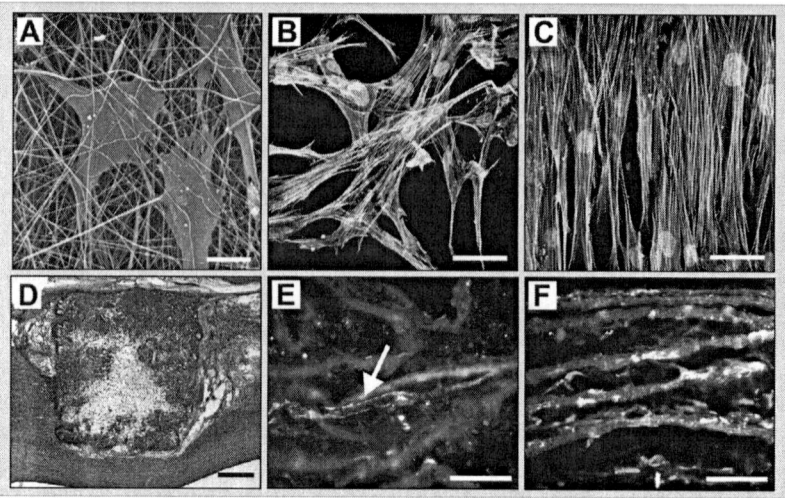

Fig. 5. **Electrospun nanofibers as a scaffold for cell culture and implantation into the spinal cord.** (A) Representative scanning electron micrographs of human mesenchymal stem cells (hMSCs) cultured on gelatin nanofibers. (B) Confocal micrographs of hMSCs grown on random and (C) aligned gelatin nanofibers. Cells were stained for F-actin cytoskeletal filaments and cell nuclei. (D) A rolled nanofiber scaffold was implanted into a hemisection of a rat spinal cord and evaluated 4 weeks after implantation. The ingrowth of tissue into an implanted rolled nanofiber scaffold stained with hematoxylin-eosin. (E) Blood vessel (RECA staining) ingrowth (arrow) into the nanofibrous scaffold. (F) The ingrowth of Schwann cells (p75 staining) into a nanofibrous scaffold. Scale bar: 20 μm (A), 50 μm (B, C, E, F) and 500 μm (D). Unpublished results.

Color image of this figure appears in the color plate section at the end of the book.

substrates for neuronal cell cultivation *in vitro*, while aligned nanofibrous substrates have proved to be effective in enhancing neurite outgrowth, axonal extension and the differentiation of stem cells into neural lineages (Cao et al. 2009). Nevertheless, the incorporation of 2D nanofibrous meshes into 3D constructs still remains a challenge, and despite numerous *in vitro* studies, there are currently only a few studies that have attempted to verify the potential of electrospun nanofibers in *in vivo* experiments.

To promote spinal cord regeneration, polyamide nanofibers modified with a bioactive peptide derived from the neuroregulatory extracellular matrix molecule tenascin-C have been implanted into hemisected spinal cord. However, histological observation revealed only slight axonal ingrowth into the implant, and the random orientation of the nanofiber layers turned out to be a critical impediment for axonal movement (Meiners et al. 2007). In our experiments, scaffolds based on layers of electrospun nanofibers were implanted into the hemisected spinal cord, where they integrated into the surrounding tissue and promoted the ingrowth of connective elements, blood vessels and neural cell processes (Fig. 5 D-F).

In contrast to spinal cord regeneration, electrospun nanofibers have been more intensively studied as tubular scaffolds for stimulating and guiding functional peripheral nerve regeneration. Bio-absorbable grafts of nonwoven micro/nanofiber mesh tubes made from, e.g., electrospun chitosan, a PLGA/PCL blend or PCL, were used in a rat model of sciatic nerve transection (Cao et al. 2009).

Self-Assembling Nanofibers

Self-assembly is another technique to fabricate scaffolds suitable for SCI treatment. Self-assembling (SA) nanofibers are composed of peptide amphiphilic molecules that spontaneously aggregate from an aqueous solution into three-dimensional networks due to multiple non-covalent interactions in the presence of a physiological salt solution or changes in pH. A range of proteins or peptides can produce such nanofibers with diameters of tens of nanometers. The peptides can be chemically designed to incorporate specific functional ligands that further enhance scaffold performance, such as peptide epitopes containing integrin receptor-binding sites, bone marrow homing proteins, insulin growth factor or stromal cell-derived factor-1 (Kubinova and Sykova 2010). Since the self-assembly is triggered by the ionic strength of the *in vivo* environment, it is possible to inject SA nanofibers into the injured nervous tissue as a liquid and simultaneously encapsulate various cells into the nanofibrous matrix.

The SA peptide RADA16-I has been demonstrated to be a scaffold that supports *in vitro* as well as *in vivo* cell growth and differentiation. When implanted into the transected spinal cord, alone or with transplanted NPCs or Schwann cells, the scaffold integrated and bridged the injured spinal cord. Transplanted cells survived 6 week after implantation, and some NPCs differentiated into neuronal lineages. Moreover, scaffolds with transplanted cells elicited improved axonal growth into the implants compared to plain scaffolds (Guo et al. 2007). In other *in vivo* experiments, RADA16-I was shown to repair the disrupted optic tract and to return functional vision (Ellis-Behnke et al. 2006) and to help to reconstruct lost tissue in the acutely injured brain (Guo et al. 2009).

Other SA nanofibers formed from peptide-amphiphile molecules containing IKVAV (isoleucin-lysin-valin-alanin-valin), an integrin receptor-binding epitope derived from laminin, have been shown to support the neuronal differentiation and neurite outgrowth of *in vitro* cultured NPCs and to inhibit the development of astrocytes (Silva et al. 2004). The injection of these IKVAV- peptide-amphiphile molecules into a mouse spinal cord lesion reduced cell death and the formation of a glial scar and increased the number of oligodendroglia at the site of injury.

Moreover, this treatment promoted the regeneration of both motor and sensory fibers through the lesion site and resulted in significant behavioral improvement (Tysseling-Mattiace et al. 2008).

Taken together, SA nanofibers have been shown in many experiments to be promising materials for neural tissue engineering as they provide a 3D environment for cell growth, are biocompatible and non-immunogenic, and integrate well into a SCI site when implanted into the lesion. On the other hand, the SA nanofiber network is held together by weak hydrophobic, ionic or van der Waal's interactions, enabling dynamic fiber de-assembling and re-assembling, and therefore this system possesses uncontrollable porosity and mechanical instability, which may limit the use of SA nanofibers in, e.g., peripheral nerve regeneration.

Key Facts about Spinal Cord Injury

- According to the National Spinal Cord Injury Statistical Center (https://www.nscisc.uab.edu/), there are approximately 12,000 new cases of SCI in the U.S. each year.
- Standard therapy of acute SCI consists of stabilization and the use of high doses of methylprednisolone; chronic SCI treatment focuses on rehabilitation, pain relief, spasticity removal, and the prevention of complications.
- Experimental therapeutic strategies include cell transplantation, neuroprotection, the promotion of neuronal regeneration, the neutralization of inhibitory factors, the regulation of gliosis and scar formation, bridging the lesion gap with permissive biomaterials and genetic engineering (Bunge 2008).
- Currently, a number of novel pharmacological as well as cell-based approaches are being tested in the clinic (Thuret et al. 2006).

Summary Points

- Systemic ultrasmall superparamagnetic iron oxide nanoparticles administration utilize their preferential uptake by monocytes/macrophages for magnetic resonance visualization of the inflammatory response after stroke or other magnetic resonance disorders.
- Superparamagnetic iron oxide nanoparticles labeling together with noninvasive magnetic resonance imaging enable following the fate of transplanted cells in the host organism and establishing the optimal conditions for transplantation.
- Biologically compatible nanofiber scaffolds can serve as a permissive bridge for axonal regeneration and the re-establishment of damaged connections.

- Electrospun nanofibrous scaffolds with oriented fibers can serve as suitable guidance conduits for nervous tissue repair.
- Self-assembling nanofibers with incorporated specific functional ligands promote the growth of cultured neural cells and improve regeneration after a spinal cord lesion.

Applications in Health and Disease

The use of cell therapy in regenerative medicine has been examined extensively to replace cells lost due to various disorders or injuries. Further progress in cell therapy leading to clinical trials requires the crucial use of non-invasive techniques for monitoring the efficacy of cell therapy and graft survival in the host organism. In regard to cell therapy, non-invasive MRI monitoring of SPIO-labeled grafted cells can allow researchers to evaluate the effect of cell therapy in patients with various disorders and brain or spinal cord injuries and consequently to establish the optimal conditions for transplantation in terms of the number of transplanted cells, the route of administration and the therapeutic time window (Fig. 6).

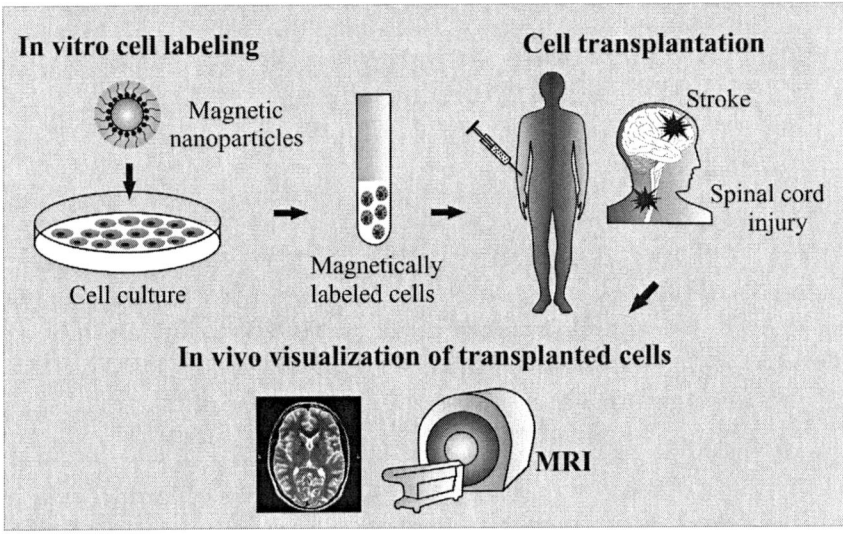

Fig. 6. Utilization of magnetic nanoparticles in CNS treatment. Prior to transplantation, the cells are labeled *in vitro* with magnetic nanoparticles. Magnetic cell labeling together with noninvasive MRI enables following the fate of transplanted cells in the targeted area and establishing the optimal conditions for transplantation.

In the case of large lesions, however, cell transplantation alone is not sufficient for tissue regeneration and thus it is necessary to bridge the lesion site with a permissive environment that fills the cavity, reduces the glial scar and enables axonal ingrowth. In addition to various types of biomaterials, nanofibers, especially self-assembling nanofibers, are suitable implantation materials whose chemical and structural properties can be tailored to promote cell adhesion, growth and differentiation or that can be coupled with various drugs, growth or other factors to improve regeneration.

Definitions

Spinal cord injury: is mostly caused by tissue compression due to vertebral disc dislocation or bone fragments. The acute tissue damage evokes a cascade of cellular and biochemical events leading to apoptosis and tissue loss. Reactive astrocytes migrate to and proliferate at the lesion site, which leads to the formation of a dense glial scar and the increased expression of extracellular matrix molecules, which have been shown to inhibit neurite outgrowth. In the chronic phase of injury, apoptosis continues together with demyelinization and the development of cystic cavities.

MRI: is a medical imaging technique that uses the property of hydrogen atoms (^1H) to return to their equilibrium state at different rates due to the different distribution of hydrogen ions in particular tissues. There are two principal relaxation processes that characterize an MR signal, T1 (spin-lattice relaxation) and T2 (spin-spin relaxation). T2 is reduced to T2* due to inhomogeneities in the magnetic field gradient. MRI contrast agents contain metal ions that shorten the T1 or T2 relaxation times of ^1H atoms in their vicinity. Paramagnetic metals, such as gadolinium, iron, and manganese, mainly affect T1 relaxation, while SPION mostly reduce T2 and T2*.

Emryonic stem cells: are pluripotent cells derived from the inner cell mass of blastocysts that can expand indefinitely and differentiate into diverse cell types, including neuronal and glial lineage phenotypes. The use of ESCs for clinical applications considerably diminishes the risk that transplanted cells could form a teratoma as well as introducing ethical or religious objections.

Neural stem/progenitor cells: are multipotent, self-renewing stem cells that can be isolated from both embryonic and adult tissues. NSPCs in the adult nervous system reside in their niches, the subventricular zone of the lateral ventricle, the subgranular zone of the dentate gyrus or the ependyma in the spinal cord. Stroke, injury or other pathological states induce increased neurogenesis and the migration of NSPCs into the damaged areas, which suggest the potential ability of the brain for self-repair. In culture, NSPCs can generate neural as well as glial cell types. Currently, a powerful new therapeutic strategy utilizes immortalized cell lines of embryonic or fetal

NSPCs as well as neuroepithelial or teratocarcinoma cell lines (Locatelli et al. 2009).

Mesenchymal stem cells: are present in adult tissue, primarily in the bone marrow, but they can be found in fat, skin, liver, peripheral blood, umbilical cord, etc.; they are easy to isolate and expand for autologous application, which gives them promising therapeutic potential in human medicine. As multipotent cells, MSCs can differentiate not only into cells of mesenchymal origin, but also into non-mesenchymal cell phenotypes.

Acknowlegments

This work was supported by CZ: GA ČR P304/11/0731, 304/07/1129, 203/09/1242; CZ: AV ČR: KAN 200520804, IAA500390902; CZ: MŠMT: 1M0538.

Abbreviations

CNS	:	central nervous system
SPION	:	superparamagnetic iron oxide nanoparticles
MRI	:	magnetic resonance imaging
SCI	:	spinal cord injury
USPION	:	ultra-small superparamagnetic iron oxide nanoparticles
MPIO particles	:	micron size paramagnetic iron oxide particles
ESCs	:	embryonic stem cells
MSCs	:	mesenchymal stem cells
NSPCs	:	neural stem/progenitor cells
OECs	:	olfactory ensheathing cells
i.v.	:	intravenous
i.a.	:	intraarterial
BrdU	:	bromodeoxyuridine
GFP	:	green fluorescent protein
NPCs	:	neural progenitor cells
MCAO	:	middle cerebral artery occlusion
BBB	:	blood brain barrier
PCL	:	poly(ε-caprolactone)
PLGA	:	poly(lactide-co-glycolic acid)
SA nanofibers	:	self-assembling nanofibers

References

Bunge, M.B. 2008. Novel combination strategies to repair the injured mammalian spinal cord. J. Spinal Cord Med. 31: 262–269.

Cao, H., T. Liu and S.Y. Chew. 2009. The application of nanofibrous scaffolds in neural tissue engineering. Adv. Drug Deliv. Rev. 61: 1055–1064.

Dunning, M.D., A. Lakatos, L. Loizou, M. Kettunen, C. ffrench-Constant, K.M. Brindle and R.J. Franklin. 2004. Superparamagnetic iron oxide-labeled Schwann cells and olfactory ensheathing cells can be traced in vivo by magnetic resonance imaging and retain functional properties after transplantation into the CNS. J. Neurosci. 24: 9799–9810.

Ellis-Behnke, R.G., Y.X. Liang, S.W. You, D.K. Tay, S. Zhang, K.F. So and G.E. Schneider. 2006. Nano neuro knitting: peptide nanofiber scaffold for brain repair and axon regeneration with functional return of vision. Proc. Natl. Acad. Sci. USA 103: 5054–5059.

Guo, J., K.K. Leung, H. Su, Q. Yuan, L. Wang, T.H. Chu, W. Zhang, J.K. Pu, G.K. Ng, W.M. Wong, X. Dai and W. Wu. 2009. Self-assembling peptide nanofiber scaffold promotes the reconstruction of acutely injured brain. Nanomed. 5: 345–351.

Guo, J., H. Su, Y. Zeng, Y.X. Liang, W.M. Wong, R.G. Ellis-Behnke, K.F. So and W. Wu. 2007. Reknitting the injured spinal cord by self-assembling peptide nanofiber scaffold. Nanomed. 3: 311–321.

Gupta, A.K., R.R. Naregalkar, V.D. Vaidya and M. Gupta. 2007. Recent advances on surface engineering of magnetic iron oxide nanoparticles and their biomedical applications. Nanomed. 2: 23–39.

Hejcl, A., P. Lesny, M. Pradny, J. Michalek, P. Jendelova, J. Stulik and E. Sykova. 2008. Biocompatible hydrogels in spinal cord injury repair. Physiol. Res. 57 Suppl. 3: S121–132.

Hejcl, A., J. Sedy, M. Kapcalova, D.A. Toro, T. Amemori, P. Lesny, K. Likavcanova-Masinova, E. Krumbholcova, M. Pradny, J. Michalek, M. Burian, M. Hajek, P. Jendelova and E. Sykova. 2010. HPMA-RGD hydrogels seeded with mesenchymal stem cells improve functional outcome in chronic spinal cord injury. Stem cells dev. 19: 1535–1546.

Hoehn, M., D. Wiedermann, C. Justicia, P. Ramos-Cabrer, K. Kruttwig, T. Farr and U. Himmelreich. 2007. Cell tracking using magnetic resonance imaging. J. Physiol. 584: 25–30.

Horák, D., M. Babič, P. Jendelova, V. Herynek, M. Trchová, K. Likavčanová, M. Kapcalová, M. Hajek and E. Sykova. 2009. Effect of different magnetic nanoparticle coatings on the efficiency of stem cell labeling. J. Magn. Magn. Mater. 321: 1539–1547.

Jendelova, P., V. Herynek, L. Urdzikova, K. Glogarova, J. Kroupova, B. Andersson, V. Bryja, M. Burian, M. Hajek and E. Sykova. 2004. Magnetic resonance tracking of transplanted bone marrow and embryonic stem cells labeled by iron oxide nanoparticles in rat brain and spinal cord. J. Neurosci. Res. 76: 232–243.

Kim, D., B.G. Chun, Y.K. Kim, Y.H. Lee, C.S. Park, I. Jeon, C. Cheong, T.S. Hwang, H. Chung, B.J. Gwag, K.S. Hong and J. Song. 2008. *In vivo* tracking of human mesenchymal stem cells in experimental stroke. Cell Transplant. 16: 1007–1012.

Kim, Y.T., J.M. Caldwell and R.V. Bellamkonda. 2009. Nanoparticle-mediated local delivery of Methylprednisolone after spinal cord injury. Biomaterials. 30: 2582–2590.

Kubinova, S., E. Sykova. 2010. Nanotechnologies in regenerative medicine. Minim. Invasive Ther. Allied Technol. 19: 144–156.

Lee, J.S., J.M. Hong, G.J. Moon, P.H. Lee, Y.H. Ahn and O.Y. Bang. 2010. A long-term follow-up study of intravenous autologous mesenchymal stem cell transplantation in patients with ischemic stroke. Stem Cells. 28: 1099–1106.

Locatelli, F., A. Bersano, E. Ballabio, S. Lanfranconi, D. Papadimitriou, S. Strazzer, N. Bresolin, G.P. Comi and S. Corti. 2009. Stem cell therapy in stroke. Cell Mol. Life Sci. 66: 757–772.

Meiners, S., I. Ahmed, A.S. Ponery, N. Amor, S.L. Harris, V. Ayres, Y. Fan, Q. Chen, R. Delgado-Rivera and A. N. Babu. 2007. Engineering electrospun nanofibrillar surfaces for spinal cord repair: a discussion. Polym. Int. 56: 1340–1348.

Saleh, A., M. Schroeter, C. Jonkmanns, H.P. Hartung, U. Modder and S. Jander. 2004. In vivo MRI of brain inflammation in human ischaemic stroke. Brain 127: 1670–1677.

Silva, G.A., C. Czeisler, K.L. Niece, E. Beniash, D.A. Harrington, J.A. Kessler and S.I. Stupp. 2004. Selective differentiation of neural progenitor cells by high-epitope density nanofibers. Science 303: 1352–1355.

Song, M., Y. Kim, Y. Kim, S. Ryu, I. Song, S.U. Kim and B.W. Yoon. 2009. MRI tracking of intravenously transplanted human neural stem cells in rat focal ischemia model. Neurosci. Res. 64: 235–239.

Sykova, E. and P. Jendelova. 2005. Magnetic resonance tracking of implanted adult and embryonic stem cells in injured brain and spinal cord. Ann. N.Y. Acad. Sci. 1049: 146–160.

Sykova, E. and P. Jendelova. 2007. Migration, fate and *in vivo* imaging of adult stem cells in the CNS. Cell Death Differ. 14: 1336–1342.

Thuret, S., L.D. Moon and F.H. Gage. 2006. Therapeutic interventions after spinal cord injury. Nature reviews 7: 628–643.

Tysseling-Mattiace, V.M., V. Sahni, K.L. Niece, D. Birch, C. Czeisler, M.G. Fehlings, S.I. Stupp and J.A. Kessler. 2008. Self-assembling nanofibers inhibit glial scar formation and promote axon elongation after spinal cord injury J. Neurosci. 28: 3814–3823.

Veiseh, O., J.W. Gunn and M. Zhang. 2010. Design and fabrication of magnetic nanoparticles for targeted drug delivery and imaging. Adv. Drug Deliv. Rev. 62: 284–304.

Walczak, P., J. Zhang, A.A. Gilad, D.A. Kedziorek, J. Ruiz-Cabello, R.G. Young, M.F. Pittenger, P.C. van Zijl, J. Huang and J.W. Bulte. 2008. Dual-modality monitoring of targeted intraarterial delivery of mesenchymal stem cells after transient ischemia. Stroke 39: 1569–1574.

Xu, X.M. and S.M. Onifer. 2009. Transplantation-mediated strategies to promote axonal regeneration following spinal cord injury. Respir. physiol. neurobiol. 169: 171–182.

Zhang, Z.G., Q. Jiang, R. Zhang, L. Zhang, L. Wang, L. Zhang, P. Arniego, K.L. Ho and M. Chopp. 2003. Magnetic resonance imaging and neurosphere therapy of stroke in rat. Ann. Neurol. 53: 259–263.

12

Delivering siRNA Using Nanoparticles to the Central Nervous System

S. Neslihan Alpay,[1,a] Bulent Ozpolat,[1,b] Hee-Dong Han,[2,a] Gabriel Lopez-Berestein,[1,c] Anil K. Sood[2,b] and Hui-Lin Pan[3]

ABSTRACT

The incidence of central nervous system (CNS) diseases is expected to increase significantly in the 21st century due to an increase in lifespan and changing population demographics. The most challenging of the CNS diseases are cancer and neurodegenerative diseases, including Alzheimer's, Parkinson's and Hungtington's disease that are characterized by age-related gradual decline in

---

[1]The University of Texas MD Anderson Cancer Center, 1 Department of Experimental Therapautics, 1515 Holcombe Boulevard Unit 422, Houston TX, USA, 77030.
[a]E-mail: skarabul@mdanderson.org
[b]E-mail: bozpolat@mdanderson.org
[c]E-mail: glopez@mdanderson.org
[2]The University of Texas MD Anderson Cancer Center, 2 Department of Gynecologic Oncology, 1155 Herman Pressler St, Unit 1362, Houston, TX 77030.
[a]E-mail: hhan@mdanderson.org
[b]E-mail: asood@mdanderson.org
[3]The University of Texas MD Anderson Cancer Center, 3 Department of Anesthesiology & PeriOperative Medicine–Research, 1515 Holcombe Boulevard, Unit 110, Houston TX, USA, 77030; E-mail: huilinpan@mdanderson.org

List of abbreviations after the text.

neurological function and neuronal death. Fundamental difficulties limiting the development of novel therapeutics include crossing the blood-brain barrier, and targeting of drugs to specific tissues or cells within the CNS. With the discovery of RNA interference (RNAi)-based therapeutics to achieve specific silencing of genes in the nervous system, the necessity to overcome such obstacles has become even more urgent. Most preclinical and clinical studies regarding delivery of small interfering RNA (siRNA) to the CNS have utilized invasive, intra-cerebral delivery of siRNA to the targeted tissue. New methods must be developed in order to facilitate delivery of therapeutically significant quantities of siRNA to the CNS via systemic route so that the full therapeutic potential of siRNA can be realized. Thus, developing safe and effective tissue-specific delivery vectors, minimizing the off-target effects, and achieving distribution in sufficient concentrations at the target tissue with minimal side effects are the major goals. In this chapter, we will discuss nanoparticle-based siRNA delivery to the CNS and their potential application in CNS diseases.

INTRODUCTION

The access to the central nervous system (CNS) is more limited compared to other organs due to the blood-brain barrier (BBB). The BBB plays the important role in the regulation of molecular exchanges at the interfaces between the blood and neural tissue or its fluid spaces (Abbott et al. 2006). The BBB is formed by cerebrovascular endothelial cells that create a barrier between the blood and brain interstitial fluid. The endothelial cells of capillaries are different from the peripheral tissues because of the close binding to each other by tight junctions. The outside of these capillary vessels is surrounded by the astrocytes that facilitate formation of these tight junctions (Liu-Synder and Webster 2006). In addition to BBB, two more barriers including the choroid plexus epithelium between the blood and ventricular cerebrospinal fluid (CSF), the blood–CSF barrier (BCSFB) and the arachnoid epithelium between the blood and subarachnoid CSF control the exchange of molecules at the interfaces between the blood and neural tissue (Fig. 1). The CSF is secreted by choroid plexuses in the lateral, third and fourth ventricles and circulates in the ventricles and subarachnoid space (Abbott et al. 2006). Neurons are usually located within the 8–20 μm from a brain capillary (Schlageter et al. 1999), although they may be positioned millimeters or centimeters away from a CSF compartment. Hence, BBB exerts the most control over the immediate microenvironment of brain cells (Abbott et al. 2006).

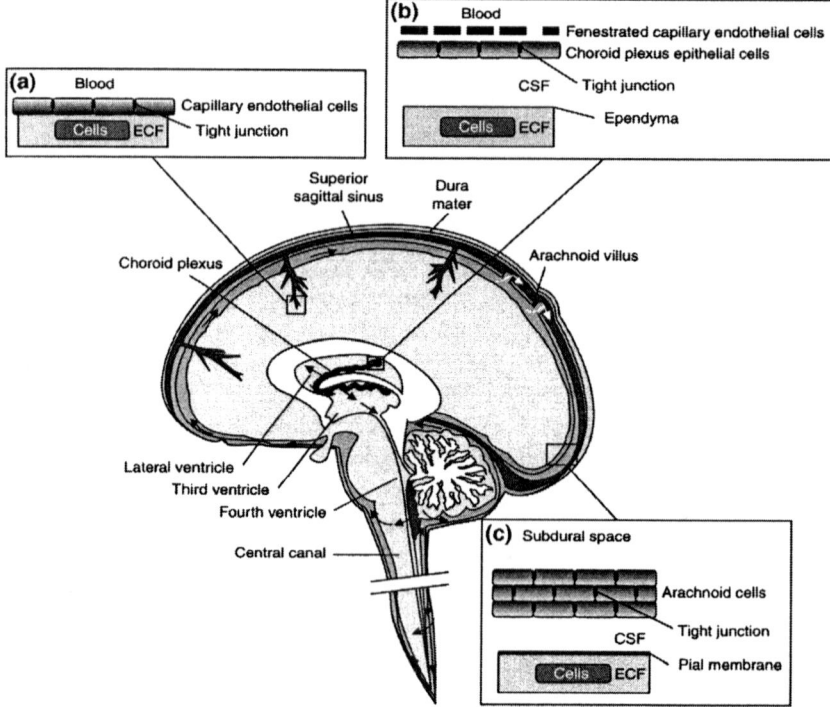

Fig. 1. Three major barriers to overcome for successful delivery of drugs to the brain. These barriers are between the blood and brain. (a) **The BBB (Blood brain barrier)**, which forms the tight junctions by endothelial cells of the cerebral capillary. The BBB is the critical path for systemic delivery of drugs to the CNS. (b) **The blood–CSF barrier (BCSFB)** is located at the choroid plexuses in the lateral, third and fourth ventricles of the brain where tight junctions are formed between the epithelial cells at the CSF-facing surface (apical surface) of the epithelium. Some drugs and solutes enter the brain principally across the choroid plexuses into the CSF, while others enter via both the BBB and BCSFB. (c) **The arachnoid barrier is** a multi-layered epithelium with tight junctions between cells of the inner layer that form an effective seal. Arachnoid villi project into the sagittal sinus through the dura and a significant amount of CSF drains into the sinus through these valvelike villi which only allow CSF movement out of the brain to blood. Transport across the arachnoid membrane is not an important route for the entry of solutes into brain (adapted from Kandel et al. 2000, with permission) (Abbott et al. 2010).

There are several mechanisms by which molecules can be transported from the systemic circulation to the CNS: 1) simple diffusion, 2) facilitated diffusion, 3) active transport and 4) paracellular diffusion. Many factors, such as the size, charge and hydrophobicity determine the capability of molecules to move through the BBB (Liu-Synder and Webster 2006). The tight junctions between endothelial cells form a physical barrier by significantly reducing passive diffusion, leading molecules to move predominantly across cells (transcellular pathway) (Fig. 2). Gases

Delivering siRNA Using Nanoparticles to the Central Nervous System 231

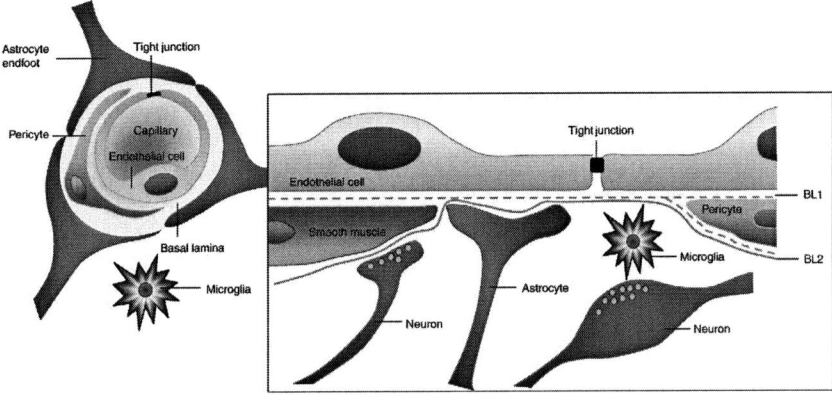

Fig. 2. **The cellular structures forming the BBB.** The cerebral endothelial cells form tight junctions at their margins, which seal the aqueous paracellular diffusional pathway between the cells. Pericytes are distributed discontinuously along the length of the cerebral capillaries and partially surround the endothelium. Both the cerebral endothelial cells and the pericytes are enclosed by and contribute to the local basement membrane, which forms a distinct perivascular extracellular matrix (basal lamina 1, BL1) different in composition from the extracellular matrix of the glial endfeet bounding the brain parenchyma (BL2). Foot processes from astrocytes form a complex network surrounding the capillaries and this close cell association is important in induction and maintenance of the barrier properties. Axonal projections from neurons onto arteriolar smooth muscle contain vasoactive neurotransmitters and peptides and regulate local cerebral blood. BBB permeability may be regulated by release of vasoactive peptides and other agents from cells associated with the endothelium. Microglia are the resident immunocompetent cells of the brain. The movement of solutes across the BBB is either passive, driven by a concentration gradient from the plasma to the brain, with more lipid-soluble substances entering most easily or may be facilitated by passive or active transporters in the endothelial cell membranes. Efflux transporters in the endothelium limit the CNS penetration of a wide variety of solutes (based on Abbott et al. 2006) (Abbott et al. 2010).

Color image of this figure appears in the color plate section at the end of the book.

such as oxygen and carbon dioxide and small lipophilic molecules can successfully diffuse across the barriers via the lipid membranes. The influx of small polar solutes, nutrients such as glucose and amino acids and the efflux of waste products are controlled by specific solute transporters in cellular membranes. Potentially toxic products derived from the diet or the environment present in the circulation can be excluded by the BBB by several members of the ATP-binding cassette (ABC) transporter family. While the BBB expresses P-glycoprotein (P-gp, ABCB1), multidrug-resistance related proteins (MRPs, ABCC family) and breast-cancer resistance protein (BCRP, ABCG2), the choroid plexus utilize ABCs and other type of transporters to control the traffic of the substances. However, most drugs with the potential to be used in treating CNS diseases are substrates for ABC transporter family (Abbott et al. 2010). Both the BBB and BCSFB constitute an important hurdle against the passage of larger molecules (i.e., peptides and proteins). However, some smaller peptides such as beta-amyloid (Aβ) that are substrates for transporters traffic in and out through neuronal membranes (Abbott et al. 2010).

The entry of larger molecules are limited by two classes of vesicular transport: 1) receptor-mediated transcytosis (RMT) or endocytosis and 2) adsorption-mediated transcytosis (AMT). RMT requires specific interaction with surface receptors before taken up. However, the AMT-mediated endocytosis requires less specific surface-charge interactions between cationic molecules and the cell membrane. A feature of the barrier layers is that they generate 'enzymatic barriers', by expressing intracellular enzymes that break down certain molecules during transit, limiting entry of many effective drugs and neurotoxic agents (Abbott et al. 2010). Collectively, these barriers protect the CNS from access by drugs, toxins, viruses, bacteria and other unwanted factors. However, they also present a great challenge for drug delivery to the CNS to treat CNS diseases (Liu-Synder and Webster 2006).

In general, molecules that have a molecular weight larger than 400–500 kDa are not transported through the BBB (Liu-Synder and Webster 2006). Among 7,000 drugs analyzed, only 5% can pass through the BBB to treat depression, schizophrenia and insomnia (Ghose et al. 1999; Liu-Synder and Webster 2006). Drugs that have been transported successfully into the brain include the hexapeptide dalargin, the dipeptide kytorphin, loperamide, tubocurarine and doxorubicin (Jain 2006). Temozolomide is one of the few chemotherapeutic agents available that can cross the BBB (i.e., glioblastoma multiforme), but its antitumor activity is limited. Therefore, new and highly effective therapeutic agents and drug delivery strategies are needed to be developed to overcome BBB for the successful treatment of brain tumors. Currently new technologies that deliver different drugs (i.e., antibodies, proteins, peptides, siRNA, small molecules, etc.) into the CNS are being investigated.

The most commonly used delivery route of drug administration for CNS is intracerebroventricular (ICV) or intrathecal (IT) injection, requiring the medication to be administered directly into the cerebrospinal fluid (Prakash et al. 2010). Although this approach is local, it provides for lower dosage compared to intravenous delivery (Prakash et al. 2010). Intrathecal injections are widely used in the clinical settings to treat acute and chronic pain conditions and also for the treatment of brain cancers and metastasis.

A Novel Therapeutic Modality for CNS Diseases

In many CNS or peripheral nervous system diseases, there is a great need for identification of novel molecular targets and effective therapeutic strategies. Recently RNAi has emerged as a promising and novel therapeutic modality that is applicable against many CNS diseases (Lingor and Bahr 2007; Mathupala 2009). RNAi (siRNA) seems to be ideal when gene suppression is expected to provide a therapeutic benefit. Development of therapeutic RNAi is now underway for brain and peripheral system tumors, infections, and neurodegenerative disorders (Gonzalez-Alegra and Paulson 2007). However, the clinical application of RNA-based therapeutics is limited due to instability of RNA and *in vivo* delivery due to degradation and multiple physiological barriers (Pardridge 2007). The potential application of RNAi in the treatment of CNS diseases will require an understanding of their interactions with biological fluids and its effective delivery across physiological barriers.

The use of the nanoparticles may provide a valuable tool to penetrate the BBB. The mechanism of the nanoparticle-mediated transport of the drugs across the BBB is not fully understood. Endocytosis seems to be the major mechanism by which endothelial cells of the brain blood capillaries take up particles. Nanoparticle encapsulation or conjugation alters the characteristics of the drug molecules and facilitate their transport across the BBB. Furthermore, nanocarriers may provide increased localization and a sustained-release system for therapeutic agents, while avoiding potential untoward side effects. The CNS delivery of anesthetics and chemotherapeutic agents incorporated into nanoparticles have been widely studied (Jain 2006). The potential applications and benefits of nanotechnologies for the treatment of both peripheral and CNS disorders are significant, and may ultimately offer novel therapeutic options for these patients (Silva 2010).

NANOTECHNOLOGY AND NANOMEDICINE

Nanotechnology is the science that is involved in the design, synthesis, characterization, and application of materials and devices less than one micron in diameter (i.e., one millionth of a millimeter). Usually nanosized objects are 100–10,000 times smaller than the size of mammalian cells (Suh et al. 2008). Nanomedicine, the application of nanotechnology to medicine, concerns the use of specifically engineered materials at this size scale to develop novel therapeutic and diagnostic modalities (Zhang et al. 2007). Nanoparticle-based drugs can be developed by choosing different biomaterials, preferably biodegradable non-toxic, creating nanoparticles with high therapeutics payloads.

siRNA DELIVERY IN NANOMEDICINE

Therapeutic Gene Silencing by Small-interfering RNA (siRNA)

Common features of RNA silencing are production of small (21–23 nucleotide) double-stranded RNAs (Ryther et al. 2005). These short double-stranded RNAs are recognized by the endonuclease, Dicer, and cleaved into two fragments called siRNA. The antisense strand of the two strands becomes associated with a complex of proteins, designated as the RNA-Induced Silencing Complex (RISC), which targets mRNA. Next, Argonaute 2 (Ago2), a RNA endonuclease within the complex cleaves the target mRNA and leads to its degradation, shutting down protein expression. The introduction of a 18–26 base pairs siRNA triggers gene silencing in cells and is expected to be more selective and more effective than longer fragments especially in tumors that have low Dicer levels. For therapeutic applications, siRNA may be preferable over long fragments such as shRNA because cells with low Dicer levels can have impaired gene silencing with shRNA, but not with siRNA. In addition, longer RNA fragments may result in greater potential for toxic effects, especially in the liver (Brummelkamp et al. 2002; Ozpolat et al. 2009). Careful selection of siRNA sequences to avoid off-target effects is an important issue and can be minimized or eliminated by avoiding certain sequence motifs, and validation of the siRNA sequences. Exclusion of partially complementary sequences and certain motifs that induce immune response and the use of the minimum effective dose of siRNA can further reduce toxicities and eliminate off-target effects. One of the most important advantages of using siRNA is that compared to antisense oligos, siRNA is 10 to 100-fold more potent for gene silencing (Ozpolat et al. 2009).

Therapeutics that are designed to engage RNA interference (RNAi) hold promise as a novel therapeutic intervention for targeted silencing of genes in patients. However, the major limitations of the use of siRNA-based therapies are their poor cellular uptake and degradation by nucleases after systemic administration (Davis et al. 2010). Consequently, the development of safe, stable, efficient and tissue-specific nanodelivery vehicles for systemic administration is the main goal for successful clinical applications of siRNA-based therapeutics for CNS diseases.

Therapeutic RNA Interference for CNS Diseases

Recently, Kumar et al. (2007) demonstrated that intravenously injected siRNAs can cross the BBB when conjugated to short peptide derived from rabies virus glycoprotein (RVG). They also fused nine arginine to this short peptide (RGV-9R) to bind to antiviral siRNAs that can selectively bind

Table 1. Some examples of siRNA deliveries to the nervous system.

Disease	Target	Type	Delivery system	Route	Models	Authors
Age-Related macular degeneration	VEGFR1	siRNA	naked	intravitreal	Phase-1 clinical trial	Kaiser et al. 2010
Drug addiction	DARPP-32	siRNA	gold nanorod	in vitro	-	Bonoui et al. 2009
Fatal viral encephalitis	RVG-9R	siRNA	siRNA/peptide complexes	intravenous	In vivo/mice	Kumar et al. 2007
JEV- and WNV-induced encephalitis	JEV and WNV-envelope	lentiviral shRNA or siRNA	Lipoplex	Intrathecal	In vivo/mice	Kumar et al. 2006
Formalin-induced nociception	NMDAR subunit NR2B	siRNA	PEI (polyethyleneimine)	Intrathecal	In vivo/mice Mice	Tan et al. 2005
DELT-induced nociception	DOR	siRNA	Lipoplex	Intrathecal	In vivo/mice	Luo et al. 2005
Mediating the antidepressant action	SERT	siRNA	naked	intraventricular	In vivo/mice	Thakker et al. 2005
Chronic neuropathic pain	P2X3	siRNA	naked /Saline	Intrathecal	In vivo/mice	Dorn et al. 2004
Neuropsychiatric disorders	EGFP or dopamine transporter (DAT)	siRNA	naked	intraventricular	In vivo/mice	Thakker et al. 2004
Increased metabolic rate and reduced body weight without changes in food intake	AGRP	siRNA	naked /Saline	Stereotactic injection (injected directly into tumor)	In vivo/ mice	Makimura et al. 2002

AGRP, agouti related peptide; DAT, dopamine transporter; DARPP-32, Adenosine 3´5´-monophosphate-regulated-phosphoprotein; DOR, delta opioid receptor; JEV, Japanese encephalitis virus; NMDAR, N-methyl-D-aspartate; P2X, pain-related cation-channel; RVG, rabies virus glycoprotein; SERT, Serotonin transporter; VEGFR1,vascular endothelial growth factor receptor 1; WNV, West Nile Virus (Tieman and Rossi 2009; Fougerolles et al. 2007).

to a specific sequence of mRNA produced by the Japanese encephalitis virus. The RVG targets acetylcholine receptor and help internalization of the peptide-siRNA complex into neuronal cells (Kumar et al. 2007). RVG-9R mediated delivery of siRNA into the neuronal cells resulted in specific gene silencing within the brain and protection of mice from the lethal viral infection. Repeated systemic administration of RVG-9R-bound siRNA did not induce inflammatory cytokines or anti-peptide antibodies. Thus, this approach of siRNA delivery offers nontoxic and safe carrier system for other potential therapeutic molecules to get across the BBB (Kumar et al. 2007).

Another nanotechnology approach to overcome the BBB for siRNA delivery is the use of gold nanoparticles (nanorods) complexed with siRNA, called nanoplexes. This type of nanocarrier was used for targeting dopaminergic signaling pathway in the brain in an *in vitro* model of BBB for a potential treatment for drug addiction therapy (Bonoiu et al. 2009; Tiemann and Rossi 2009). Using gold-nanoplexes, expression of key proteins (DARPP-32, ERK, and PP-1) of the dopaminergic pathway was successfully silenced with no observed cytotoxicity. Therefore, these nanoplexes appear to be safe and highly effective for brain-specific delivery of siRNA for therapy of other brain diseases (Bonoiu et al. 2009).

Applications and Current Challenges for *in vivo* siRNA

The delivery of any small molecule drug or siRNA for gene silencing in the CNS may consequently have some short-term and long-term side effects. The major advantages of siRNA over small molecule drugs are; 1) sequences can be rapidly designed for highly specific inhibition of the target of interest; 2) undrugable targets (i.e., those lacking crystallographic structures) can be targeted and, 3) targets can include any molecular class (Sah 2006; Prakash et al. 2010). On the other hand, there are some problems with the use of short sequences of siRNA molecules for *in vivo* applications. These include in stability of siRNA in biological solutions, difficulties *in vivo* delivery of siRNA, off-target effects and induction of non-specific immune system.

Because of the instability of siRNA in biological fluids, naked siRNA delivery will lead to rapid degradation, ineffective cellular uptake and poor target tissue distribution. Potential approaches to overcome these limitations will be discussed in the following section.

SiRNAs have short sequences and target 19–21 bp of the gene of interest. Therefore, it is possible that siRNA can bind to other untargeted genes with similar sequences and cause some off-target effects. Even a few nucleotide similarities with untargeted gene can cause silencing by both the sense and antisense strands of siRNA. This may cause some important side effects because of unintentional inhibition of many genes. Studies utilizing *in silico* prediction have shown that the probability for off-target effect in the human genome is quite significant (Qui et al. 2005; Prakash et al. 2010). Addition of specific chemical modifications on the 5'- end of the siRNA has been used to effectively eliminate most of the off-target effects (Soutscheck et al. 2004; Song et al. 2003). Therefore, it is critical to design a highly specific siRNA sequence that will target only the gene of interest without inhibiting the functions of other genes due to off-target effects of siRNA that may happen by binding to other gene targets with similar sequences.

Recently, several algorithms have been developed to design siRNAs with the highest efficacy and least amount of off-target effects. Most of the existing tools and algorithms apply Tuschl's and Reynolds' rules together with rational design rule, followed with Basic Local Alignment Search Tool (BLAST) search to eliminate non specific bindings and "off-target effect" for the effective siRNA design (Prakash et al. 2010).

Another key factor influencing the *in vivo* application of siRNA is the potential induction of non-specific immune reactions. Several studies indicate that the use of siRNA especially long dsRNA molecules can trigger the innate immune system, involving activation of interferon-response. Interferon is an important defense mechanism against viral infections, which can be activated by double- stranded siRNA due to viral infection mimicry. This interferon response leads to activation of interferon responsive genes (IRG) and the activation of dsRNA-dependent protein kinase (PKR), which causes phosphorylation of eIF2α, resulting in inhibition of global protein synthesis (Forte et al. 2005). Recent studies demonstrated that the innate immune system is not activated by siRNA molecules shorter than 30bp, thus providing an opportunity for the use of RNAi technology as a potential therapeutic tool (Forte et al. 2005).

NANOPARTICLES

In recent years remarkable progress has been made in the use of nanoparticles to enhance *in vivo* efficacy of many drugs. Some of the nanocarriers have been shown to display potential attractive features for systemic delivery such as increased circulation time, slow release, intracellular uptake, enhanced uptake to target cells and tissues. The majority of these nanocarriers are being tested in preclinical studies and some have gone to clinical trials or have been approved for clinical use.

Selection of Particles (challenges, biocompatibility and specificity)

Current challenges for *in vivo* siRNA delivery are largely dependent on the development of systemic delivery systems that can efficiently deliver siRNA molecules into tumor tissues and target cells (Ryther et al. 2005; Ozpolat et al. 2009). The major limitations of the use of siRNA-based therapies relate to poor cellular uptake, degradation by nucleases and rapid renal clearance following systemic administration (Bumcrot et al. 2006; Ozpolat et al. 2009). Due to its large molecular weight (~ 13 kDa) and polyanionic nature (~ 40 negative phosphate groups), naked siRNA does not freely cross the cell membranes. Negatively charged cell membranes prevent efficient intracellular delivery of siRNA fragments, which also have a negatively charged backbone, leading to electrostatic repulsion. Although many siRNA carriers have been reported for *in vitro* applications, these delivery systems are usually inappropriate for *in vivo* use due to poor safety profile and unsatisfactory tissue delivery (Bumcrot et al. 2006; Ozpolat et al. 2009). Although several siRNA based therapies have entered into clinical trials, most of these utilize local delivery such as via the intravitreal or intranasal routes. Therefore, development of novel, safe, stable, effective and tumor-specific nanoparticles for systemic administration remains an unmet goal for successful clinical applications of siRNA-based cancer therapeutics (Ozpolat et al. 2009).

Systemic administration for RNAi is an attainable goal. The ideal systemic delivery system for siRNA is expected to provide robust gene silencing, be biocompatible, biodegradable and nonimmunogenic, and bypass rapid hepatic or renal clearance. Furthermore, an ideal delivery system should be able to target siRNA specifically into the tumor by interacting with tumor-specific receptors. The biodistribution, safety, and uptake of particles by cells and tissues can vary depending on the surface charge, size and hydrophobicity of the particle. Negatively charged particles are cleared faster than positively charged particles. Particles bigger than 100 nm in diameters are taken up by reticuloendothelial system in liver, spleen, lung and bone marrow, while smaller nanoparticles have a prolonged circulation time. In addition, hydrophilic coating using polyethylene glycol (PEG) or a nonanionic surfactant enhances circulation time of the nanocarriers (Ozpolat et al. 2009).

Nanoparticles for siRNA Delivery and Preparation of Particles

Nanocarriers are usually made of biodegradable nanomaterials such as natural or synthetic lipids and polymers [e.g., poly lactic co-glycolic acid (PLGA), poly lactic acid (PLA), polyethilenimine (PEI), chitosan, atellocollagen, etc.], quantum dots, or iron oxide magnetic nanoparticles;

many of these have been investigated for effective and safe siRNA delivery (Ozpolat et al. 2009).

Table 1. Summary of sizes of the particles that have been used for siRNA delivery (Ozpolat et al. 2009).

Category of particle	Delivery system	Type	Natural versus synthetic	Toxicity
Lipid complex	Cationic liposomes	siRNA	Synthetic	Lung Toxicity
	Neutral liposomes	siRNA	Synthetic	Biodegradable-nontoxic
	Lipoplexes	siRNA	Synthetic	
	Stable nucleic acid-lipid particles	siRNA	Synthetic	
Conjugated polymers	Polymer-functional peptides	siRNA	Synthetic	
	Polymer-lipophilic molecules	siRNA	Synthetic	
	Polymer-PEG	siRNA	Synthetic	
Cationic polymers	Chitosan	siRNA/shRNA	Natural	Biodegradable-nontoxic
	Atellocollagen	siRNA/shRNA	Natural	Biodegradable-nontoxic
	Pegylated	siRNA/shRNA	Synthetic	Cytotoxicity
	Cyclodextrin	siRNA/shRNA	Synthetic	

Liposomal nanocarriers for siRNA delivery

Lipid-based nanoparticles include liposomes, micelles, emulsions and solid lipid nanoparticles. The ability of liposomes to deliver their payload to target sites (i.e., drugs, antigens, proteins and nucleotides) has made these particles the most popular and successful method for delivery of therapeutic agents. Liposomes are uni- or multilamellar vesicles consisting of a phospholipid bilayer with hydrophilic and/or aqueous inner compartment. For robust siRNA delivery, using lipid-based systems, lipid composition, siRNA-to-lipid ratio and particle size should be optimized. The circulating half-life of liposomes can be prolonged by the addition of neutral, hydrophilic polymers such as PEG to the outer surface (Yu et al. 2005; Ozpolat et al. 2009). An extended circulation half-life allows for sustained availability to take advantage of the enhanced permeability of tumor vasculature, resulting in increased delivery to target sites (Maeda. 2001; Ozpolat et al. 2009).

Cationic liposomes Amongst nonviral delivery systems for oligonucleotides (e.g., plasmid DNA, antisense oligonucleotides and siRNA/shRNA), cationic lipids have been traditionally the most popular and widely used delivery systems. Cationic lipids, such as 1,2-dioleoyl-3 trimethylammonium-propane (DOTAP) and N-[1-(2,3- Dioleoyloxy)propyl]-N,N,N-trimethyl-ammonium methyl sulphate (DOTMA) form complexes with negatively charged DNA and siRNA resulting in high *in vitro* transfection efficiency. Cationic liposomes, while efficiently taking up nucleic acids, have had limited success for *in vivo* gene downregulation, perhaps because of their stable intracellular nature and resultant failure to release siRNA contents. The effectiveness of cationic liposomes as potential therapeutic molecules has been limited, in part, by their toxicity (Dokka et al. 2000; Spagnou et al. 2004; Ozpolat et al. 2009). The *in vivo* use of cationic liposomes in animal models elicited dose-dependent toxicity and pulmonary inflammation (Dokka et al. 2000; Spagnou et al. 2004; Ozpolat et al. 2009). However, neutral and negatively charged liposomes were not found to exhibit lung toxicity. Overall, although cationic lipid-based delivery systems offer some advantages as a potential siRNA delivery system, potential for lung and other toxicities may require alternative preparations for safety (Dokka et al. 2000; Spagnou et al. 2004; Lv et al. 2006; Ozpolat et al. 2009). However Different lengths of hydrocarbon chains can also influence the cytotoxicity of cationic lipids (Gewirtz 2007). Therefore, careful selection of lipids and formulation strategies may help reduce or eliminate toxicity and potential adverse effects.

Neutral nanoliposomes One of the most important advances in the siRNA delivery field has been the development of neutral 1,2-dioleoyl-sn-glycero-3-phosphatidylcholine (DOPC) based nanoliposomes (mean size 65 nm) (Ozpolat et al. 2009). These nanoliposomes can deliver siRNA *in vivo* into tumor cells 10- and 30-fold more effectively than cationic liposomes (DOTAP) and naked siRNA, respectively (Gewirtz 2007). Overall, these data support our hypothesis that neutral DOPC-nanoliposomes effectively deliver siRNA into tumor cells and can be combined with other conventional anti-cancer therapies, such as chemotherapy, to enhance the efficacy of conventional drugs. The neutral nanoliposomal (DOPC) carriers seem to overcome barriers for developing siRNA-therapeutics for the treatment of cancers other than brain tumors (Ozpolat et al. 2009).

Polymeric nanocarriers for siRNA delivery

Biodegradable cationic polymers include; 1) natural (e.g., chitosan, atelocollagen, and cationic polypeptides) and 2) synthetic (e.g., polyethylimine (PEI), poly-L-lysine (PLL), cyclodextrin). Furthermore, PLA, PLGA (Shinde et al. 2007) based nanoparticles, quantum dots and magnetic iron oxide particles (Tan et al. 2007) have also been studied for siRNA delivery with some promising results. These carriers may be useful for *in vivo* applications because of their safety. Cationic polymers rapidly form a complex with negatively charged siRNA (Ozpolat et al. 2009).

Chitosan as a novel nanocarriers for siRNA delivery

Chitosan is a biocompatible, non-toxic and biodegradable positively charged polysaccharide material, making it a perfect candidate for siRNA delivery. The amine group of chitosan is protonated at lower pH, producing a polycationic polymer that enables strong electrostatic binding of oligonucleotides. Chitosan nanoparticles have been shown to be highly effective and safe for *in vivo* delivery of siRNA when administered in mice through intranasal or intravenous administration (Howard et al. 2006; Pille et al. 2006; Ozpolat et al. 2009; Han et al. 2010).

The most important factors that enhance chitosan's cellular uptake and cell specificity for *in vitro* and *in vivo* applications include molecular weight, chemical modification and degree of deacetylation of chitosan particles, acidic or basic nature of the environment, presence of serum and the ratio of chitosan to siRNA (reviewed by Rudzinski and Abinabhavi 2010). In addition to novel formulation and optimization, modification of chitosan structure, and attachment of functional groups or ligands are highly useful methods to improve the cellular uptake, enhance targeting potential and stability of the carrier in *in vivo* conditions (i.e., blood) and intracellular release from lysosomes (Mao et al. 2010; Han et al. 2010).

Applications of chitosan-mediated siRNA delivery to CNS

The muscarinic acetylcholine receptors (mAChRs) in the spinal cord for important for regulating nociceptive transmission. Because there are no highly specific agonists and antagonists for mAChRs subtypes, it has been difficult to define which mAChRs subtypes are involved in the analgesic effects produced by spinally administered cholinergic agents. We therefore used small-interference RNA (siRNA) targeting specific mAChRs subtypes to determine the contribution of M_2, M_3, and M_4 subtypes in the spinal cord to muscarinic analgesia in rats. Our study demonstrates that chitosan nanoparticles can be used for efficient delivery of siRNA into the neural tissues to specifically knockdown the target genes *in vivo*. Our findings provided important evidence that M_2 and M_4, but not M_3, subtypes contribute to nociceptive regulation by mAChRs at the spinal level (Cai et al. 2009).

Axons of neuronal cells are not able to regenerate after injury, partly because of the activity of neurite outgrowth inhibitors in the CNS. Axons in the adult vertebrate CNS of neuronal cells exhibit are not able to regenerate after injury, partly because of the activity of presence of neuritis outgrowth inhibitors in the CNS. Mittnacht et al. (2009) have tried to enhance functional recovery and axon regrowth using siRNA by transiently inhibiting the intracellular signaling pathway of central components into which various repulsive inputs converge. Analysis of a variety of chitosan-siRNA nanoparticles using gel-retardation assays demonstrated high stability of those complexes disrupted only in the presence of competing agents. Experiments utilizing fluorescent-labeled siRNA showed significant uptake of chitosan-siRNA nanoparticles into different variety of cells in the nervous system. Taken together, these data highlight the potential application of chitosan in siRNA-mediated gene knockdown to promote neuronal regeneration after spinal cord injury (Mittnacht et al. 2009).

Recently, we have developed tumor-targeting version of Arg-Gly-Asp (RGD) peptide-labeled chitosan nanoparticle (RGD-CH-NP) as a novel tumor targeted delivery system for siRNA which was highly effective in *in vivo* in animal models of ovarian cancer models in mice (Han et al. 2010). RGD cyclic peptide specifically binds to αvβ3 integrin receptors, which are specifically expressed in tumor associated angiogenic endothelium. αvβ3 is also expressed in some tumors including ovarian cancer. Therefore highly effective tumor-targeting nanocarriers can be produced by coating them with RGD peptides. In summary, the transfection efficiency of chitosan-based delivery systems can be adjusted by specific need and targeted tissues (Han et al. 2010).

Concluding Remarks and Future Direction

Recent improvement in the understanding of cancer biology and tumor microenvironment open new opportunities to overcome challenges in the treatment of cancer and will lead to development of novel diagnostic and molecularly targeted therapies. For instance, discovery of the discontinuous alignment and increased gaps between endothelial cells as well as expression of specific receptors in the tumor-associated vascular bed provided a window of opportunity for enhanced delivery of tumor targeted nanoparticle-based therapies. Also, discovery of RNA interference (RNAi) which is a powerful method for specific gene silencing led to design of siRNA-based promising novel therapeutic strategies. A major challenge with the use of siRNA is their intracellular delivery to specific tissues and organs such as CNS and peripheral nerves that express the target gene. Although this promising technology initially suffered from lack of safe and effective systemic delivery systems for *in vivo* applications, neutral lipid-based nanoliposomes or chitosan-based nanocarriers have emerged as highly effective and safe systemic delivery systems. However, as we learn about the new biomaterials and their behavior in biological systems, designing and developing more effective nanocarriers for diagnostic and therapeutic applications become more achievable. Data accumulated so far on the safety of systemic administration will allow for phase I clinical studies in the near future. However, pharmacokinetic, pharmacodynamic parameters and safety profiles of various delivery systems following *in vivo* delivery of siRNA should be evaluated carefully in future studies. Understanding and evaluating safety considerations associated with the use and application of nanotechnologies for BBB delivery is also a critical step to their clinical use.

In conclusion, nanocarriers hold great potential for successful and safe *in vivo* delivery of siRNA-based therapeutics into hard to deliver tumor tissues by systemic administration and expected to be one of the highest-impact contributions to clinical applications in neuroscience. Applying various nanotechnology platforms for siRNA-therapeutics seems to be the solution to overcoming challenges with regard to delivery and specifically targeting and therapeutically silencing genes of interest in cancer cells in tumors and the other diseases of CNS.

In the future, multiple challenges including safety, reproducibility, purity, particle size distribution, charge, porosity, loading capacity, biocompatibility and toxicity, fabrication cost will determine the commercialization of nanoparticle-based therapeutics. Nevertheless, it appears that overcoming these issues could make nanoparticle-based therapeutics an integral part of mainstream medicine and the pharmaceutical industry.

Key Facts

- The incidence of CNS diseases is expected to increase significantly in the 21st century due to an increase in lifespan and changing population demographics.
- The discovery of RNA interference (i.e., siRNA)-based gene silencing in 1998 resulted in the award of Nobel Prize in Physiology or Medicine to Drs. Andrew Fire and Craig Mello in 2004. An siRNA-based mechanism silences genes in a sequence-specific manner. Very small amounts of double-stranded RNA (dsRNA) complementary to a gene of interest could silence the expression of the gene by degradation of the mRNA.
- As a result of its potency and selectivity, RNAi has rapidly become standard methodology in biological laboratories to study gene function. Scientists can exploit this phenomenon to silence or eliminate activity of overexpressed genes critical to cancer growth and pathogenesis of other disease for therapeutic strategy.

Summary Points

Neurological diseases and targets where need for new treatments exist

- The most challenging of the CNS diseases are cancer and neurodegenerative diseases, including, Alzheimer's, Parkinson's and Hungtington's diseases. The major problems in treating these diseases include difficulties in the systemic delivery of therapeutics across the blood-brain barrier, and targeting of drugs to specific tissues or cells within the CNS. In general, molecules that have a molecular weight larger than 400–500 kDa cannot pass through the BBB. Among 7,000 drugs analyzed in one study, only 5% can pass through the BBB.
- The most widely used technique for CNS drug delivery is intracerebroventricular (ICV) and intrathecal (IT) injection wherein the medication is administered directly into the cerebrospinal fluid. In many CNS or peripheral nervous system diseases, novel molecularly targeted therapeutic strategies and delivery systems that can cross BBB is needed.

RNAi and why this approach is needed

- In many CNS or peripheral nervous system diseases, there is a great need for identification of novel molecular targets and effective therapeutic strategies. Recently, RNAi has emerged as a promising and novel therapeutic modality that is applicable against many CNS disease. RNAi seems to be ideal when gene suppression is expected to provide a therapeutic benefit. Development of therapeutic RNAi is now underway for brain and peripheral system tumors, infections, and neurodegenerative disorders.
- The introduction of a synthetic siRNA of nucleotides triggers gene silencing in cells and is expected to be more effective than longer fragments especially in tumors that have low Dicer levels. mRNA silencing can be achieved by small (21–23 nucleotide) double-stranded RNAs. Careful selection of siRNA sequences to avoid off-target effects is an important issue and can be minimized or eliminated by avoiding certain sequence motifs, and validation of the siRNA sequences. Exclusion of partially complementary sequences and certain motifs that induce immune response and the use of the minimum effective dose of siRNA can further reduce toxicities and eliminate off-target effects.

Current limitations with delivery

- Therapeutics that are designed to engage RNA interference holds promise as a novel therapeutic intervention for targeted silencing of genes in patients. However, the major limitations of the use of siRNA-based therapies are their poor cellular uptake and degradation by nucleases after systemic administration. Consequently, development of safe, stable, efficient and tissue-specific nanodelivery vehicles for systemic administration is the main goal for successful clinical applications of siRNA-based therapeutics for CNS diseases.

- Most preclinical and clinical studies regarding delivery of small interfering RNA (siRNA) to the CNS have utilized invasive, intra-cerebral delivery of siRNA to the targeted tissue. Methods need to be developed to facilitate delivery of therapeutically significant quantities of siRNA to the CNS via systemic route. The systemic route is the most preferred, noninvasive route of administration for RNAi. Therefore, the need for an efficient target tissue-specific siRNA delivery vehicle is evident.

Delivery options and why they are attractive

- One of the possibilities for overcoming the BBB is drug delivery to the CNS using nanoparticles. Nanocarriers are usually made of biodegradable nanomaterials such as natural or synthetic lipids and polymers [e.g., poly lactic co-glycolic acid (PLGA), poly lactic acid (PLA), polyethilenimine (PEI), chitosan, atellocollagen, etc.], quantum dots, or iron oxide magnetic nanoparticles.

Current state-of-science

- Chitosan is a biocompatible, non-toxic and biodegradable positively charged polysaccharide. Chitosan nanoparticles have been shown to be highly effective and safe for *in vivo* delivery of siRNA when administered in mice through intranasal and intravenous administration.
- Chitosan nanoparticles can be used to deliver siRNA into the neural tissues to specifically knock down the target genes.

Future directions

- Developing safe and effective tissue-specific delivery vectors, minimizing the off-target effects, and achieving distribution in sufficient concentrations of the payload at the target tissue with minimal side effects will lead to major modifications in the therapeutic window of siRNA. The ideal systemic delivery system for siRNA is expected to provide robust gene silencing, be biocompatible, biodegradable and nonimmunogenic, and bypass rapid hepatic or renal clearance. Furthermore, an ideal delivery system should be able to target siRNA specifically into the tumor by interacting with tumor-specific receptors.

Abbreviations

ABC	:	ATP-binding cassette
Ago2	:	argonaute 2
AMT	:	adsorptive-mediated transcytosis
BBB	:	blood-brain barrier
BCSFB	:	blood cerebrospinal fluid barrier
CNS	:	central nervous system
CSF	:	cerebrospinal fluid
DARPP-32	:	dopamine- and adenosine 3':5'- monophosphate-regulated phosphoprotein
DAT	:	dopamine transporter
DOPC	:	1,2-Dioleoyl-sn-glycero-3-phosphatidylcholine
DOR	:	delta opioid receptor
DOTAP	:	1,2-dioleoyl-3methylammonium-propane
DOTMA	:	N-[1-(2,3-Dioleoyloxy) propyl]-N,N,N-trimethyl-ammonium methyl sulphate
ICV	:	intracerebroventricular
IT	:	intrathecal
JEV	:	Japanese encephalitis virus
mAChRs	:	muscarinic acetylcholine receptors
MRP	:	multidrug-resistance related proteins
NMDAR	:	N-methyl-D-aspartate
PE1	:	polyethylleneimine
P2X	:	pain-related cation-channel EGFP
PEG	:	polyethylene glycol
PEI	:	polyethilenimine
P-GP	:	P-glycoprotein
PLA	:	polylactic acid
PLGA	:	poly lactic co-glycolic acid
RISC	:	RNA-induced silencing complex
RMT	:	receptor-mediated transcytosis
RVG	:	rabies virus glycoprotein
SERT	:	serotonin transporter
siRNA	:	small-interfering RNA
VEGFR1	:	vascular endothelial growth factor receptor 1
WNV	:	West Nile Virus

Acknowledgements

Portions of this work were supported by the NIH (P50 CA083639, P50 CA098258, CA128797, RC2GM092599, U54 CA151668), the Ovarian Cancer Research Fund, Inc. (Program Project Development Grant, the Zarrow Foundation, the Kim Medlin Fund, and the Laura and John Arnold Foundation.

References

Abbott, N.J., L. Ronnback and E. Hansson. 2006. Astrocyte- endothelial interactions at the blood-brain barrier. Nat. Neurosci. 4: 41–53.

Abbott, N.J., A.A.K. Patabendige, D.E.M. Dolman, S.R. Yusof and D.J. Begley. 2010. Structure and function of the blood–brain barrier. Neurobiology of Disease 37: 13–25.

Bonoiu, A.C., S.D. Mahajan, H. ding, I. Roy, K.T. Yong, R. Kumar, R. Hu, E.J. Bergey, S.A. Schwartz and P.N. Prasad. 2009. Nanotechnology approach for drug addiction therapy: gene silencing using delivery of gold nanorod-siRNA nanoplex in dopaminergic neurons. Proc. Natl. Acad. Sci. USA 106: 5546–5550.

Bumcrot, D., M. Manoharan, V. Koteliansky and D.W. Sah. 2006. RNAi therapeutics: A potential new class of pharmaceutical drugs. Nat. Chem. Biol. 2: 711–719.

Brummelkamp et al. T.R., R. Bernards, R. Agami. 2002. Stable suppression of tumorigenicity by virus-mediated RNA interference. Cancer Cell 2: 243–247.

Cai, Y.Q., S.R. Chen, H.D. Han, A.K. Sood, G. Lopez-Berestein, H.L. Pan. 2009. Role of M2, M3, and M4 muscarinic receptor subtypes in the spinal cholinergic control of nociception revealed using siRNA in rats. J. Neurochem. 111(4): 1000–1010.

Davis, M.E., J.E. Zuckerman, C.H.J. Choi, D. Seligson, A. Tolcher, A.A. Christopher, Y. Yen, J.D. Heidel and A. Ribas. 2010. Evidence of RNAi in humans from systemically administered siRNA via targeted nanoparticles. Nature 464: 1067–1071.

Dokka, S., D. Toledo, X. Shi, V. Castranova and Y. Rojanasakul. 2000. Oxygen radical-mediated pulmonary toxicity induced by some cationic liposomes. Pharm. Res. 17: 521.

Dorn, G., S. Patel, G. Wotherspoon, M. Hemmings-Mieazczak, J. Barclay, F.J. Natt, P. Martin, S. Bevan, A. Fox, P. Ganju, W. Wishart and J. Hall. 2004. Nucleic Acids Res. 32(5): e49.

Forte, A., M. Cipollaro, A. Cascino and U. Galderisi. 2005. Curr. Drug Targets 6: 21–29.

Fougerolles, A., H.P. Vornlocher, J. Maraganore and J. Lieberman. 2007. Interfering with disease: A progress report on siRNA-based therapeutics. Nat Rev Drug Discov. 6: 443–453.

Gewirtz, A.M. 2007. On future's doorstep: RNA interference and the pharmacopeia of tomorrow. J. Clin. Invest. 117: 3612–3614.

Ghose, A.K., W.N. Viswanadhan and J.J. Wendoloski. 1999. A knowledge-based approach in designing combinatorial or medicinal chemistry libraries for drug discovery.1. A qualitative and quantative characterization of known drug databases. J. Comb. Chem. 1 (1): 55–68.

Gonzalez-Alegra, P. and H. Paulson. 2007. Technology Insight: therapeutic RNA interference-how far from the neurology clinic? Nature Clinical Practice Neurology. 3: 394–40.

Han, H.D., L.S. Mangala, J.W. Lee, M.M. Shahzad, H.S. Kim, D. Shen, E.J. Nam, E.M. Mora, R.L. Stone, C. Lu, S.J. Lee, J.W. Roh, A.M. Nick AM, G. Lopez-Berestein and A.K. Sood. 2010. Targeted gene silencing using RGD-labeled chitosan nanoparticles.Clin. Cancer Res. 16(15): 3910–3922.

Howard, K.A., U.L. Rahbek, X. Liu, C.K. Damgaard, Sz. Glud, M. Anderson, M.B. Hovgaard, A. Schmitz, J.R. Nyengaard, F.B. esenbacher and J. Kiems. 2006. RNA interference in vitro and in vivo using a novel chitosan/siRNA nanoparticle system. Mol. Ther. 14: 476–484.

Jain, K.K. 2006. Role of nanotechnology in developing new therapies for diseases of nervous system. Nanomedicine 1(1): 9–12.

Kaiser, P.K., R.C.A. Symons, S.M. Shah, E.J. Quinlan, H. Tabandeh, D. V.Do, G. Reisen, J.A. Lockridge, B. Short, R. Guerciolini and Q.D. Nguyen. 2010. RNAi-based treatment for neovascular age-related macular degeneration by siRNA-027. Am. J. Ophthalmol. 150: 33–39.

Kumar, P., H. Wu, J.L. McBride, K.-E. Jung, M.H. Kim, B.L. Davidson, S.K. Lee, P. Shankar and N. Manjunath. 2007. Transvascular delivery of small interfering RNA to the central nervous system. Nature 448: 39–43.

Kumar, P., S.K. Lee, P. Shankar and N. Manjunath. 2006. A single siRNA suppresses fatal encephalitis induced by two different flavavirus. PLoS Med. 3(4): e96.

Lingor, P. and M. Bahr. 2007. Targeting neurological disease with RNAi. Mol. Biosyst. 3(11): 773–780.

Liu-Synder, P. and T.J. Webster. 2006. Designing drug-delivery systems for the nervous system using nanotechnology: opportunities and challenges. Expert Rev. Med. Devices 3(6): 683–687.

Lv, H., S. Zhang, B. Wang, S. Cui and J. Yan. 2006. Toxicity of cationic lipids and cationic polymers in gene delivery. J. Control Release 114: 100–9.

Luo, M.C., D.Q. Zhang, S.W. Ma, Y.Y. Huang, S.J. Shuster, F. Porreca and J. Lai. 2005. An efficient intrathecal delivery of small interfering RNA to the spinal cord and peripheral neurons. Mol. Pain. 1: 29.

Maeda, H. 2001. The enhanced permeability and retention (EPR) effect in tumor vasculature: the key role of tumor-selective macromolecular drug targeting. Adv. Enzyme Regul. 41: 189–207.

Makimura, H., T.M. Mizuno, J.W. Mastaitis, R. Agami and C.V. Mobbs. 2002. Reducing hypothalamic AGRP by RNA interference increases metabolic rate and decreases body weight without influencing food intake. BMC Neurosci. 3: 18.

Mao, S., W. Sun and T. Kissel. 2010. Chitosan-based formulations for delivery of DNA and siRNA. Adv. Drug Deliv. Rev. 62(1): 12–27.

Mathupala, S.P. 2009. Delivery of small-interfering RNA (siRNA) to the brain. Expert Opin. Ther. Pat. 19(2): 137–140.

Mittnacht, U., H. Hartman, M. Andersen, S. Hein, K. Howard, J. Kjems and B. Schlosshauer. 2009. Chitosan nanoparticles for si-RNA-mediated gene knockdown to promote neuronal regeneration.

Ozpolat, B., A.K. Sood and G. Lopez-Berestein. 2009. Nanomedicine based approaches for the delivery of si RNA in cancer. Internal Med. 1: 44–53.

Pardridge, W.M. 2007. Blood-brain barrier delivery. Drug Discov. Today. 12(1-2): 54–61.

Prakash, S., M. Malhotra and V. Rengaswamy. 2010. Nonviral siRNA delivery for gene silencing in neurodegenerative diseases. pp. 211–229. In: Wei-Ping Min and T. Ichim. [eds.] RNA interference, Methods in Molecular Biology. Springer Science+ Business Media, LLC.

Pille, J.Y., H. Li and E. Blot, J.R. Bertrand, L.L. Pritchard, P. Opolon, A. Maksimenko, H. Lu, J.P. Vannier, J. Soria, C. Malvy and C. Soria. 2006. Intravenous delivery of anti-RhoA small interfering RNA loaded in nanoparticles of chitosan in mice: safety and efficacy in xenografted aggressive breast cancer. Hum. Gene Ther. 17: 1019–26.

Qui, S., C.M. Adema and T. Lane. 2005. A computational study of off-target effects of RNA interference. Nucleic Acids Res. 33: 1834–1847.

Schlageter, K.E., P. Molnar, G.D. Lapin and D.R. Groothuis. 1999. Microvessel organization and structure in experimental brain tumors: microvessel populations with distinctive structural and functional properties. Microvasc. Res. 58: 312–328.

Song, E., S.K. Lee, J. Wang, N. Ince, N. Ouyang, J. Min, J. Chen, P. Shankar and J. Liberman. 2003. RNA interference targeting Fas protects mice from fulminant hepatitis. Nat Med. 9(3): 347–351.

Southcheck, J., A. Akinc, B. Bramlage, K. Charisse, R. Constien, M. Donaghue, et al. 2004. Therapeutic silencing of an endogenous gene by systemic administration of modified siRNAs. Nature 432: 173–178.

Rudzinski, W.E., T.M. Aminabhavi. 2010. Chitosan as a carrier for targeted delivery of small interfering RNA. Int. J. Pharm. 399(1-2): 1–11.

Ryther, R.C., A.S. Flynt, J.A. Phillips and J.G. Patton. 2005. siRNA therapeutics: big potential from small RNAs. Gene Ther. 12: 5–11.

Sah, W.Y. 2006. Therapeutic potential of RNA interference for neurological disorders. Life Sci. 79: 1773–1780.

Shinde, R.R., M.H. Bachmann, Q. Wang, R. Kasper and C.H. Contag. 2007. PEG-PLA/PLGA nanoparticles for in-vivo RNAi delivery. California: NSTI Nano tech.

Silva, G.A. 2010. Nanotechnology applications and approaches for neuroregeneration and drug delivery to the central nervous system. Ann. N. Y. Acad. Sci. 1199: 221–30.

Spagnou, S., A.D. Miller and M. Keller. 2004. Lipid carriers of siRNA: differences in the formulation, cellular uptake and delivery with plasmid DNA. Biochemistry 43: 13348–56.

Suh, W.Y., K.S. Suslick, G.D. Stucky and Yoo-Huh Suh. 2008. Nanotechnology, nanotoxicology and neuroscience. Prog. Neurobiol. 87: 133–170.

Tan, P.H., L.C. yang, H.C. Shih, K.C. Lan and J.T. Taiwan. 2005. Gene knockdown with intrathecal siRNA of NMDA receptorNR2B subunit reduces formalin-induced nociceptin in the rat. Gene Ther. 12(1): 59–66.

Tan, W.B., S. Jiang and Y. Zhang. 2007. Quantum-dot based nanoparticles for targeted silencing of HER2/neu gene via RNA interference. Biomaterials 28: 1565–71.

Thakker, D.R., F. Natt, D. Husken, H. van der Putten, R. Maier, D. Hoyer and J.F. Cryan. 2005. si-RNA mediated knockdown of the serotonin transporter in the adult mouse brain. Mol. Psychiatry. 10(8): 782–789.

Tiemann, K. and J.J. Rossi. 2009. RNAi-based therapeutics–current status challenges and prospects EMBO Molecular Medicine 1(3): 142–151.

Yu, D., P. Peng, S.S. Dharap, et al. 2005. Antitumor activity of poly (ethylene glycol)-camptothecin conjugate: the inhibition of tumor growth *in vivo*. J. Control Release 110: 90–102.

Zhang, L., F.X. Gu, J.M. Chan, A.Z. Wang, R.S. Langer and O.C. Farokhzad. 2007. Clin Pharmacol. Ther. 83(5): 761–769.

13

Targeting Drug Delivery to the Brain via Transferrin Anchored Nanoparticles

Jiang Chang[1,a] *and Didier Betbeder*[1,b]

ABSTRACT

Drug delivery to the brain is one of the most challenging research areas in pharmaceutical sciences due to the presence of the blood-brain barrier (BBB). This barrier represents an insurmountable obstacle for a large number of drugs, including anticancer agents, antibiotics, peptides, oligonucleotides and macromolecular drugs. The BBB presents a physical barrier owing to the presence of tight junctions between endothelial cells, and also a metabolic barrier restricting the entry of undesired substances. Nanotechnology can help in the transfer of drugs across the BBB. Recently, researchers have been building liposomes and nanoparticles in order to gain access through the BBB. Brain targeting can be achieved using specific anchored ligands on the surface of these vectors. Some over-expressed receptors on the brain capillary endothelial cells,

---

[1]EA 4483, IMPRT, IFR 114, Faculté de Médecine pôle recherche, Université de Lille 2, 59000 Lille, France.
[a]E-mail: jiang.chang@univ-lille2.fr
[b]E-mail: dbetbeder@aol.com

List of abbreviations after the text.

such as transferrin, insulin and others, have been chosen as targets to improve delivery of drugs to the brain. Based on the "Trojan horse" strategy, the drugs are loaded in the nanoparticles. This drug delivery system will then be recognized and mediated by the receptors at the BBB and finally be transferred to the brain. This chapter summarizes the research and development of drug delivery and targeting of brain tissue using nanoparticles to target transferrin receptor for crossing BBB.

Keywords: Nanoparticles, transferrin, antibody, targeting, drug, receptor, blood-brain barrier, glioma

INTRODUCTION

Most diseases of the nervous system, cancers, and infections of the brain cannot be effectively treated due to the presence of the blood-brain barrier (BBB) that restricts the entry of undesired substances to the brain. For example, more than 98% of all potential new drugs for the treatment of central nervous system (CNS) disorders do not cross the BBB. The BBB is situated at the interface of blood and brain and its primary function is to maintain the homeostasis of the brain. The human BBB has a total blood vessel length of approximately 650 km and an estimated surface area of approximately 20 m^2. It is considered the most important barrier to solutes reaching the brain (Pardridge 1999). In recent years, pharmaceutical companies have focused on the development of small drug molecules as therapeutic moieties. In general, small molecules should be lipid-soluble and have a molecular weight below 400–600 Dalton to be able to cross the BBB in therapeutically effective quantities (Pardridge 1999). The BBB is formed by a network of closely sealed endothelial cells in the brain's capillaries (Fig. 1), and it expresses a high level of proteins that pump foreign molecules away from the brain. Therefore, even small drug molecules cannot cross the BBB in sufficient quantities without brain drug-targeting strategies. Meanwhile, more and more larger molecules have been generated by biotechnological means which constitute promising alternatives for the treatment of diseases of the CNS. These include proteins (neurotrophins) (Pardridge 2002) and genes (neprylysin gene) for Alzheimer's Disease, antisense therapy for Huntington's Disease and monoclonal antibodies for diagnostic purposes and treatment of brain metastasis of breast cancer tumors. These macromolecules cannot cross the BBB without using targeting and delivery strategies. Receptors on the brain's capillary endothelial cells for insulin, insulin-like growth factor

252 Nanomedicine and the Nervous System

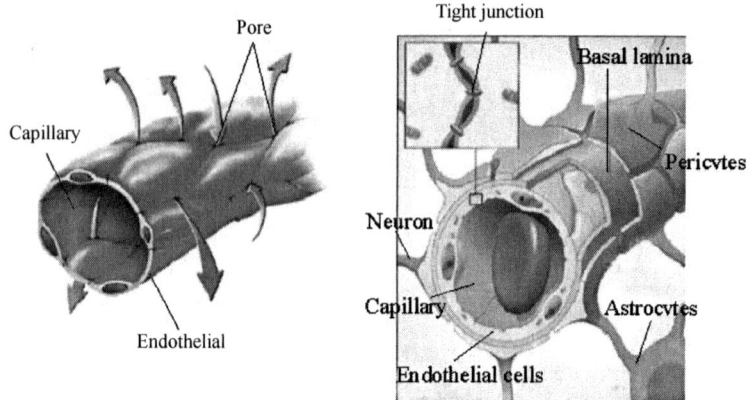

Fig. 1. General capillary (left) and brain capillary (right). Brain capillaries are made of endothelial cells sealed with tight junction surrounded by astrocytes and pericytes.

Color image of this figure appears in the color plate section at the end of the book.

(IGF), leptin, low-density density lipoprotein (LDL) and transferrin (Tf), allow some molecules, such as glucose, transferrin and insulin (which are necessary for brain homeostasis) to cross the BBB via specific pathways (Roberts et al. 1993). Ligand-receptor mediated drug delivery has received major attention in the past few years owing to the potential of site-specific crossing of the BBB therefore allowing delivery to the brain of drugs bound to a specific ligand. Among the ligands used to cross the BBB, transferrin (Tf) has been widely applied in the active targeting of anticancer agents, proteins and genes, as well as to proliferating malignant cells in the brain which also over-express transferrin receptors (Prior et al. 1990). The applications of nanoparticles (NPs) in biomedicine, as controlled drug delivery and drug targeting vehicles, have been widely studied and rapidly advanced in the last twenty years (Cao 2004). In this chapter, we present an overview of the studies using nanoparticulate vectors to target the brain with the transferrin receptor as a target (Fig. 2).

Key Facts of Blood-brain Barrier and its Drug Delivery System by Nanoparticles

- The Blood Brain Barrier preserves brain homeostasis.
- The Blood brain barrier prevents the entry of unwanted substances and restricts the entry of potential harmful compounds.
- The blood–brain barrier is formed by the endothelial cells that line cerebral microvessels, these endothelial cells are highly linked together by tight junctions blocking para-cellular passage by small molecules.

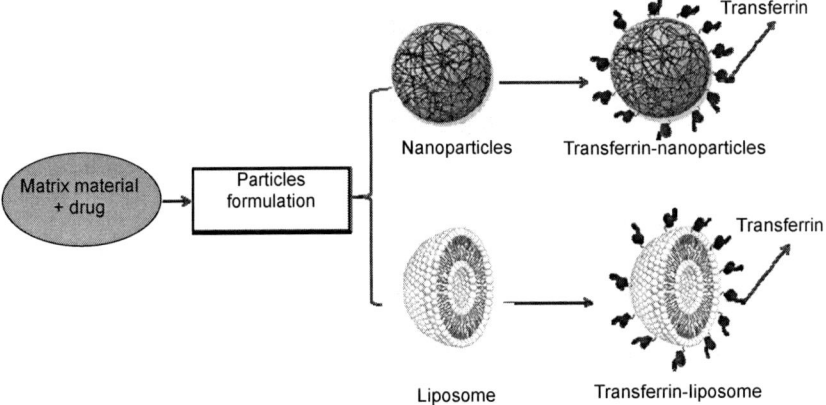

Fig. 2. **Transferrin-nanoparticles/liposome targeting drug delivery system.** As typical drug delivery system, drug is loaded in nanoparticles/liposome matrix during particles formulation, transferrin as targeting ligand is connected on the particle surface.

- To reach the brain nanoparticles must be stealthy, therefore avoiding the immune cells, and be targeted to blood brain endothelial receptors.
- Drug targeting to the brain can be achieved by nanoparticles, however more research is needed to improve brain targeting and efficacy.

NANOPARTICLES AND LIPOSOMES AS TROJAN HORSES

Nanoparticles are characterized as objects in which at least one dimension has approximately a size range of 1 nm to 100 nm ([ISO/TS 27687, 2008]). This specification refers to core terms such as the nanoscale and nanoobjects, which include nanoparticles, nanoplates, nanofibers, nanotubes, nanorods, nanowires and quantum dots (Table 1, Fig. 3). Nanoparticles are promising tools for drug delivery, medical imaging, and as diagnostic sensors. However, the biodistribution of nanoparticles after intravenous administration is characterized by their rapid elimination from the bloodstream and their accumulation in the tissues of the mononuclear phagocyte system (MPS), especially the liver, spleen and bone marrow (Bazile et al. 1992). Thus, the most effective therapeutic use of drugs associated with nanoparticles is currently limited to the treatment of several liver diseases (Hay 1994). Nanoparticles for nanomedicine applications are formulated from a variety of materials and are engineered to carry an array of substances in a controlled and targeted manner. The nanocarriers are made of biocompatible and biodegradable materials such as synthetic proteins, peptides, lipids, polysaccharides,

Table 1. Examples of nanoparticles and their characteristics.

Nanoparticle	Characteristics
Micelle	Formed when amphiphiles are placed in water. Consist of an inner core of assembled hydrophobic segments and an outer hydrophilic corona serving as a stabilizing interface between the core and the external aqueous environment
Liposome	Spherical particle formed by at least one lipid bilayer enclosing an aqueous compartment
Polymeric nanoparticles	Prepared from synthetic or natural polymers
Lipid nanocapsulates	Composed of an oily liquid core surrounded by a solid or semi-solid shell of surfactant
Quantum dots	Nanoscale crystals of semiconductor material that glow or fluoresce when excited by a light source such as a laser
Dendrimers	Three-dimensional, nanoscale core-shell structures
Magnetic nanoparticles	Superparamagnetic iron oxide particles; display large magnetic moments in a magnetic field
Inorganic nanoparticles	Silica, titanium, or alumina nanoparticles Metallic nanoparticles ex: iron, gold, silver nanoparticles
Carbon nanotubes	Fullerene-related structures which consist of rolled grapheme sheets. Typically a few nanometers in diameter and several micrometers to centimeters long

This table listed the nanoparticles of different forms and their characteristics.

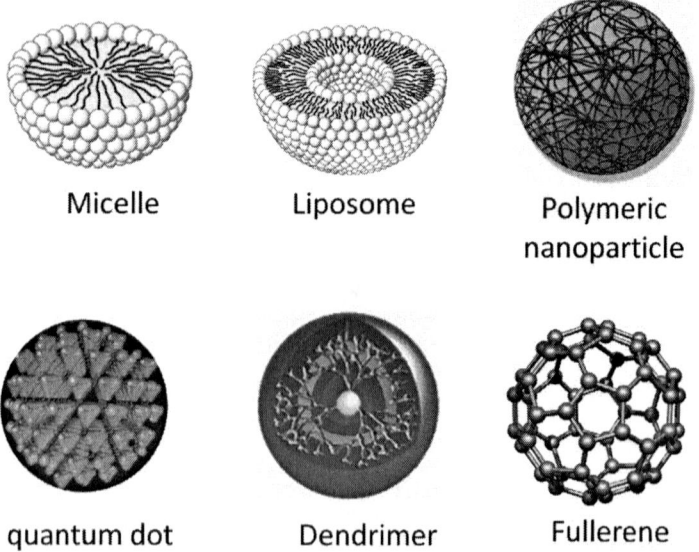

Micelle   Liposome   Polymeric nanoparticle

quantum dot   Dendrimer   Fullerene

Fig. 3. Different types of nanoparticles. There are different type of articial nanoparticles for drug delivery system.

Color image of this figure appears in the color plate section at the end of the book.

biodegradable polymers and fibers. They are used as "Trojan horses", the name of this strategy being derived from Greek mythology. Among others, Homer's *Iliad* recounts the story of Troy after a long siege by the Greeks. The Greeks built a hollow, wooden horse and offered it to the Trojans in the guise of a peace offering to their enemy. The Trojans took the large wooden horse inside their city walls and celebrated their victory. Early the next morning, Greek soldiers hidden in the horse emerged from its hollow belly, killed all the Trojan sentries and opened the city gates to their army, which successfully captured the city. Nanoparticles used as drug delivery systems consist of the active principle which is loaded into the NPs or adsorbed on its surface. The association of drugs can be made during the polymerisation process (the manufacture of the NPs), by adsorption onto pre-manufactured nanoparticles, or both (Mohanraj 2006).

The benefits of improving drug delivery by using nanoparticles are:

- Feasibility of incorporation of both hydrophilic and lipophilic substances
- Protection of drug from degradation
- High drug carrier capacity
- Improvement of the drug solubility
- Potential for sustained release
- Improvement of the bioavailability of drugs
- Improvement of the therapeutic properties of drugs
- Lowering of drug toxicity
- Target the specific site of action
- Intra-cellular delivery of the drug

The rapid clearance of circulating nanoparticles from the bloodstream coupled with their high uptake by liver and spleen has thus far been overcome by reducing the particle size and by making the particle surface hydrophilic with polymers such as polyethylene glycol (PEG), dextrane, poloxamers and poloxamines (Gaur et al. 2000). To reach the brain itself, nanoparticles injected intravenously must overcome different barriers. The first of these is the mononuclear phagocytic system (MPS). The MPS is a part of the immune system that consists of the phagocytic cells located in reticular connective tissue. The cells are primarily monocytes and macrophages, and they accumulate in the liver (more particularly the Kupffer cells), spleen and lymph nodes. These cells eliminate damaged cells and particles circulating in the blood, so they must be avoided by the NPs in order to arrive at the BBB. Researchers have attempted to produce "stealthy" nanoparticles in order to avoid blood opsonins whose role is to increase MPS clearance. Opsonins are any molecule that acts as a binding enhancer for the process of phagocytosis, for example by coating the surface of a particle. Main opsonins are antibodies or complement molecules. The most important are IgG and C3b. Long circulating nanoparticles or liposomes can be obtained by modifying their surface properties with dysopsonic polymers such as poly (ethylene glycol) (PEG) (Avgoustakis et al. 2003). These hydrophilic polymers can prevent the opsonin-nanoparticle interaction, which is the first step in the recognition of NPs by the immune system.

The second obstacle to the brain is the BBB. In order to cross the BBB and target the brain itself, these stealthy nanoparticles can be grafted with ligands specific to the BBB and allowing endocytosis and transcytosis (Andrieux and Couvreur 2009; Chang et al. 2009). Circulating nanoparticles in the blood must cross the BBB in order to diffuse within brain parenchyma and target specific cells such as neurons or tumor cells such as glioma derived from glial cells.

RECEPTOR MEDIATED TARGETING

The goal of this approach is to deliver therapeutics across the BBB into the brain parenchyma using vesicular transport systems. Receptor-mediated transcytosis (RMT, Fig. 4) allows the transport of peptides, for example endogenous nanoparticles such as LDL across the BBB (Dehouck et al. 1997). Upon binding, the receptor-ligand complex is internalized into the endothelial cell and the ligand is transcytosed using the vesicular pathway. Insulin, transferrin, insulin-like growth factors and other endogenous peptides employ the vesicular trafficking machinery of the brain's endothelial cells to transport substances between the blood and the

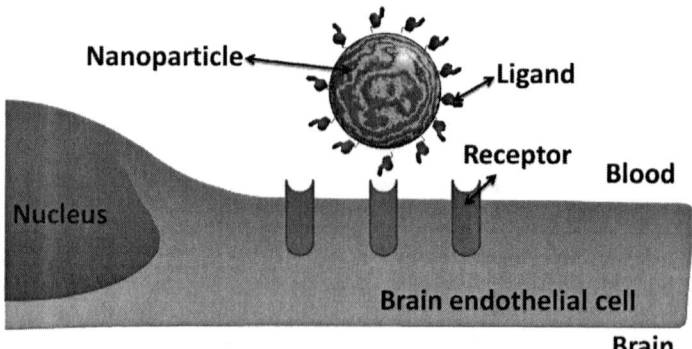

Fig. 4. Receptor mediated targeting of nanoparticles. The targeted nanoparticles are functionalized with an outer layer of receptor-specific ligands. After binding the receptor-ligand complex is endocytosed and potentially transcytosed to the brain.

Color image of this figure appears in the color plate section at the end of the book.

brain (Wang et al. 2009). Transferrin, or antibodies recognizing transferrin receptors are generally used to target nanoparticles to cells over-expressing transferrin receptor such as in the BBB.

TRANSFERRIN

Transferrin (Tf) (Fig. 5) is a single chain, 80 kDa protein and the naturally occurring ligand for the Tf receptor. The transferrin receptor (TfR) assists iron uptake into vertebrate cells through a cycle of endo- and exocytosis of transferrin. The TfR has been found in red blood cells, monocytes, the BBB and brain. In plasma and other extracellular fluids, Tf is present as a mixture of iron-free (apo-Tf), monoferric Tf, and diferric Tf (*holo*-Tf) forms. The relative abundance of each form depends on the concentrations of iron and Tf. In human serum, the concentration of transferrin is about 2.5 mg/ml (35 mM) with 30% containing iron (Fe). The primary function of plasma Tf is to act as an Fe carrier in the plasma and interstitial fluids of the body. In serving this function it accepts Fe released from cells and transports it to other cells where the Fe is taken up by TfR-mediated processes. The active step in this cycle is uptake by TfR-containing cells, and the distribution of plasma Fe to the various cells and tissues of the body is determined largely by their content of TfR. Targeting approaches via Tf ligand can involve conjugating drug molecules themselves or drug loading complexes permitting the targeting of cells expressing transferrin receptors. The endogenous ligand, Tf, is exploited in the targeting of drugs.

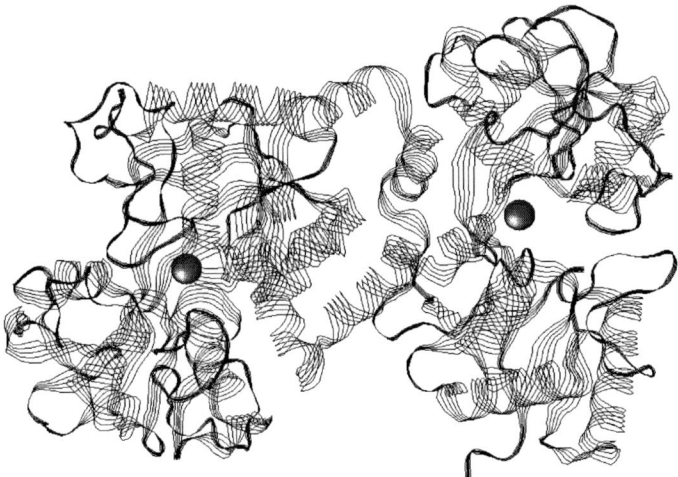

Fig. 5. Structure of *holo*-transferrin (●: Fe3+). Holo-transferrin is loaded with 2 iron atoms, one in each lobe. Internet site: http://pubpages.unh.edu/~ndc/

BRAIN TARGETING USING TRANSFERRIN OR ANTIBODIES TARGETED TO TRANSFERRIN RECEPTOR

The crossing of the BBB using a targeted or a non targeted nanoparticle has been demonstrated both *in vitro* and *in vivo* (Fenart et al. 1999; Huwyler et al. 1996; Jallouli et al. 2007). Evidence of nanoparticle transfer by transcytosis across the BBB on physiological models where endothelial cells are linked by tight junctions has been observed in different studies. The mechanisms involved in transferrin transcytosis through the BBB on physiological *in vitro* BBB models remain controversial as either the clathrin or caveolae pathway might be involved, depending on the model used (Chang et al. 2009; Fenart and Cecchelli 2003; Visser et al. 2004). A study performed on an *in vitro* BBB model using a *co*-culture of endothelial cells and astrocytes showed that 90-nm Tf-NPs enter the endothelial cells using a cholesterol-dependent pathway (Fig. 4). Moreover, the uptake of these NPs was increased 20-fold by Tf targeting (targeting was inhibited with excess Tf suggesting that these Tf-NPs enter the cells via the TfR). One study further suggests that the Tf-NPs can specifically interact with the BBB via the TfR and are highly endocytosed via the caveolae pathway (Chang et al. 2009).

Recently, Mahajan and colleagues observed a significant enhancement in the ability of quantum rods (QR)-Tf-Saquinavir to cross the BBB, along with a marked decrease in HIV-1 viral replication in the peripheral blood monocytes (Mahajan et al. 2010). There have been several studies on liposome for brain delivery by targeting TfR. Pegylated liposomes were used for delivery of the antineoplastic agent daunomycin to rat brain. OX26 is a monoclonal antibody (mAb) that recognizes the rat anti-transferrin receptor. OX26 antibodies were coupled to the terminal end of a PEG-conjugated linker lipid coating 85 nm liposomes to form immunoliposomes (antibody-directed liposomes) (Huwyler et al. 1996). The authors observed that coupling of thirty OX26 antibodies per liposome resulted in optimal brain delivery. Owing to the high loading capacity of these liposomes the authors concluded that the conjugation of OX26 mAb to PEG-liposomes greatly increases the carrying capacity of daunomycin by up to 4 logarithmic orders in magnitude. Gosk et al. studied the uptake after *in situ* perfusion of OX26 conjugated to PEG-coated liposomes by brain capillary endothelial cells (BCECs) in 18 day-old rats (Gosk et al. 2004). The uptake of PEG liposome conjugated OX26 by BCECs after a 15-minute perfusion was approximately 16 times higher than that of non-immune IgG2a (Ni-IgG2a). OX26 and OX26-conjugated liposomes selectively distributed to BCECs, leaving choroid plexus epithelium, neurons, and glia unlabeled. Ni-IgG2a and unconjugated liposomes did not reveal any labeling of BCECs. The failure to label neurons and glial shows that OX26 and OX26-conjugated liposomes did not pass through BCECs. These results reflect the complexity of this approach.

Nanoparticle-based formulations have several advantages compared with liposome formulations such as stability that can enable potent drug delivery across the BBB.

In vivo studies using Tf conjugated to nanoparticles of poly(lactide)-D-α-Tocopheryl polyethylene glycol succinate (PLA-TPGS) showed that imaging/therapeutic agents could be delivered across the BBB in rats (Gan and Feng 2010). Moreover, human serum albumin nanoparticles (HSA NPs) coupled to transferrin or transferrin receptor monoclonal antibodies (OX26 or R17217) exhibited significant loperamide transport across the BBB into the brain and produced strong anti-nociceptive effects (Ulbrich et al. 2009). In rats, poly (ethyleneglycol)-poly (epsilon-caprolactone) (PEG-PCL) polymersomes conjugated with OX26 (~100 nm) loaded peptide (NC-1900) improved the scopolamine-induced learning and memory impairments in a water maze task via intravenous administration (Pang et al. 2008). The inhibition of the caspase-3 enzyme is reported to increase neuronal cell survival following cerebral ischemia. Studies in rats showed that OX26-chitosan-PEG was able to reach the brain of healthy animals and deliver anticaspase peptide after intravenous injection, suggesting a potential for the treatment of ischemia (Aktas et al. 2005). When given intravenously, the caspase-3 inhibitor-loaded chitosan nanospheres, conjugated with an anti-mouse transferrin receptor monoclonal antibody, were rapidly transported across the BBB without being measurably taken up by liver or spleen, and dose-dependently decreased infarct volume, neurological deficit, and ischemia induced caspase-3 activity in mice subjected to focal cerebral ischemia/reperfusion (Karatas et al. 2009).

Applications to Areas of Health and Disease

The treatment of neurodegenerative diseases (such as Parkinson's, Alzheimer's, and Huntington's) and of brain tumors is hampered by poor drug delivery across the BBB. The development of new drugs for treating brain diseases has not kept pace with progress in the molecular neurosciences, primarily because the majority of new drugs discovered do not cross the BBB. BBB drug targeting technology can be built around a knowledge base of specific endogenous transporters within the brain capillary endothelium, which forms the BBB *in vivo*.

TARGETING GLIOMA

There are four basic criteria for choosing targeting ligand for cancer therapeutic applications: (1) the receptor is overexpressed by tumor cells, (2) the receptor participates as a principle component in the progression of the disease, (3) the receptor is stable in its present form upon the tumor cell surface, and (4) the receptor is expressed by a large percentage of tumor cells and a large variety of tumors. While BBB-targeted vector-therapeutic conjugates may be able to cross the BBB, generally they also need to target a specific population of cells in the brain. As transferrin receptor is overexpressed at the BBB and in glioma, the same ligand can potentially be used to target both the BBB and tumor cells in the brain.

Glioma is the most common primary malignant brain tumor and represents a morphologically diverse group of tumors. These tumors are rapidly proliferating and leave patients with a mean survival of only 12 months after diagnosis. Although the interest for treating this devastating disease has intensified the research in drug discovery and drug delivery to the brain, the prognosis for patients with malignant brain tumor remains poor. Using this strategy Ren and colleagues showed that Tf functionalized PEG-PLA nanoparticles could cross BBB and penetrate glioma in a rat model (Ren et al. 2010). Moreover, prolonged average survival time of glioma seeded rats up to 88.37% was obtained with camustin-loaded Tf-PLA NPs (Kang et al. 2009). Furthermore, paclitaxel loaded transferrin PLA-PEG nanoparticles increased markedly anti-tumoral activity of paclitaxel to brain glioma cells (Pulkkinen et al. 2008).

Summary

- The delivery of macromolecules and non-lipophilic compounds to the brain is severely limited by the tightly apposed capillary endothelial cells that form the blood-brain barrier (BBB).
- Brain capillary endothelial cells possess specific receptor-mediated transport mechanisms that potentially can be exploited as a means to deliver therapeutic molecules to the brain across the BBB.
- By modifying the surface of drug loaded nanoparticles with targeting ligands, specially, transferrin and transferrin receptor antibodies, it is possible to improve drug delivery to the brain and thence therapeutic effect.
- The research of specific BBB ligands is necessary to improve effective targeting of brain tissue by nanomedicines.

Abbreviations

BBB	:	Blood-brain barrier
BCECs	:	Brain capillary endothelial cells
CNS	:	Central nervous system
IGF	:	Insulin-like growth factor
LDL	:	Low-density lipoprotein
LNC	:	Lipid nanocapsules
mAb	:	Monoclonal antibody
MPS	:	Mononuclear phagocytic system
NPs	:	Nanoparticles
PCL	:	Poly (epsilon-caprolactone)
PEG	:	Poly (ethyleneglycol)
PLA	:	Poly (lactide)
PLGA	:	Poly (D, L-lactide-co-glycolide)
QR	:	Quantum rods
RMT	:	Receptor mediated transcytosis
Tf	:	Transferrin
TfR	:	Transferrin receptor

References

[ISO/TS 27687, 2008] International Organization for Standardization/Technical Specification 27687. Nanotechnologies–terminology and definitions for nano-objects–nanoparticle, nanofibre and nanoplate. 2008. International Organization for Standardisation.

Aktas, Y., M. Yemisci, K. Andrieux, R.N. Gürsoy, M.J. Alonso, E. Fernandez-Megia, R. Novoa-Carballal, E. Quiñoá, R. Riguera, M.F. Sargon, H.H. Celik, A.S. Demir, A.A. Hincal, T. Dalkara, Y. Capan and P. Couvreur. 2005. Development and brain delivery of chitosan-PEG nanoparticles functionalized with the monoclonal antibody OX26. Bioconjug. Chem. 16: 1503–1511.

Andrieux, K. and P. Couvreur. 2009. Polyalkylcyanoacrylate nanoparticles for delivery of drugs across the blood-brain barrier. Wiley Interdiscip. Rev. Nanomed. Nanobiotechnol. 1: 463–474.

Avgoustakis, K., A. Beletsi, Z. Panagi, P. Klepetsanis, E. Livaniou, G. Evangelatos and D.S. Ithakissios. 2003. Effect of copolymer composition on the physicochemical characteristics, *in vitro* stability, and biodistribution of PLGA-mPEG nanoparticles. Int. J. Pharm. 259: 115–127.

Bazile, D.V., C. Ropert, P. Huve, T. Verrecchia, M. Marlard, A. Frydman, M. Veillard and G. Spenlehauer. 1992. Body distribution of fully biodegradable [14C]-poly(lactic acid) nanoparticles coated with albumin after parenteral administration to rats. Biomaterials 13: 1093–1102.

Cao, G. 2004. Nanostructures & Nanomaterials: Synthesis, Properties & Applications. Imperial College Press, London, UK.

Chang, J., Y. Jallouli, M. Kroubi, X.B. Yuan, W. Feng, C.S. Kang, P.Y. Pu and D. Betbeder. 2009. Characterization of endocytosis of transferrin-coated PLGA nanoparticles by the blood-brain barrier. Int. J. Pharm. 379: 285–292.

Dehouck, B., L. Fenart, M.P. Dehouck, A. Pierce, G. Torpier and R. Cecchelli. 1997. A new function for the LDL receptor: transcytosis of LDL across the blood-brain barrier. J. Cell Biol. 138: 877–889.

Fenart, L., A. Casanova, B. Dehouck, C. Duhem, S. Slupek, R. Cecchelli and D. Betbeder. 1999. Evaluation of effect of charge and lipid coating on ability of 60-nm nanoparticles to cross an *in vitro* model of the blood-brain barrier. J. Pharmacol. Exp. Ther. 291: 1017–1022.

Fenart, L. and R. Cecchelli. 2003. Protein transport in cerebral endothelium. *In vitro* transcytosis of transferrin. Methods Mol. Med. 89: 277–290.

Gan, C.W. and S.S. Feng. 2010. Transferrin-conjugated nanoparticles of poly(lactide)-D-alpha-tocopheryl polyethylene glycol succinate diblock copolymer for targeted drug delivery across the blood-brain barrier. Biomaterials 31: 7748–7757.

Gaur, U., S.K. Sahoo, T.K. De, P.C. Ghosh, A. Maitra and P.K. Ghosh. 2000. Biodistribution of fluoresceinated dextran using novel nanoparticles evading reticuloendothelial system. Int. J. Pharm. 202: 1–10.

Gosk, S., C. Vermehren, G. Storm and T. Moos. 2004. Targeting anti-transferrin receptor antibody (OX26) and OX26-conjugated liposomes to brain capillary endothelial cells using in situ perfusion. J. Cereb. Blood Flow. Metab. 24: 1193–1204.

Hay, R.J. 1994. Liposomal amphotericin B, AmBisome. J. Infect. 28 Suppl 1: 35–43.

Huwyler, J., D. Wu and W.M. Pardridge. 1996. Brain drug delivery of small molecules using immunoliposomes. Proc. Natl. Acad. Sci. USA 93: 14164–14169.

Jallouli, Y., A. Paillard, J. Chang, E. Sevin and D. Betbeder. 2007. Influence of surface charge and inner composition of porous nanoparticles to cross blood-brain barrier *in vitro*. Int. J. Pharm. 344: 103–109.

Kang, C., X. Yuan, Y. Zhong, P. Pu, Y. Guo, A. Albadany, S. Yu, Z. Zhang, Y. Li, J. Chang and J. Sheng. 2009. Growth inhibition against intracranial C6 glioma cells by stereotactic delivery of BCNU by controlled release from poly(D,L-lactic acid) nanoparticles. Technol. Cancer Res. Treat 8: 61–70.

Karatas, H., Y. Aktas and Y. Gursoy-Ozdemir, E. Bodur, M. Yemisci, S. Caban, A. Vural, O. Pinarbasli, Y. Capan, E. Fernandez-Megia, R. Novoa-Carballal, R. Riguera, K. Andrieux, P. Couvreur and T. Dalkara. 2009. A nanomedicine transports a peptide caspase-3 inhibitor across the blood-brain barrier and provides neuroprotection. J. Neurosci. 29: 13761–13769.

Mahajan S.D., I. Roy, G. Xu, K.T. Yong, H. Ding, R. Aalinkeel, J. Reynolds, D. Sykes, B.B. Nair, E.Y. Lin, P.N. Prasad and S.A. Schwartz. 2010. Enhancing the delivery of anti retroviral drug "Saquinavir" across the blood brain barrier using nanoparticles. Curr. HIV Res. 8: 396–404.

Mohanraj, V.J. and Y. Chen. 2006. Nanoparticles—A Review. Tropical Journal of Pharmaceutical Research 5: 561–573.

Pang, Z., W. Lu, H. Gao, K. Hu, J. Chen, C. Zhang, X. Gao, X. Jiang and C. Zhu. 2008. Preparation and brain delivery property of biodegradable polymersomes conjugated with OX26. J. Control Release 128: 120–127.

Pardridge, W.M. 1999. Blood-brain barrier biology and methodology. J. Neurovirol. 5: 556–569.

Pardridge, W.M. 2002. Blood-brain barrier drug targeting enables neuroprotection in brain ischemia following delayed intravenous administration of neurotrophins. Adv. Exp. Med. Biol. 513: 397–430.

Prior, R., G. Reifenberger and W. Wechsler. 1990. Transferrin receptor expression in tumours of the human nervous system: relation to tumour type, grading and tumour growth fraction. Virchows. Arch. A. Pathol. Anat. Histopathol. 416: 491–496.

Pulkkinen, M., J. Pikkarainen, T. Wirth, T. Tarvainen, V. Haapa-aho, H. Korhonen, J. Seppälä and K. Järvinen. 2008. Three-step tumor targeting of paclitaxel using biotinylated PLA-PEG nanoparticles and avidin-biotin technology: Formulation development and *in vitro* anticancer activity. Eur. J. Pharm. Biopharm. 70: 66–74.

Ren, W.H., J. Chang, C.H. Yan, X.M. Qian, L.X. Long, B. He, X.B. Yuan, C.S. Kang, D. Betbeder, J. Sheng and P.Y. Pu. 2010. Development of transferrin functionalized poly(ethylene glycol)/poly(lactic acid) amphiphilic block copolymeric micelles as a potential delivery system targeting brain glioma. J. Mater. Sci. Mater. Med. 21: 2673–2681.

Roberts R.L., R.E. Fine and A. Sandra. 1993. Receptor-mediated endocytosis of transferrin at the blood-brain barrier. J. Cell Sci. 104 : 521–532.

Ulbrich, K., T. Hekmatara, E. Herbert and J. Kreuter. 2009. Transferrin- and transferrin-receptor-antibody-modified nanoparticles enable drug delivery across the blood-brain barrier (BBB). Eur. J. Pharm. Biopharm. 71: 251–256.

Visser, C.C., S. Stevanovic, L. Heleen Voorwinden, P.J. Gaillard, D.J. Crommelin, M. Danhof and A.G. De Boer. 2004. Validation of the transferrin receptor for drug targeting to brain capillary endothelial cells *in vitro*. J. Drug Target. 12: 145–150.

Wang, Y.Y., P.C. Lui and J.Y. Li. 2009. Receptor-mediated therapeutic transport across the blood-brain barrier. Immunotherapy 1: 983–993.

14

Therapeutic-loaded Lipid Nanostructures and Brain Diseases

Maria Luisa Bondì[1,*] *and Emanuela Fabiola Craparo*[2]

ABSTRACT

Central Nervous System (CNS) diseases represent the largest and fastest growing area of unmet medical need since an alarming increase in brain disease incidence is going on. Despite major advances in neuroscience, many potential therapeutic agents are denied access to the CNS because of the existence of a physiological low permeable barrier, the Blood-Brain Barrier (BBB). To obtain an improvement of drug CNS performance, sophisticated approaches such as nanoparticulate systems are rapidly developing. In particular, in this chapter, the most recent data demonstrating the potential of lipid nanostructures, such as Solid Lipid Nanoparticles (SLN) and Nanostructured Lipid Carriers (NLC), to transport drugs successfully into the brain

---

[1]Istituto per lo Studio dei Materiali Nanostrutturati (ISMN), Consiglio Nazionale delle Ricerche, via Ugo La Malfa, 153, 90146 Palermo, Italy; E-mail: marialuisa.bondi@ismn.cnr.it

[2]Dipartimento di Scienze e Tecnologie Molecolari e Biomolecolari (STEMBIO), Università di Palermo, via Archirafi, 32– 90123 Palermo, Italy; E-mail: emanuela.craparo@unipa.it

*Corresponding author

List of abbreviations after the text.

for the treatment of CNS diseases including Alzheimer's and Parkinson's diseases, cancer, mood disorder, AIDS, and bacterial infections, are summarised. Their use as drug delivery systems is associated with many advantages that include an excellent storage stability, a relatively easy production without the use of any organic solvent, the possibility of steam sterilization and lyophilization, and large scale production. Moreover, SLN and NLC are obtained by using physiologically well-tolerated ingredients already approved for pharmaceutical applications in humans and show low toxicity when administered. Because of their small size, these systems may be injected intravenously and avoid the uptake of macrophages of mononuclear phagocyte system (MPS). Moreover, their lipophilic features lead them to CNS by an endocytotic mechanism, overcoming the BBB.

INTRODUCTION

In the last decade, an emerging interest has been growing towards brain targeting of therapeutic agents (Denora et al. 2009; Alam et al. 2010; Craparo et al. in press). In fact, the delivery of drugs to brain diseases is limited by the presence of the Blood-Brain Barrier (BBB). The BBB is a physical and physiologic barrier separating blood from the extracellular fluid in the brain parenchyma. It regulates the influx and efflux of biological molecules between the bloodstream and the brain, either by paracellular or transcellular pathways, thus protecting the brain from systematically circulating potentially cytotoxic agents and maintaining its homeostasis (Pardridge 2005). An understanding of the specific mechanisms of the brain capillary endothelium has led to the development of various strategies to enhance the penetration of drugs into the brain tissue. The pharmaceutical technology contribution to the brain diseases area has been found of paramount importance. Several targeted drug delivery systems (TDDS) and strategies have been developed for the sole purpose of an effective drug deposition to the Central Nervous System (CNS). Among these, lipid nanostructures capable of penetrating into brain capillary endothelial cells and cerebral cells have shown promising potential in brain therapy.

Lipid nanostructures are nanosized systems ranging in size from 1 to 500 nm and are classified into two main groups: Solid Lipid Nanoparticles (SLN) and Nanostructured Lipid Carriers (NLC); moreover, they have a great potential to carry lipophilic or hydrophilic drugs, and diagnostics (Kaur et al. 2008). Size, surface charge, morphology and matrix composition play a fundamental role for their *in vivo* fate and all these features can be modulated accordingly to obtain suitable modified release profile, biological activity and compatibility. Lipid nanoparticles are administrable for all routes even for intravenous route being smaller than the size of the smallest vessels. Their use as TDDS is associated with many advantages that include biocompatibility/biodegradability, an excellent storage stability, a relatively easy production without the use of any organic solvent, the possibility of steam sterilization and lyophilization, and large scale production. The surface modification with hydrophilic polymers improves their pharmacokinetic behaviour with an increase of the mean residence time in bloodstream (Beduneau et al. 2007). In order to optimize their biodistribution, these vectors can be surface decorated by site-specific biomolecules recognizing target tissue. These latter biomolecules focus on nanocarriers able to deliver specifically drugs to cerebral tissues by active transport mechanisms (Koziara et al. 2004).

THE BBB AND TRANSPORT OF NANOPARTICLES ACROSS IT

Normal BBB

The normal BBB is a highly effective physical and physiological barrier that regulates the CNS homeostasis and thereby controls the delivery of drugs to the brain. It results from the unique properties of endothelial cells (Olivier and de Oliviera 2007): i) the tight junctions that connect adjacent endothelial cells and physically restrict solute flux between the blood and the brain (Yang 2010); ii) an elaborate system of transport proteins that allows the selective influx transport of hydrophilic solutes and macromolecules; iii) a metabolic barrier which serves as a biotransformation and detoxification systems and (iv) a negligible pinocytotic activity. To summarize, the complex structure of the BBB, the presence of high levels of efflux transport proteins, including P-gp and Multidrug Resistance Protein-1 (MRP-1), and the expression of metabolic enzymes pose hurdles for drug-brain entry. Then, the BBB impedes the access of a large number of diagnostic and therapeutic agents (belonging to classes such as antibiotics, anti-inflammatory and antineoplastic agents, and a variety of CNS–active drugs, especially neuropeptides) into the brain parenchyma in physiological and, often, in pathological conditions. One of the possibilities of bypassing this barrier relies on specific properties of nanoparticulate vectors designed to interact with BBB-forming cells at a molecular level, as a result of which the transport of drugs or other molecules (such as nucleic acids, proteins or imaging agents) could be achieved without interfering with the normal function of the brain.

BBB Alteration in Pathology

In various brain pathologies, including stroke, septic encephalopathy or neurodegenerative diseases, the alterations of the BBB permeability results in leakage of normally restrained plasma components and contribute to neuro-inflammation and neuronal damage (Olivier and de Oliviera 2007). It is still controversial whether the BBB permeability increase results from the opening of passageways through brain endothelial cells or from tight junction degradation, and little is known on the degree of permeabilization in inflammatory brain pathologies, but its actual impact on nanoparticulate passive diffusion is likely to be low.

In the case of brain tumours, the BBB integrity is locally compromised by the absence of thigh junctions, allowing for tumour core penetration and retention of drug macromolecules or nanoparticulate systems otherwise excluded from normal brain (Olivier and de Oliviera 2007). This phenomenon is known as the enhanced permeability and retention-EPR-effect of tumours.

NOVEL APPROACHES FOR BRAIN TARGETING

The field of novel drug delivery has fully emerged and has come into existence as an ideal approach of drug targeting to brain (Alam et al. 2010). It includes mainly the use of small colloidal particles due to advantages such as sub-cellular size and the possibility to employ biocompatible and biodegradable materials for their realization (Kaur et al. 2008). Thus, the pharmaceutical strategies based on nanotechnology represent a valid approach to achieve brain drug release. Various pharmaceutical nanocarriers, such as lipid nanoparticles, polymeric nanoparticles, micelles and some others were developed to achieve brain drug release (Denora et al. 2009; Gabathuler 2010; Craparo et al. in press; Fig. 1).

For the majority of these systems, a long circulating time does not ensure an effective BBB overcome either in physiological or pathological condition and, therefore, additional carrier features are needed. In this respect, exploitation of specific carrier mediated transport systems on the BBB can be considered as a major strategy for delivering drug-loaded colloidal carriers as well as drugs to the CNS (Gabathuler 2010). As an example it could be cited the development of brain-targeted SLN by coating the surface with thiamine, that facilitate binding and/or association of SLN with BBB thiamine transporters (Lockman et al. 2003).

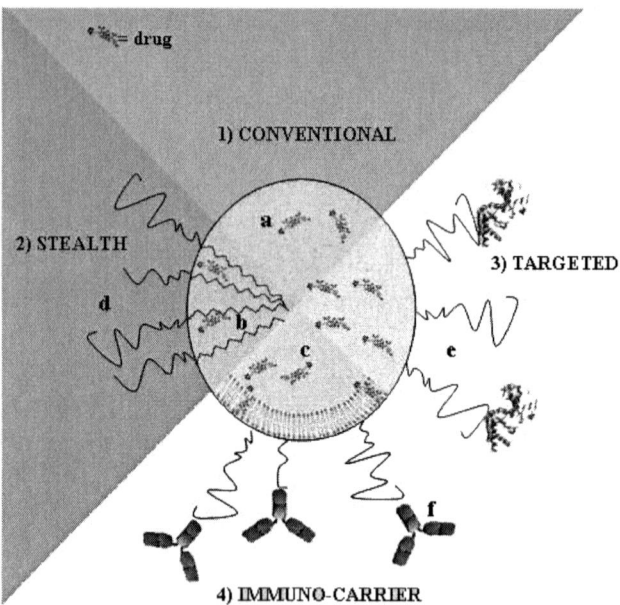

Fig. 1. Pharmaceutical nanocarriers. The schematic structure of pharmaceutical nanoparticulate carriers (a–polymeric or lipid nanoparticles; b–polymeric micelles, and c–liposomes). 1–Conventional nanocarrier; 2–long-circulating nanocarrier (d–surface-attached protecting polymer, usually PEGs, allowing for prolonged circulation of the nanocarrier in the bloodstream); 3–targeted nanocarrier (e–specific targeting ligand attached to the carrier surface); 4–immunocarrier (f–specific monoclonal antibody linked to the carrier surface). Adapted with permission from (Craparo et al. in press). Copyright (2010) Blackwell.

Color image of this figure appears in the color plate section at the end of the book.

LIPID NANOSTRUCTURES

Lipid-based nanostructures, such as SLN, are an important class of colloidal systems in the area of modified drug delivery technology (Joshi and Müller 2009). These are generally made up of solid lipids and natural surfactants. More recently, modification of SLN, the so-called NLC, have been obtained by mixing spatially different lipids or blending solid lipids with liquid lipids, which show a higher drug loading capacity and a longer term stability during storage than SLN. Important advantages deriving from the use of either SLN or NLC include the possibility to control drug release and targeting, and to increase drug stability. Moreover, these do not have toxicity or biodegradability problems, being obtained from physiological and biodegradable/biocompatible lipids (triglycerides, fatty acids, steroids and waxes, etc.), solid at body temperature, such as trimyristin, cetyl palmitate, glyceryl behenate, stearic acid, caprylic/capric triglyceride. The most common stabilizers employed are ionic and non-ionic molecules, such as lecithins, polysorbates, poloxamers, derivatized fatty acids and their combinations (Denora et al. 2009; Joshi and Muller 2009).

Lipid nanostructures have been investigated as possible carriers for modified drug delivery and targeting after intravenous administration or as oral drug carriers to overcome administration problems of drugs that are instable in the gastrointestinal tract or are inadequately absorbed (Yang et al. 1999; Joshi and Müller 2009). Furthermore, the commercial feasibility to produce SLN on a large scale at low cost and excellent reproducibility performing high pressure homogenization (HPH) method is not less important (Souto and Muller 2007; Almeida and Souto 2007; Denora et al. 2009; Bondì and Craparo 2010). SLN and NLC can also be formulated by using the warm oil-in-water (o/w) micro-emulsion and the solvent emulsification/evaporation or diffusion techniques (Joshi and Müller 2009; Fig. 2). The warm o/w micro-emulsion technique (Fig. 2A) is favourable when substances are unstable to the high mechanical stress produced by HPH. However, using this method, huge amounts of surfactants and co-surfactants have to be used. In the solvent emulsification/evaporation (Fig. 2B), an organic solvent immiscible with water must be used to dissolve lipids. Either the hot and the cold homogenization techniques (Fig. 2C) have in common an initial step, consisting of the lipid melting at approximately 5–10°C above the mixture melting point, although larger particles sizes and a broader size distribution are observed in cold homogenized sample, compared to the hot homogenization.

SIZE AND SURFACE PROPERTIES

Chemical-physical characteristics (mainly size and surface properties) of lipid nanostructures could be easily modified in order to modulate the body distribution hence increasing bioavailability into CNS of the loaded drug. In fact, since nanocarriers for the body defence system usually represent foreign particles, they become easily opsonised and removed by phagocytosis from the circulation prior to complete their function. This process depends mainly on particle size, charge and surface properties of the nanocarriers. In general, size not exceeding a maximum diameter of 200 nm, neutral surface charge, sphericity and an adequate deformability are crucial peculiarities to ensure prolonged blood circulation and a relatively low rate of mononuclear phagocyte system (MPS) uptake. Moreover, chemical modification of nano-carrier surface with certain hydrophilic polymers, such as poly(ethylene glycols) (PEGs) and/or the

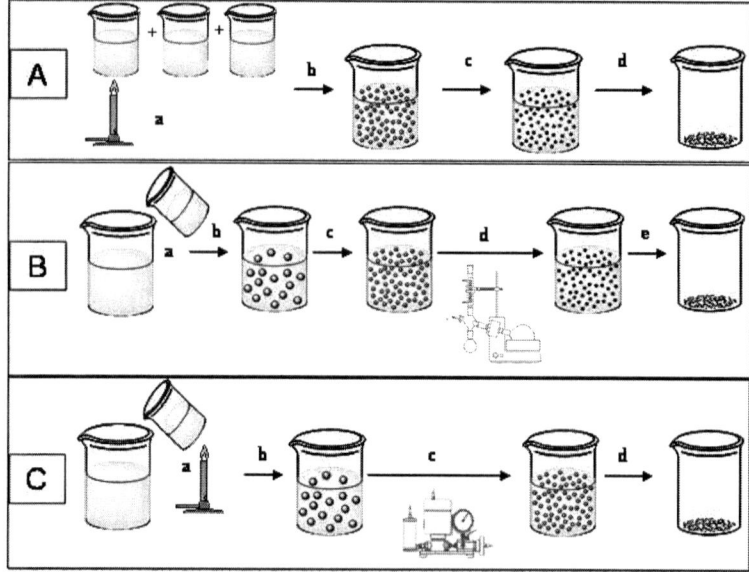

Fig. 2. SLN preparation techniques. Schematic diagrams for the preparation of SLN and NLC. A) o/w microemulsion technique: a) heating the lipid to around 10°C its melting point, b) addition of surfactant aqueous solution(s) to the melted lipid; c) dispersion of the warm microemulsion in cold water (2–3°C); d) purification and freeze-drying procedure. B) solvent emulsification/evaporation technique: a) dissolution of the lipid in the organic solvent immiscible in water; b) addition of the organic solution in an aqueous phase containing the surfactant(s); c) sonication of the obtained emulsion; d) evaporation by rotavapor to remove organic solvent; e) purification and freeze-drying procedure. C) hot HPH technique: a) heating the lipid to around 10°C its melting point; b) addition of the melted lipid in a hot aqueous phase containing the surfactant(s) and formation of the primary emulsion by vigorous stirring with a high-speed stirrer; c) homogenization in a heated homogenizer for several homogenization cycles; d) purification and freeze-drying procedure. (Unpublished material of the authors).

use of a surfactant (such as Polysorbate and Epikuron), represents the most frequent way to stabilize them sterically by a polymer-mediated protection and to impart the *in vivo* longevity to drug carriers (Juillerat-Jeanneret 2008; Park et al. 2008). PEG chains can be physically adsorbed on the particle surface or covalently bonded to fully formed lipid nanocarriers by using PEGylated lipids as starting material (Hu et al. 2007; Denora et al. 2009).

In order to increase the accumulation specifically in the required pathological site, nanocarriers may be actively targeted by incorporating surface grafted recognition moieties that impart an affinity for cellular receptors or components that are present on and/or up-regulated by specific cells (Byrne et al. 2008).

SLN BRAIN RETENTION AND ACCUMULATION

Generally, when SLN reach the blood stream, these may avoid the macrophage uptake due to their small size and surface characteristics. In addition, thanks to their lipophilic features, these systems could reach the CNS overcoming the BBB by an endocytosis or a transcytosis mechanism which occurs in the endothelial cells lining the brain blood capillaries, or SLN could also permeate the tight junctions between the endothelial cells (Denora et al. 2009; Joshi and Müller 2009). The absorption of a plasmatic protein, the Apoliprotein E, onto nanoparticle surface is supposed to be responsible for the brain SLN uptake by mediating its adherence to the endothelial cells of the BBB (Blasi et al. 2007; Joshi and Müller 2009). The transport mechanisms of free drugs or drug-loaded SLN across the BBB are reported in Fig. 3.

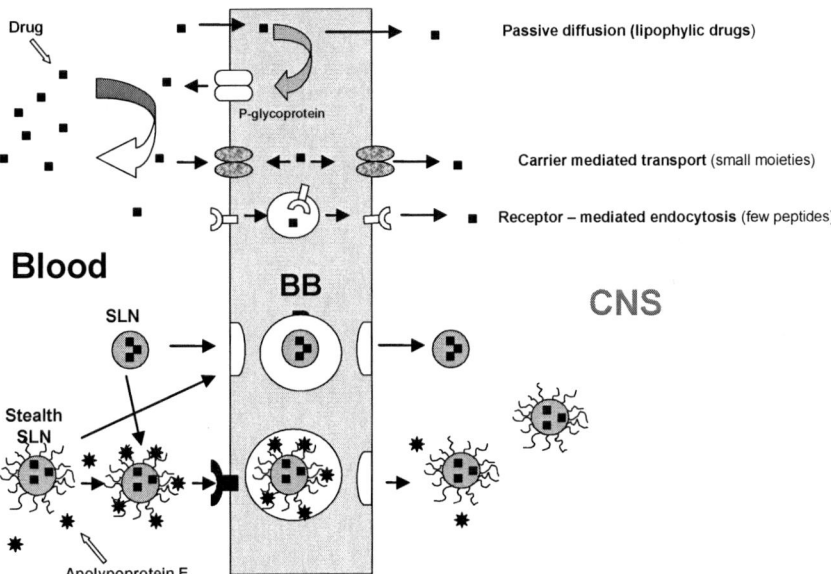

Fig. 3. Transport mechanisms across the BBB. The transport mechanisms of free drugs or drug-loaded SLN and NLC across the BBB. (Unpublished material of the authors).

Moreover, retention of SLN in brain–blood capillaries with absorption to capillary walls could occur and create a higher drug concentration gradient leading to enhanced drug transport across the endothelial cells.

ADMINISTRATION ROUTES

A majority of CNS drugs and brain drug delivery systems are administered systemically, most likely via intravenous (i.v.) injection because of the ease of application and avoidance of the first pass effect (Sarin 2009). Following systemic administration, nanoparticles can follow either receptor-mediated transcytosis (RMT) or adsorptive-mediated transcytosis (AMT) to get into the brain. Intranasal drug administration has been investigated recently as an alternative to i.v. administration due to the large surface area available for drug absorption, porous endothelial membrane, high blood flow, the avoidance of drug loss in first pass metabolism and mainly because it allows direct access of drugs from the nasal cavity to the brain via the olfactory epithelium by circumventing the BBB (Yang 2010). This route offers a new means for the long-term non-invasive management of chronic neurological disorders.

SLN AND NEURODEGENERATIVE DISORDERS

With an increase in lifespan and changing population demographics, the incidence of CNS diseases is expected to increase significantly in the 21st century. The most challenging of the CNS diseases are neurodegenerative such as Alzheimer's and Parkinson's diseases, characterized by age-related gradual decline in neurological function, often accompanied by neuronal death, which have been well described in terms of disease mechanisms and pathology (Thomas and Mansoor 2009). However, successful treatment strategies for neurodegenerative diseases have so far been limited.

Advances in novel colloidal systems can open a new perspective for effective brain delivery of a vast variety of drugs. Some successful examples of drug incorporation into lipid based nanoparticles are listed below.

Bromocriptine, an anti-parkinson agent, was encapsulated into NLC, and the *in vivo* efficacy of these formulations was tested on 6-hydroxydopamine hemilesioned rats which are models for Parkinson's disease, by intraperitoneal injection (Joshi and Muller 2009). The results indicated that Bromocriptine-loaded NLC have rapid onset of action as compared to that of drug solution. Furthermore, the anti-parkinson effect was retained for longer duration in case of NLC as compared to that of drug solution. This study demonstrated indirectly that higher brain levels of Bromocriptine were achieved after administration as NLC. However, suitable pharmacokinetic studies would be required to validate this observation.

Bondì and coworkers described the preparation of SLN loaded with ferulic acid (FA), a phenolic compound with a significant antioxidant activity potentially useful in the treatment of Alzheimer's disease (Bondì et al. 2009). FA-loaded SLN caused a reduced ROS production in cells, more evident than for cells treated with free FA, demonstrating that FA-loaded SLN possessed a higher protective activity than free FA against oxidative stress induced in neurons and suggested that these nanoparticles were excellent carriers to transport FA into the cells (Fig. 4).

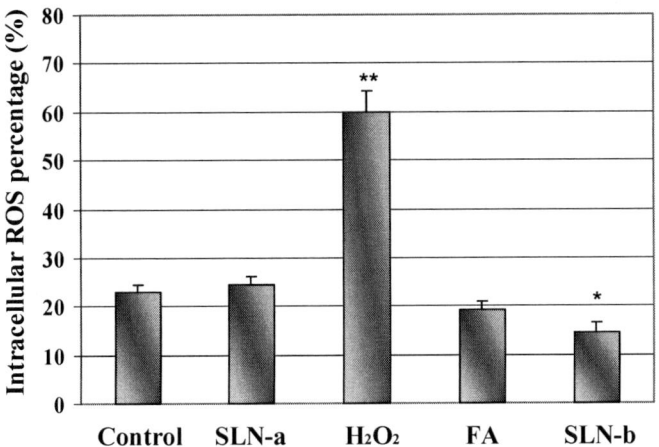

Fig. 4. **Effect of drug entrapment into SLN on oxidative stress.** ROS measurement in LAN5 cells untreated (control) and after incubation with FA (28 μM), H_2O_2 (positive control), empty SLN (sample SLN-a), and FA-loaded SLN (sample SLN-b) corresponding to 28 μM of FA. Data are the mean ± SD of three separate experiments, expressed as percentage of control value (* $p<0.05$, ** $p<0.02$). Adapted with permission from (Bondì et al. 2009). Copyright (2009) Bentham Science.

More recently, riluzole-loaded SLN with great potential as drug delivery systems for amyotrophic lateral sclerosis (ALS) were produced (Bondì et al. 2010). These particles showed a greater efficacy than free riluzole on rats that were immunised using the experimental allergic encephalomyelitis (EAE), and a smaller accumulation in not target organs of trapped riluzole than free one. The concentration of riluzole (expressed as μg riluzole·g^{-1} tissue weight) in brain, serum, liver, spleen, kidneys and lungs after intraperitoneal administration of free drug or blended into SLNs in rats 7th day after immunization is summarized in Figs. 5a and 5b.

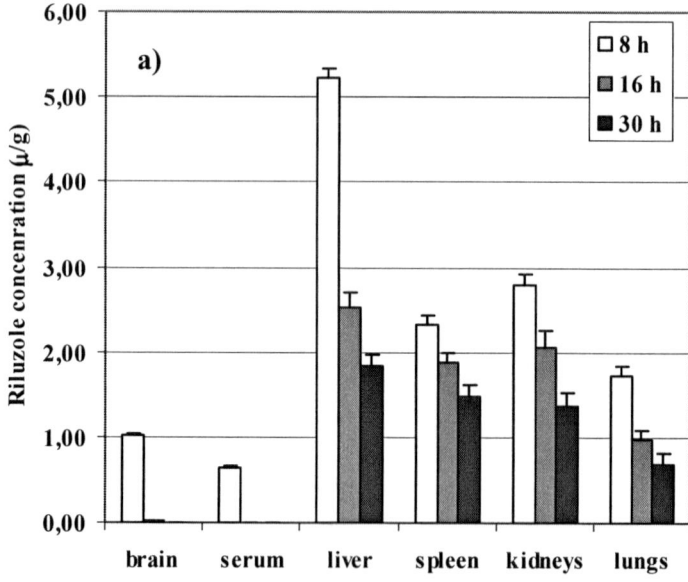

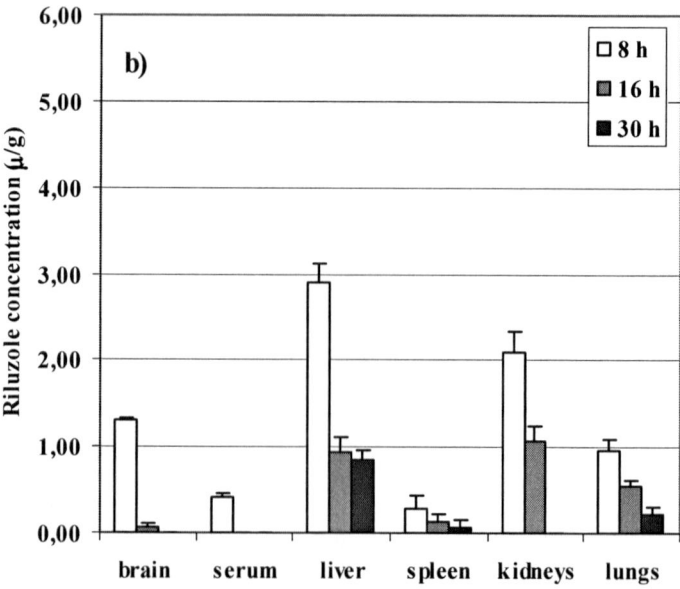

Fig. 5. Riluzole biodistribution. The riluzole levels (µg·g^{-1} tissue weight) in brain, serum, liver, spleen, kidneys and lung after a single intraperitoneal administration at 8 mg·kg^{-1} equivalent drug of (a) the free riluzole, (b) the riluzole-loaded SLN (mean ± SD and n = 4). Adapted with permission from (Bondì et al. 2010). Copyright (2010) Future Medicine.

SLN FOR BRAIN TUMOURS

Brain tumours constitute a profound and unsolved clinical problem although significant strides have been made in the treatment of many other cancer types (Koo et al. 2006; Brioshi et al. 2010). Despite recent improvements in surgical and adjuvant therapy for brain tumours, the multimodality approach currently used (surgery followed by adjuvant therapy such as radiation therapy, chemotherapy and photodynamic therapy) has limiting factors and does not produce a meaningful improvement in patient outcome (Koo et al. 2006). Conventional chemotherapy has shown a poor outcome due to the low permeability of most anti-cancer agents through the BBB. Moreover, it has dose-related side effects owing to non specific biodistribution of drugs. Passive and active brain targeting with lipid nanoparticulate carriers has emerged as a promising tool for brain chemotherapy due to the nanoparticle advantages such as the ability to improve the therapeutic index of drugs by preferential localization into the brain, lower distribution into healthy tissues and extended release rate, and the evidence for their ability to cross the BBB (Craparo et al. in press).

Yang and colleagues described the incorporation of campthotecin, an alkaloid with a peculiar anticancer mechanism of action, into stearic acid-based SLN and the body distribution of intravenously injected campthotecin-loaded SLN in mice (Yang et al. 1999). The result of this study showed the induction of a higher maximum concentration (>180%) and a better profile of the area under the concentration (AUC)/dose curve in the brain, heart and RES after i.v. administration of campthotecin-loaded SLN stabilized with Poloxamer 188 than the free solution.

Paclitaxel is an active chemotherapeutic agent used against malignant gliomas and brain metastases but, because of its systemic side effects, its clinical use is highly limited. Moreover, it does not cross the BBB and it is a substrate for MRP and P-gp. SLN loaded with paclitaxel have been investigated both *in vitro* and *in vivo* (Koziaka et al. 2004). The obtained SLN were smaller than 100 nm and the incorporation of paclitaxel into SLN has no effect on particle size. By *in vitro* cytotoxicity tests using two different cell lines on paclitaxel-loaded SLN as well as with a commercially available Taxol® formulation containing paclitaxel, it was demonstrated that the paclitaxel delivery by the SLN was efficient to overcome drug efflux at the BBB. The brain uptake studies in an *in situ* rat brain perfusion model suggested the significantly increased paclitaxel brain uptake due to the use of SLN.

Temozolomide (TMZ), an alkylating agent that can be administered orally, was also incorporated into SLN (Huang et al. 2008). The AUC/dose of TMZ-loaded SLN i.v. injected in healthy rabbits demonstrated much

Table 1. Summary of lipid nanostructures used as brain drug delivery and targeting.

Brain disease	Drug	Targeting ligand	Relevant results	Ref.
—	—	thiamine	Association of thiamine-targeted SLN with the BBB thiamine transporter, and accumulation at the BBB, increasing brain uptake during perfusion time frames in BALB/c mice.	Lockman et al. 2003
Parkinson	Bromocriptine	—	Bromocriptin-loaded NLC had rapid onset of action compared to that of drug solution in 6-hydroxylamine hemilesioned rats, and its effect was retained for longer duration.	Joshi and Muller 2009
Alzheimer	FA	—	FA-loaded SLN caused a reduced ROS production in cells and possessed a higher protective activity than free FA.	Bondi et al. 2009
Amyotrophic lateral sclerosis (ALS)	Riluzole	—	Riluzole entrapped into SLNs that reached the brain was 3-fold greater than free riluzole, 16h postinjection, and showed a smaller accumulation in the RES organs. In addition, rats treated with riluzole-loaded SLNs showed clinical signs of EAE later than those treated with free riluzole.	Bondi et al. 2010
Cancer	Campthotecin	—	The administration of campthotecin-loaded SLN stabilized with Poloxamer 188 induced of a higher maximum concentration and a better profile of the area under the concentration (AUC)/dose curve in the brain, heart and RES after i.v. administration than the free solution.	Yang et al. 1999
Cancer	Paclitaxel	—	Paclitaxel delivery by the SLN overcame drug efflux at the BBB. The brain uptake studies in an *in situ* rat brain perfusion model suggested the significantly increased paclitaxel brain uptake due to the use of SLN.	Koziaka et al. 2004
Cancer	Temozolomide	—	The AUC/dose of TMZ-loaded SLN i.v. injected in healthy rabbits demonstrated much higher and longer than those obtained with free drug solution, especially in brain and in RES organs.	Huang et al. 2008
Cancer	Doxorubicin	—	The mean peak plasma concentrations of doxorubicin with non-stealth and stealth SLN were higher than that with the doxorubicin solution by a factor of 5 and 7, respectively. Moreover, it was demonstrated that pegylated doxorubicin-loaded SLN reach the brain in larger amounts than non-stealth SLN.	Brioshi et al. 2010
Cancer	DO-FudR	—	The drug brain concentration (intended as FUdR or DO-FUdR) was higher at each time point for the group treated with the SLN formulations with respect to FUdR solution.	Wang et al. 2002

Cancer	VEGF-AS-ODN	—	The treatment with VEGF-AS-ODN-loaded SLN induced a downregulation of VEGF expression in neoplastic cells, effectively reaching both central and peripheral regions of the implanted tumours.	Brioshi et al. 2010
AIDS	D4T, delaviridine and saquinavir	—	The incorporation into SLN enhanced the BBB permeability of all these drugs.	Kuo and Su 2007
AIDS	Atazanavir	—	A higher drug cell uptake was due to the drug incorporation into SLN.	Chattopadhyay et al. 2008
Malaria	Quinine dihydrochloride	Tf	Enhancement of brain uptake of quinine dihydrochloride into Tf-targeted SLNs, as demonstrated by the recovery of a higher percentage of the dose from the brain after in vivo administration of Tf-coupled SLNs compared with unconjugated SLNs or drug solution.	Gupta et al. 2007
Psychosis	Clozapine	—	Positively charged clozapine SLN enhanced the bioavailability of clozapine from 3.1- to 4.5-fold on intraduodenal administration. The AUC of clozapine SLN showed higher uptake in RES organs and brain after intravenous administration than clozapine suspension.	Manjunath and Venkateswarlu 2005

Note: This table was never published before.

higher and longer than those obtained with free drug solution, especially in brain and in RES organs; moreover, the AUC ratio between the drug-loaded SLN and drug solution in the brain was the highest among the tested organs.

Anthracyclin antibiotics such as doxorubicin have general anticancer properties that include interaction with DNA by intercalation and DNA strand breakage. However, the chemical properties and the dose-related characteristics systemic side effects, such as cardiomyopathy and congestive heart failure, limited their use for the treatment of brain tumours. Non-stealth and stealth SLN were prepared with an average size between 60–100 nm in diameter (Brioshi et al. 2010). The pharmacokinetics and tissue distribution of doxorubicin were studied after i.v. administration of equivalent doses of commercial doxorubicin solution and doxorubicin incorporated SLN to conscious (healthy) rats. The mean peak plasma concentrations of doxorubicin with non-stealth and stealth SLN were higher than that with the doxorubicin solution by a factor of 5 and 7, respectively. In all rat tissues, except the brain, the amount of doxorubicin was always lower after the injection of the two samples than after the injection of the commercial solution. It was also demonstrated that pegylated doxorubicin-loaded SLN reach the brain in larger amounts than non-stealth SLN and that the brain drug concentrations increase proportionally to the percentage of stealth agent used in the formulation. However, the therapeutic potential of the SLN-based chemotherapy has not yet been investigated in tumor model.

Recently, Wang et al. have reported the synthesis of the lipophilic prodrug 3′,5′-dioctanoyl-5-fluoro-2′-deoxyuridine (DO–FUdR), to overcome the limited access of the drug 5-fluoro-2′-deoxyuridine (FUdR), and its incorporation into SLN (DO–FUdR SLN) (Wang et al. 2002). The drug brain concentration (intended as FUdR or DO–FUdR) was higher at each time point for the group treated with the SLN formulations with respect to FUdR solution. These results indicated that DO–FUdR SLN had a good (2 times the free drug) brain targeting efficiency *in vivo*, demonstrating, also for this drug, that the incorporation into SLN can improve the ability of the drug to penetrate through the BBB (Kaur et al. 2008; Fernandes et al. 2010). However, the pro-drug approach can be very useful in the development of SLN for brain drug targeting. In fact, not all the therapeutic molecules possess suitable characteristics to be incorporated efficiently into a lipid matrix. Moreover, as a result of the low particle uptake available at the BBB surface, SLN drug content may become a critical issue for the clinical success of this strategy.

Histological main characteristics of malignant brain tumours include extensive vascularization, which is related to the expression and up-regulation of the vascular endothelial growth factor A (VEGF-A).

An interesting study regarded the administration of VEGF antisense oligonucleotides (VEGF-AS-ODN) after incorporation into SLN, in order to down-regulate VEGF expression *in vivo* in the intracerebral rat glioma model (Brioschi et al. 2010). It was found that only the treatment with VEGF-AS-ODN-loaded SLN induced a down-regulation of VEGF expression in neoplastic cells, effectively reaching every part (both central and peripheral regions) of the implanted tumours.

SLN FOR BRAIN INFECTIONS

CNS is an important reservoir for the human immunodeficiency virus (HIV) and hence targeting of anti-HIV/ acquired immunodeficiency syndrome (AIDS) drugs to the brain has become a significant goal for drug therapy. The BBB prevents access of anti-HIV/AIDS drugs to the brain due to the tight endothelial cell junctions of the brain capillaries and the presence of efflux transporters on the cell surface (Gupta and Jain 2010). In a recent study, Kuo and Su demonstrated the enhanced transport of D4T, delaviridine and saquinavir across the BBB *in vitro* thanks to the use of either polymeric nanoparticles or SLN (Kuo and Su 2007). These investigated systems enhanced permeability of the drugs with higher permeabilities being attributed to smaller particle sizes.

In a further study, the brain targeting potential of SLN was also reported for the anti-HIV/AIDS drug, atazanavir (Chattopadhyay et al. 2008). In particular, spherical SLN with a high encapsulation efficiency of atazanavir (≥90%) were successfully prepared and characterized, and results from cell viability studies coupled with higher drug cell uptake suggested that these SLN can effectively deliver atazanavir to human brain endothelial cells, *in vitro*.

Quinine was the first effective treatment for malaria that is a mosquito-borne infectious disease caused by Plasmodium falciparum. Severe malaria can affect the brain and the rest of the CNS, leading to changes in the level of consciousness, convulsions and paralysis. Recently, Transferrin (Tf)-conjugated SLN with a great potential in the treatment of brain diseases such as cerebral malaria, were fabricated (Gupta et al. 2007). It was demonstrated by *in-vivo* performance studies that intravenous administration of quinine dihydrochloride solution resulted in much higher concentrations of drug in the serum than with quinine-loaded SLN. Moreover, decoration of SLN with Tf significantly enhanced the brain uptake of quinine which was shown by the recovery of a higher percentage of the dose from the brain following administration of Tf-coupled SLN compared with unconjugated SLN or drug solution.

SLN AND PSYCHIATRIC DISORDERS

Schizophrenia is a long-term mental illness that causes people to have false perceptions of the senses, see the world in a different way from the majority, or to withdraw socially and psychologically. Antipsychotic medication can reduce these symptoms in most people and improve their functioning.

Clozapine is an atypical antipsychotic medication prescribed for the treatment of psychosis associated with resistant schizophrenia. After incorporation in stearylamine-based SLN, this drug was successfully carried into the brain on intravenous and intraduodenal administration (Manjunath and Venkateswarlu 2005). In particular, stearylamine-containing clozapine SLN were found to give significantly higher plasma levels and AUC as compared to clozapine-loaded SLN without stearylamine and clozapine suspension. Furthermore, biodistribution studies indicated that the administration of stearylamine-containing clozapine SLN resulted in significantly higher amount of clozapine in brain as compared to that of clozapine-loaded SLN without stearylamine and clozapine suspension.

Application to Areas of Health and Diseases

In recent years, the amazing growth of Central Nervous System (CNS) diseases has generated enormous research efforts in an attempt to develop new therapeutic strategies. However, the main interest has been focused on the discovery of new therapeutic molecules rather than developing new approaches and systems to target actives to the brain. Lipid nanostructures, e.g., Solid Lipid Nanoparticles (SLN) and Nanostructured Lipid Carriers (NLC), may represent promising carriers towards the brain tissue of a vast variety of therapeutic agents, including analgesics, antitubercular, anticancer, antiageing, antianxiety, neuroleptics, antibiotics and antiviral agents, since their prevalence over other formulations in terms of toxicity, production feasibility, and scalability is widely documented in the literature (Kaur et al. 2008; Joshi and Muller 2009). The suitability of these carriers has been proven because they are either already on the market or they have been used in clinical phases/human studies.

Today, there is also an increasing awareness of maintaining personal health by balanced nutrition, natural bio-products and the intake of nutraceutical supplements but many of these supplements have bioavailability problems. For many compounds, a nanoparticulate delivery system enhancing oral absorption is highly suited to create new improved and innovative products. Lipid nanoparticles are already described as a delivery system for food bioactive compounds (Weiss et al. 2008). Moreover, these are applied to the cosmetic market, thanks to the feasibility to large scale production that is one of the most important prerequisite for the development of a pharmaceutical product (Souto and Müller 2007).

Key Facts for Lipid Nanoparticles Formulations

- All the components of the formulation should be safe, affordable, non-toxic and biodegradable.
- The manufacturing process should be simple, affordable and easy to scale up. It must exclude organic solvents or potentially toxic ingredients.
- Nanoparticles should have the proper size, shape and surface characteristics.
- The nanoparticles should be stable with respect to size, surface morphology and size distribution.
- Nanoparticle formulation should be stable on storage and be able to regenerate the nanoparticles after administration.
- Nanoparticle formulations should protect the incorporated drug from degradation and prevent premature release of its payload.

Definitions

Mood disorder: It is the term designating a group of diagnoses in the Diagnostic and Statistical Manual of Mental Disorders classification system where a disturbance in the person's mood is hypothesized to be the main underlying feature.

Biodistribution: It is a method of tracking where compounds of interest travel in an experimental animal or human subject.

Pharmacokinetics: It is a branch of pharmacology dedicated to the study of the mechanisms of absorption and distribution of an administered drug, the rate at which a drug action begins and the duration of the effect, the chemical changes of the substance in the body and the effects and routes of excretion of the metabolites.

Pharmacodynamics: It is the study of the biochemical and physiological effects of drugs on the body, the mechanisms of drug action and the relationship between drug concentration and effect.

Mononuclear phagocyte system (MPS): It is a part of the immune system that consists of the phagocytic cells located in reticular connective tissue. The cells are primarily monocytes and macrophages, and they accumulate in lymph nodes and the spleen.

Blood Brain Barrier (BBB): It is a highly effective physical and physiological barrier that regulates the CNS homeostasis and thereby controls the delivery of drugs to the brain. It consists of a monolayer of polarized capillary endothelial cells that line the cerebral microvessels and surrounding perivascular elements make up the BBB, separating the brain from the rest of the body with tight junctions.

Multidrug resistance-associated protein 1 (MDR-1): It is a member of the superfamily of ATP-binding cassette (ABC) transporters. ABC proteins transport various molecules across extra- and intra-cellular membranes.

Immunogenicity: It is the ability of a particular substance, such as an antigen or epitope, to provoke an immune response in the body of a human or animal.

Antigenicity: It is the ability of a chemical structure (referred to as an Antigen) to bind specifically with certain products of adaptive immunity: T cell receptors or Antibodies.

Opsonin: It is any molecule that acts as a binding enhancer for the process of phagocytosis.

Reactive oxygen species (ROS): They are chemically-reactive molecules containing oxygen and they have important roles in cell signalling. However, during times of environmental stress (e.g., UV or heat exposure), ROS levels can increase dramatically. This may result in significant damage to cell structures. This cumulates into a situation known as oxidative stress.

Amyotrophic lateral sclerosis (ALS): It is a progressive, neurodegenerative disease caused by the degeneration of motor neurons, the nerve cells in the CNS that control voluntary muscle movement.

Experimental Allergic Encephalomyelitis (EAE): It is an animal model of brain inflammation. It is an inflammatory demyelinating disease of the Central Nervous System (CNS). It is widely studied as an animal model of the human CNS demyelinating diseases, and is also the prototype for T-cell-mediated autoimmune disease in general.

Summary Points

- BBB restricts the entry of foreign substances such as drugs and neuropharmaceutical agents into the brain tissue, making the CNS treatment ineffective.
- The conventional drug delivery systems which release drug into general circulation fail to deliver drugs effectively to brain and are therefore not very useful in treating certain diseases that affect CNS including Alzheimer's and Parkinson's disease, cancer, mood disorder, AIDS, and bacterial infections.
- There is a need to develop and design approaches which specifically target the brain in a better and effective way.

- Advances in novel colloidal systems for drug delivery have progressed rapidly in recent years, and therapeutically efficient formulations have been already developed for drugs that were unable to cross the BBB. Among all, SLN and NLC may represent promising carriers. Their prevalence over other formulations in terms of toxicity, production feasibility, and scalability is widely documented in the literature.
- These nanoparticles are showing a great potential as drug carriers for the treatment of several brain disorders thanks to many advantages associated with their use, and are becoming an alternative to the present surgical and conventional methods.

Abbreviations

CNS	:	Central nervous system
BBB	:	Blood-brain barrier
SLN	:	Solid lipid nanoparticles
NLC	:	Nanostructured lipid carriers
TDDS	:	Targeted drug delivery systems
MPS	:	Mononuclear phagocyte system
P-gp	:	P-glycoprotein
MRP-1	:	Multidrug resistance protein-1
HPH	:	High pressure homogenization
RMT	:	Receptor-mediated transcytosis
AMT	:	Adsorptive-mediated transcytosis
PEGs	:	Poly(ethylene glycols)
EPR	:	Enhanced permeability and retention effect
FA	:	Ferulic acid
ALS	:	Amyotrophic lateral sclerosis
EAE	:	Experimental allergic encephalomyelitis
AUC	:	Area under the concentration
TMZ	:	Temozolomide
DO–FUdR	:	3′,5′-dioctanoyl-5-fluoro-2′-deoxyuridine
FUdR	:	5-fluoro-2′-deoxyuridine
VEGF-A	:	Vascular endothelial growth factor A
AS-ODN	:	Antisense oligonucleotides
HIV	:	Human immunodeficiency virus
AIDS	:	Acquired immunodeficiency syndrome
Tf	:	Transferrin
ROS	:	Reactive oxygen species

References

Alam, M.I., S. Beg, A. Samad, S. Baboota, K. Kohli, J. Ali, A. Ahuja and M. Akbard. 2010. Strategy for effective brain drug delivery. Eur. J. Pharm. Sci. 40: 385–403.
Almeida, A.J. and E. Souto. 2007. Solid lipid nanoparticles as a drug delivery system for peptides and proteins. Adv. Drug Del. Rev. 59: 478–490.
Beduneau, A., P. Saulnier and J.P. Benoit. 2007. Active targeting of brain tumors using nanocarriers. Biomaterials 28: 4947–4967.
Blasi, P., S. Giovagnoli, A. Schoubben, M. Ricci and C. Rossi. 2007. Solid lipid nanoparticles for targeted brain drug delivery. Adv. Drug Del. Rev. 59: 454–477.
Bondì, M.L., G. Montana, E.F. Craparo, P. Picone, G. Capuano, M. Di Carlo and G. Giammona. 2009. Ferulic acid-loaded lipid nanostructures as drug delivery systems for Alzheimer's disease: preparation, characterization and cytotoxicity studies. Current Nanosci. 5: 26–32.
Bondì, M.L., E.F. Craparo, G. Giammona and F. Drago. 2010. Brain-targeted solid lipid nanoparticles (SLNs) containing riluzole: preparation, characterisation and biodistribution. Nanomedicine 5: 25–32.
Bondì, M.L. and E.F. Craparo. 2010. Solid lipid nanoparticles for applications in gene therapy: a review of the state of the art. Expert Opin. Drug Del. 7: 7–18.
Brioschi, A.M., S. Calderoni, G.P. Zara, L. Proano, M.R. Gasco and A. Mauro. Solid lipid nanoparticles for brain tumors therapy: state of the art and novel challenger. pp. 193–223 In: H.S. Sharma [Eds.] 2010. Progress in Brain Research.
Byrne, J.D., T. Betancourt and L. Brannon-Peppas. 2008. Active targeting schemes for nanoparticle systems in cancer therapeutics. Adv. Drug Del. Rev. 60: 1615–1626.
Chattopadhyay, N., J. Zastre, H.L. Wong, X.Y. Wu and R. Bendayan. 2008. Solid lipid nanoparticles enhance the delivery of the HIV protease inhibitor, atazanavir, by a human brain endothelial cell line. Pharm. Res. 25: 2262–2271.
Craparo, E.F., M.L. Bondì, G. Pitarresi and G. Cavallaro. Nanoparticulate systems for drug delivery and targeting to the central nervous system. CNS Neuroscience & Therapeutics (in press) doi:10.1111/j.1755-5949.2010.00199.x
Denora, N., A. Trapani, V. Laquintana, A. Lopedota and G. Trapani. 2009. Recent advances in medicinal chemistry and pharmaceutical technology-strategies for drug delivery to the brain. Current Topics in Medicinal Chemistry 9: 182–196.
Fernandes, C., U. Soni and V. Patravale. 2010. Nano-interventions for neurodegenerative disorders, Pharmacological Research 62: 166–178.
Gabathuler, R. 2010. Approaches to transport therapeutic drugs across the blood–brain barrier to treat brain diseases. Neurobiology of Disease 37: 48–57.
Gupta, U. and N.K. Jain. 2010. Non-polymeric nano-carriers in HIV/AIDS drug delivery and targeting. Adv. Drug Del. Rev. 62: 478–490.
Gupta, Y., A. Jain and S.K. Jain. 2007. Transferrin-conjugated solid lipid nanoparticles for enhanced delivery of quinine dihydrochloride to the brain. J. Pharm. Pharmacol. 59: 935–940.
Hu, Y., J. Xie, T.Y. Wah and C.-H. Wang. 2007. Effect of PEG conformation and particle size on the cellular uptake efficiency of nanoparticles with the HepG2 cells. J. Control Release 118: 7–17.
Huang, G., N. Zhang, X. Bi and M. Dou. 2008. Solid lipid nanoparticles of temozolomide: Potential reduction of cardial and nephric toxicity. Int. J. Pharm. 355: 314–320.
Joshi, M.D. and R.H. Müller. 2009. Lipid nanoparticles for parenteral delivery of actives. Eur. J. Pharm. Biopharm. 71: 161–172.
Juillerat-Jeanneret, L. 2008. The targeted delivery of cancer drugs across the blood-brain barrier: chemical modifications of drugs or drug-nanoparticles? Drug Discov. Today 13: 1099–1106.
Kaur, I.P., R. Bhandari, S. Bhandari and V. Kakkar. 2008. Potential of solid lipid nanoparticles in brain targeting. J. Control Release 127: 97–109.

Koo, Y.-E.L., G.R. Reddy, M. Bhojani, R. Schneider, M.A. Philbert, A. Rehemtulla, B.D. Ross and R. Kopelman. 2006. Brain cancer diagnosis and therapy with nanoplatforms. Adv. Drug Del. Rev. 58: 1556–1577.

Koziara, J.M., P.R. Lockman, D.D. Allen and R.J. Mumper. 2004. Paclitaxel nanoparticles for the potential treatment of brain tumors. J. Control Rel. 99: 259–269.

Kuo, Y.C. and F.L. Su. 2007. Transport of stavudine, delavirdine, and saquinavir across the blood–brain barrier by polybutylcyanoacrylate methyl methacrylate sulfopropyl methacrylate and solid lipid nanoparticles. Int. J. Pharm. 340: 143–152.

Lockman, P.R., M.O. Oyewumi, J.M. Koziara, K.E. Roder, R.J. Mumper and D.D. Allen. 2003. Brain uptake of thiamine-coated nanoparticles. J. Control Release 93: 271–282.

Manjunath, K. and V. Venkateswarlu. 2005. Pharmacokinetics, tissue distribution and bioavailability of clozapine solid lipid nanoparticles after intravenous and intraduodenal administration. J. Control. Release 107: 215–228.

Olivier, J.-C. and M.P. de Oliviera. 2007. Nanoparticulate systems for central nervous system drug delivery. pp. 281–290. In: M. Deleers, Y. Pathak, Y.D. Thassu. [eds.] Nanoparticulate Drug Delivery Systems. Informa Healthcare, new York, USA.

Pardridge, W.M. 2005. Molecular biology of the blood-brain barrier. Mol. Biotechnol. 30: 57–70.

Park, J.H., S. Lee, J.H. Kim, K. Park, K. Kim and I.C. Kwon. 2008. Polymeric nanomedicine for cancer therapy. Progress in Polymer Science 33: 113–137.

Sarin, H. 2009. Recent progress towards development of effective systemic chemotherapy for the treatment of malignant brain tumors. J. Transl. Med. 7: 77.

Souto, E.B. and R.H. Müller. 2007. Lipid nanoparticles (Solid Lipid Nanoparticles and Nanostructured Lipid Carriers) for Cosmetic, Dermal, and Transdermal applications. pp. 213–233. In: M. Deleers, Y. Pathak, Y.D. Thassu [eds.] Nanoparticulate Drug Delivery Systems. Informa Healthcare, new York, USA.

Thomas, M.B. and M.A. Mansoor. 2009. Challenges and opportunities in CNS delivery of therapeutics for neurodegenerative diseases. Expert Opin. Drug Deliv. 6: 211–25.

Wang, J.X., X. Sun and Z.R. Zhang. 2002. Enhanced brain targeting by synthesis of 3',5'-dioctanoyl-5-fluoro-2'-deoxyuridine and incorporation into solid lipid nanoparticles. Eur. J. Pharm. Biopharm. 54: 285–290.

Weiss, J., E.A. Decker, D.J. McClements, K. Kristbergsson, T. Helgason and T. Awad. 2008. Solid lipid nanoparticles as delivery systems for bioactive food components. Food Biophisics 3: 146–154.

Yang, S., J. Zhu, Y. Lu, B. Liang and C. Yang. 1999. Body distribution of camptothecin solid lipid nanoparticles after oral administration. Pharm. Res. 16: 751–757.

Yang, H. 2010. Nanoparticle-Mediated Brain-Specific Drug Delivery, Imaging, and Diagnosis. Pharm. Res. 27: 1759–1771.

15

Nanocarriers for the Treatment of Brain Tumours

Pauline Resnier,[1,a] *Anne Clavreul*[1,b] *and Catherine Passirani*[1,c,*]

ABSTRACT

Many strategies have been developed to deliver chemotherapeutic agents into the brain, especially to treat gliomas. Among them, nanostructures have appeared like new promising drug carriers thanks to their small size (<1µm) and generally lipid components that should improve the passage of the blood-brain barrier (BBB). Firstly, local strategies by stereotaxy like bolus injection and convection enhanced delivery (CED) have the advantage to bypass the hurdle of BBB. However, these techniques are invasive and repeated administrations are excluded. On the other hand, nanocarriers, thanks to their stability in bloodstream and their long-circulating properties, have been developed for intravenous strategies by passive or active targeting. The passive strategy permits to target tumours thanks to the increased permeability of

---

[1]INSERM U646, Université d'Angers, IBS-CHU ANGERS, 4 rue Larrey, 49933 Angers Cedex 9 France.
[a]E-mail: pauline.resnier@etud.univ-angers.fr
[b]E-mail: anne.clavreul@univ-angers.fr
[c]E-mail: catherine.passirani@univ-angers.fr
*Corresponding author

List of abbreviations after the text.

tumour vessels while the active strategy uses ligands that promote directly the interaction with cancer cells or indirectly facilitate the passage through the BBB. More recently, these nanostructures have appeared to be an efficient solution to deliver DNA or siRNA for gene therapy.

Key Facts of Nanocarriers in Brain Tumour Therapy

- Drug-loaded nanocarriers hold great promise in the field of brain tumour therapy. They enable the control of drug characteristics such as solubility, vascular circulation time, and specific site-targeted delivery.
- Four principal types of nanocarriers are developed for brain tumour treatment: dendrimers, micelles, liposomes and nanoparticles that can be magnetic, polymer or lipid/polymer.
- Nanocarriers are divided in generations (G) that characterise their stabilised structure (G1), stealth (G2) or targeting (G3) properties. Stabilised nanocarriers (G1) provide controlled drug release but necessitate a local injection. Stealth nanocarriers (G2) escape the immune system recognition which enhances their longevity in the blood and their accumulation into the tumour *via* the EPR effect, leading to passive targeting. Functionalized nanocarriers (G3) enable active targeting based on the attachment of ligands at the surface of the carrier to bind specifically on over-expressed receptors of tumour cells or organs.
- Systemic (intravenous or intraperitoneal) and direct (stereotaxic bolus or CED) delivery to brain tumours have been used to administer drug-loaded nanocarriers.
- Gene-loaded nanocarriers are recently developed to increase or decrease the expression of a target gene and are opening a new field of research in glioma therapy.

INTRODUCTION

Brain Tumours

Primary brain tumours include tumours of the brain parenchyma, meninges, cranial nerves, and other intracranial structures (such as the pituitary and pineal glands). The most common and malignant primary brain tumours are Grade IV astrocytomas, called glioblastomas, which are characterised by the aggressive invasion and diffuse infiltration of tumour cells into the surrounding brain tissue. According to CBTRUS (Central Brain Tumour Registry of the United States), glioblastomas accounted for about

20% of the estimated 44,500 new primary brain tumours diagnosed in the United States in 2005. The conventional therapy of glioblastomas consists of surgery followed by fractionated radiotherapy with concomitant and adjuvant chemotherapy with temozolomide (Stupp et al. 2009). Despite this treatment, glioblastomas generally recur at the site of initial treatment. Median survival is around 16 months and less than 10% of patients survive for more than five years.

Numerous chemotherapeutic drugs affecting cell division or DNA synthesis have been tested to improve the prognosis of glioblastomas (Muldoon et al. 2007). However, the ineffectiveness of these drugs for the treatment of brain tumours is generally not linked to drug potency, but mainly to three hurdles:

1) The presence of the blood-brain barrier (BBB) which regulates central nervous system homeostasis and controls the delivery of molecules in the brain (Fig. 1). This restriction is carried out by endothelial cell-tight junctions, which are characterised by an absence of fenestration and a reduction of pinocytotic vesicles. In addition, the BBB is composed of specific anatomical elements that also limit this passage of, for example, astrocytic end-feet, pericytes and extracellular matrices. Furthermore, this normal barrier has an efflux activity due to the presence of effective efflux transporters on surface cells (Deeken and Löscher 2007).
2) Diffusion in the brain parenchyma is very weak, partially due to the high level of intercellular fluid pressure in tumours. Targeting tumour cells infiltrating the brain tissue surrounding the tumour is consequently rather difficult which leads to frequent GB recurrence.
3) The brain tissue is highly sensitive, so only limited doses of therapeutic agents can be employed.

Applications to Areas of Health and Disease

Many strategies have emerged to overcome these three obstacles. Among these, nanotechnology has been developed to protect drugs and improve their biodistribution and therapeutic index. These non-viral carriers can ideally offer optimal, controlled release and specific targeting that could limit unwanted side effects on healthy cells and deliver a therapeutic concentration of drugs only to cancerous cells. Chemotherapy is already a major point to treat cancer. The use of drug-loaded nanocarriers is expected to ameliorate the prognostic of patients hoping a curative treatment.

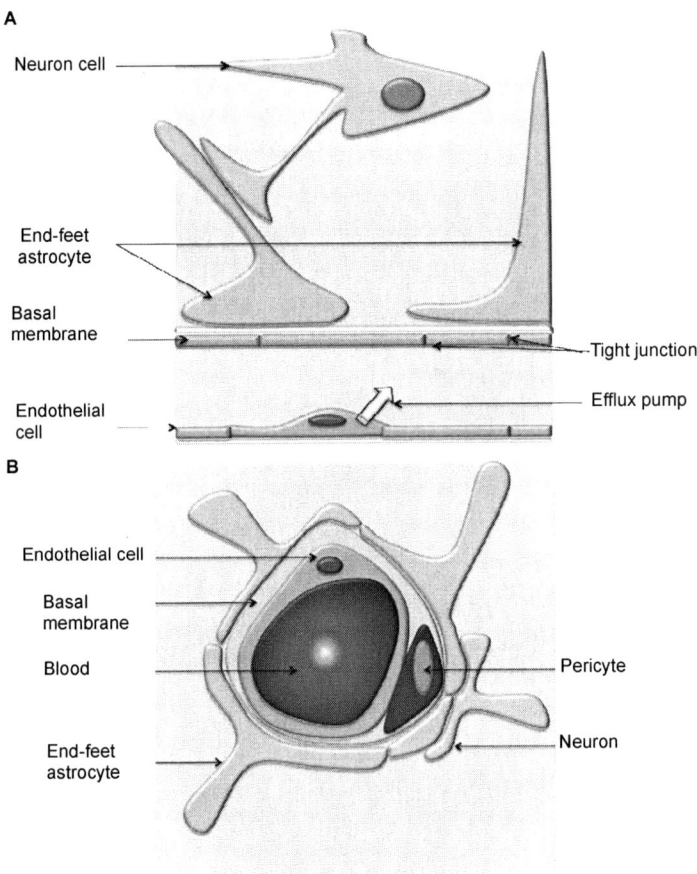

Fig. 1. Schematic representation of the blood-brain barrier (A: longitudinal and B: transversal). In the brain, tight junctions bind endothelial cells; they are surrounded by pericytes and basal membrane. Moreover, end-feet of astrocytes create a second layer that surrounds basal membrane, increases the non-permeability of the BBB and allows neuron protection.

Color image of this figure appears in the color plate section at the end of the book.

NANOCARRIERS

Nanocarriers regroup some models of nanostructures recognisable by their organization, size and lipid or polymer components. They are divided in generations (G) that characterise their stabilised structure (G1), stealth (G2) or targeting (G3) properties (Fig. 2). The principal nanocarriers described for the treatment of gliomas are dendrimers, micelles, liposomes and different types of nanoparticles.

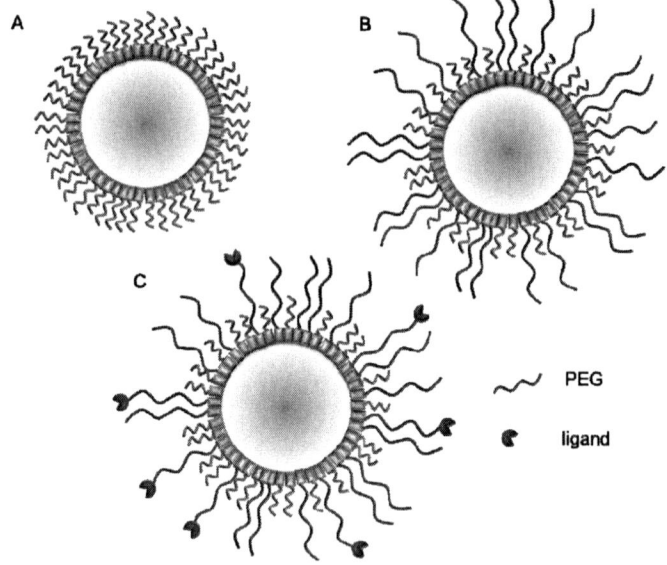

Fig. 2. Representation of nanocarrier generations (G). A: G1, B: G2 and C: G3. Nanocarriers are classified into three generations (G). The first generation (G1) corresponds to vectors stabilised with surfactant for drug encapsulation. G2 corresponds to stealth nanoparticles; they have long-circulating properties and can be used for systemic delivery. G3 are nanocarriers funtionalised with ligands to specifically target tumour cells or organs.

Dendrimers

Dendrimers are hierarchical, highly-branched polymers that form a spherical three-dimensional structure. These "macromolecules" are built with the repetitive assembly of a unique monomer around a core. This confers a narrow degree of polydispersity and low nanometer range size. The number of layers represents the generation of a dendrimer (Fig. 3A). Sizes are less than 4-5 nm for earlier generations (G0–G5) and more than 10 nm for most complex dendrimers (G9–G10). This specific shape can react with nanoscopic reagents like drugs as well as with DNA or antibodies, and form stable complexes. These very small sizes allow an improved passage through natural barriers, this being an essential point for systemic injection for the brain. Moreover, dendrimers are characterised by multivalency that confers a high capacity of reaction with all surfaces in a "molecular Velcro" behaviour (Nanjwade et al. 2009). Dendrimers are often considered like "proteins" because they can adopt a "native" or "denatured" conformation depending on polarity, ionic strength and pH of the solvent.

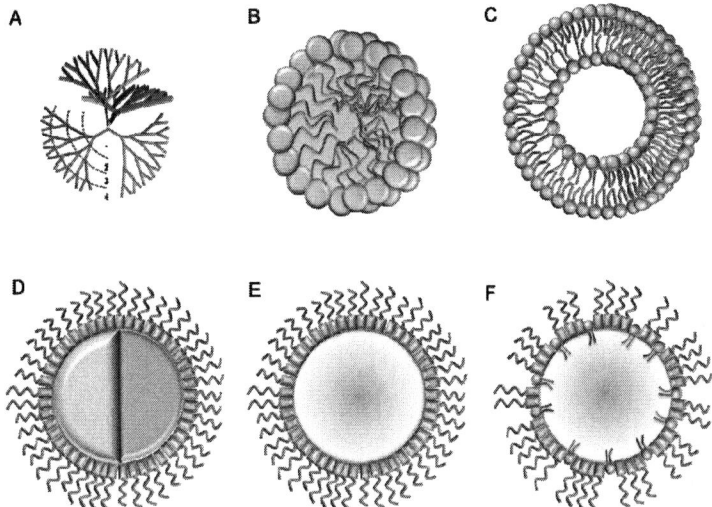

Fig. 3. Representation of various nanostructures used in brain tumor therapy. Four different types of nanocarriers are mainly developed in brain tumour therapy: (A) dendrimer: highly-branched polymers; (B) micelle: diblock polymers self-assembled; (C) liposome: bilayer of phospholipids forming a liquid aqueous core; (D) nanoparticles: polymer shell that can contain magnetic core (left), or solid polymer core (right); (E) solid lipid nanoparticle: polymer and lipid shell surrounding a solid lipid core; and (F) lipid nanocapsule: polymer and lipid shell surrounding a liquid lipid core.

Polymer Micelles

Block copolymers that have amphiphilic properties can spontaneously form polymer micelles in aqueous media with a size of between 50 and 100 nm (Fig. 3B). These are characterised by a core-shell structure whose core serves as a nanocontainer for hydrophobic drugs as anticancer drugs. Micelles provide a good alternative for poor water-soluble drugs but also for gene or siRNA thanks to electrostatic interactions. Advantages of micelles are an easy process of fabrication, efficient drug loading and a fairly narrow size distribution. They also present good thermodynamic stability in physiological solution that induces slow dissociation and controlled drug release (Nishiyama and Kataoka 2006).

Liposomes

Liposomes are nanocarriers whose structure is similar to the lipid membrane of cells (Fig. 3C) (Ulrich 2002). They are composed of amphiphilic molecules, found in biological membranes (like phospholipids, cholesterol, sphingomyelin...) that spontaneously assemble into bilayers due to their poor solubility and their configuration. Liposomes are divided into two major sub-groups: multilamellar vesicles (0.1 to 10 µm) and unilamellar ones. Small unilamellar vesicles are less than 100 nm in size; large ones are from 100 nm to 500 nm and giant ones are around 1 µm. Hydrophobic and hydrophilic drugs can be encapsulated simultaneously or singly inside liposomes. This property allows lipoplexes to be created, due to electrostatic interaction between nucleic acids (DNA or siRNA) and lipids. Under pathological conditions, such as in malignant tumours, liposomes (< 100 nm) can passively pass through the lipid membrane by fusion or by receptor-mediated endocytosis. Thanks to their size, hydrophobic and hydrophilic character, biocompatibility, biodegradability, low toxicity and immunogenicity, liposomes are promising systems for drug delivery.

Nanoparticles

Nanoparticles regroup two types of nanocarriers that differ in their intern organisation: nanospheres that are characterised by a matrix structure; and nanocapsules that have a vesicular structure with an inner core surrounded by a shell. This shell can be composed of polymer and/or lipid with various proportions and sizes from 10 nm and 200–500 nm. Nanoparticles are very interesting because they are characterised by their physical stability due to the presence of polymers, good protection of incorporated, labile drugs from degradation, and good release control.

Magnetic nanoparticles

In this case, the shell surrounds a magnetic core. Magnetic nanoparticles (MNPs) have traditionally been used for disease imaging, but recent advances have opened the door for drug delivery. This group includes metallic, bimetallic and superparamagnetic iron oxide nanoparticles (SPIONs). The latter are widely favoured because they have an inoffensive toxicity profile and a reactive surface. SPIONs are mainly made with magnetite (Fe_3O_4) and maghemite (Fe_2O_3), which form the magnetic core usually surrounded by polymer components to enhance their biocompatibility (Fig. 3D). These nanoparticles are used not only for their chemical stability, but also for their possibilities in the use of targeted drug delivery. In fact, the particular physiochemical profile of SPIONs can provide specific targeting, as with Combidex®, used for imaging of lymph node metastasis (Harisinghani et al. 2003).

Polymer nanoparticles

Polymer nanoparticles are usually formulated using hydrophobic synthetic polymers or copolymers such as polylactide/polyglycolide (PLA, PLGA), polyacrylates, polycaprolactones or natural polymers such as albumin, gelatin, alginate, collagen, and chitosan.... (Fig. 3D). PLA and PLGA have been the most investigated nanoparticles for drug delivery (Langer 1997). The degradation of these polymers can be altered by changing the block copolymer composition and molecular weight. Hence, the release of an encapsulated therapeutic agent can be adjusted to last from days to months. Polymer nanoparticles are mainly formulated using emulsion/ solvent evaporation or solvent-displacement techniques (Jain 2000). Using these methods, a variety of therapeutic agents including low molecular weight lipophilic or hydrophilic drugs and high molecular weight DNA or antisense oligonucleotides can be encapsulated (Prabha and Labhasetwar 2004).

Lipid nanocarriers

Solid lipid nanoparticles (SLNs) represent an alternative colloidal matrix carrier system with a solid lipid core inside which the hydrophobic drug is dispersed or dissolved (Fig. 3E). The whole is surrounded by a monolayer of phospholipids and polymers (Brioschi et al. 2009). They have a size around 100 nm. High drug loading, good stability and reproducibility characterise SLNs. Moreover, controlled release of the therapeutic agent can be achieved for up to several weeks in the case of specific formulations.

Lipid nanocapsules (LNCs), like SLNs, are built by combining the use of poly(ethylene glycol) (PEG) and lipid monolayer in the shell (Fig. 3F and 4) in order to surround and stabilise an oily liquid core. This core can encapsulate lipophilic drugs. LNCs have the same advantages as SLNs:

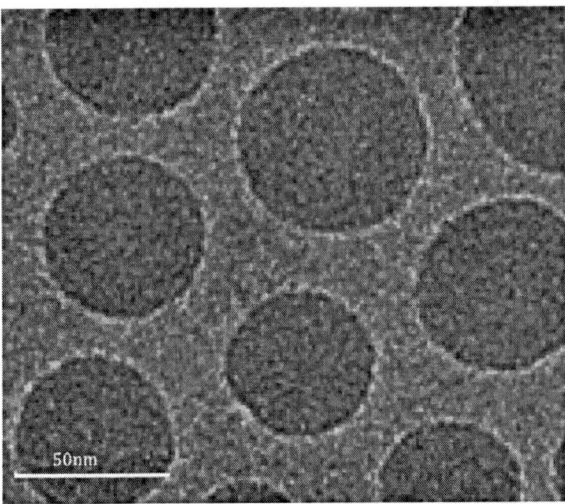

Fig. 4. Micrograph of one type of nanocarriers: LNC. Micrograh realised on 50 nm LNC by Transmission Electron Microscopy (TEM) (SCIAM, ANGERS).

good stability, good protection and control of drug delivery. Moreover, the use of solutol® in their composition inhibits drug efflux pumps which actively remove chemotherapeutic drugs from the brain (Huynh et al. 2009). SLNs and LNCs present some advantages for drug delivery because they are formulated without organic solvents, are well tolerated, are biodegradable (depending on polymer used) and possess a high degree of availability for targeting thanks to their high level stability. SLN/LNC systems can represent an innovative way to administer molecules into the brain as well as intravenously.

NANOCARRIERS IN BRAIN TUMOUR THERAPY

Chemotherapy is usually essential to obtain cancer remission. Nanotechnology constitutes a new strategy to deliver chemotherapy into the brain and consequently to improve treatment. It can also be used for immunotherapy or radiotherapy. Systemic or local delivery to the brain has been used to administer these drug-loaded nanocarriers and various approaches have been evaluated for their selective targeting of brain tumours.

Local Delivery (see examples and references in Table 1)

In order to avoid systemic side effects, local injections of chemotherapeutic agents can be performed using two different methods. The first method is the stereotaxy bolus which generates a simple diffusion of the drugs into the brain (Fig. 5A). Several assays with drug-loaded nanocarriers have shown promising results in brain tumour therapy. However, a major problem is the slow movement of nanocarriers within the brain due to limited diffusion coefficients and back-flow of the injection because of the closely-packed arrangement of cells in both the grey and white matter microenvironments (Nicholson et al. 1998).

To overcome these problems, a second stereotatic technique called CED (convection-enhanced delivery) has emerged (Fig. 5B). This method is based on the convection of a fluid by maintaining a pressure gradient during interstitial infusion, achieving a greater dose and volume of distribution compared to diffusion alone. Passive CED has been the subject of many studies in the field of nanotechnology in combination with a multitude of molecules including ferrocifen, doxorubicin, topotecan, and synthetic retinoid Am80. Radiotherapy by passive targeting has also been tested by CED.

Different approaches of active CED have emerged such as the functionalisation of the drug-loaded nanocarrier surface with ligands or antibodies to specifically target the tumour endothelium and tumour

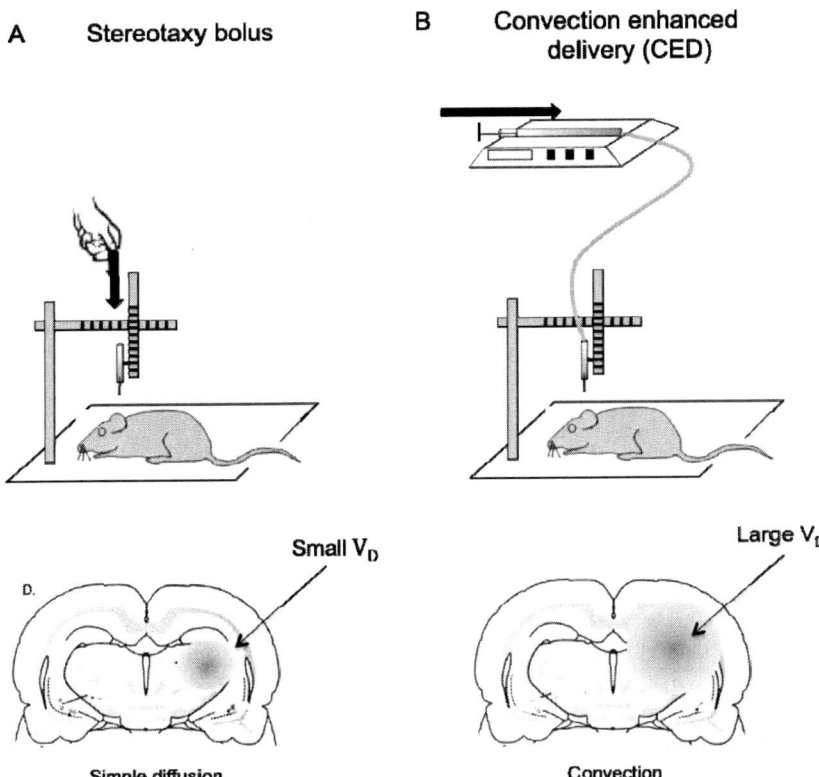

Fig. 5. Schematic representation of nanocarrier local delivery approaches. Local injection is performed by: (A) stereotaxy bolus that consists of simple diffusion in the brain parenchyma, and induces a weak volume of distribution (V_D); (B) stereotaxy convection-enhanced delivery (CED) that consists in delivering a large volume of solution by the application of a constant pressure gradient, inducing a large V_D.

associated antigens (TAAs). Currently, most tests are based on the use of antibodies against the TAA epidermal growth factor receptor (EGFR) or its mutant EGFRvIII which is expressed in 25% of glioblastomas but not in the normal brain. Another approach of active targeting is the use of stem cells as cellular carriers for nanoparticles, knowing the specific tropism of these cells for brain tumours. Roger et al. (2010) showed that polylactide nanoparticles and LNCs could be efficiently internalised into mesenchymal stem cells without cell viability and differentiation being affected. Furthermore, these nanoparticle-loaded cells were able to migrate towards an experimental human glioma model and their therapeutic effect was evidenced *in vitro* and *in vivo* (unpublished results).

Table 1. Local delivery of nanocarriers in experimental glioma models.

nanocarriers	loaded molecule	injection method	targeting	targeting molecule	Glioma model	result	ref
platinum NPs	/	bolus	passive	/	rat C6	diminution of the tumor size and volume	Lopez et al. 2010
PLA NPs	BCNU (Carmustine)	bolus	active	transferin	rat C6	significantly prolonged survival	Kang et al. 2009
LNCs	ferrocifen	CED	passive	/	rat 9L	significantly prolonged survival	Allard et al. 2009
PEG liposomes	doxorubicine	CED	passive	/	human U87MG + U251	significantly prolonged survival	Kikuchi et al. 2008
liposomes (no PEG)	topotecan	CED	passive	/	human U87MG	significantly prolonged survival	Grahn et al. 2009
polymer micelles	doxorubicin	CED	passive	/	rat 9L	significantly prolonged survival	Inoue et al. 2009
Polymer micelles	synthetic retinoid Am80	CED	passive	/	human U87MG	significantly prolonged survival	Yokosawa et al. 2010
LNCs	ferrocifen + radiotherapy	CED	passive	/	rat 9L	significantly prolonged survival	Allard et al. 2010
LNCs	paclitaxel + radiotherapy	CED	passive	/	rat 9L	significantly prolonged survival	Vinchon-Petit et al. 2010
dendrimers	methotrexate	CED	active	mAb EGFR (cetuximab)	rat F98	significantly prolonged survival	Wu et al. 2006
boronated dendrimers (PAMAM)	/	bolus and CED	active	EGF	rat F98	significantly prolonged survival	Yang et al. 2009a

boronated dendrimers	/	CED	active	mAb EGFR and EGFRVIII (cetuximab + L8A4)	rat F98	significantly prolonged survival	Yang et al. 2009b
MNPs	/	CED	active	mAb EGFRvIII (L8A4)	human U87MG (mutant EGFR vIII)	speicific targeting	Hadjipanayis et al. 2010

NPs: Nanoparticles; PLA: Polylactide; BCNU: 1,3-bis(2-chloroethyl)-1 nitrosourea; LNCs: Lipid nanocapsules; CED: Convection enhanced delivery; PEG: Polyethylene glycol; mAb: Monoclonal antibody; EGF: Epidermal growth factor; EGFR: EGF receptor; PAMAM: poly(amidoamine); MNPs: Magnetic NPs.

All these local strategies showed promising results; however, they do not provide practical solutions to everyday life nor the possibility of adjustment and adaptation for each patient.

Systemic Delivery (see examples and references in Table 2)

The greatest advantages of using systemic, drug-loaded nanocarrier delivery are their non-invasive nature and the possibility to allow repeated injections over short periods. The most commonly used method for systemic administration of nanocarriers into the body is the intravenous route. After intravenous injection, the nanocarriers are very quickly eliminated from the bloodstream due to the activation of the immune system by these foreign particles. The phagocytic mechanism of the mononuclear phagocyte system cells is the first cause advanced for this phenomenon. To improve half-life of nanoparticles in blood, stealth nanoparticles (G2) have been developed according to different parameters such as size, zeta potential, and surface modification with hydrophilic PEG (Vonarbourg et al. 2006). The improvement of the half-life allows passive targeting of a tumour *via* the phenomenon of EPR (enhanced permeability and retention) (Maeda et al. 2000) (Fig. 6A). The occurrence of this phenomenon at the tumour area is due to the organisation, structure and properties of neoangionenesis tumour vessels, these being more permeable than conventional vessels. In addition, cancer cells are constantly growing and therefore more readily accept the entry of molecules or external structures as nanocarriers. Finally, there is a lack of lymphatic drainage in tumoural environment. This causes a "natural" accumulation of nanostructures in tumours.

Drug-loaded liposomes, micelles and nanoparticles were used for passive targeting which led to significantly prolonged survival of C6 glioma bearing rats. To improve passive targeting, "semi-active" strategies were used. For example, the addition of surfactants such as polysorbate 80 in the formulation of nanoparticles is known to allow receptor-mediated endocytosis by the brain capillary endothelial cells after absorption of apolipoproteins B and E. In a number of rat studies, polysorbate 80- nanocarriers carrying doxorubicine, gemcitabine, or indomethacin crossed the BBB and reached therapeutic areas in the brain, thereby extending survival in glioma models. In another strategy, MNPs, which can be concentrated within the tumour tissue by local application of an external magnetic field, have also been developed and thereby allow non-targeted tissue and organs to remain unaffected.

As with local strategy, active targeting strategies consisting of grafting specific molecules onto the surface of nanocarriers to facilitate their transport across the BBB and then target brain gliomas have also been explored (Fig. 6B). Transferrin (TF) ligand that binds to the TF receptor,

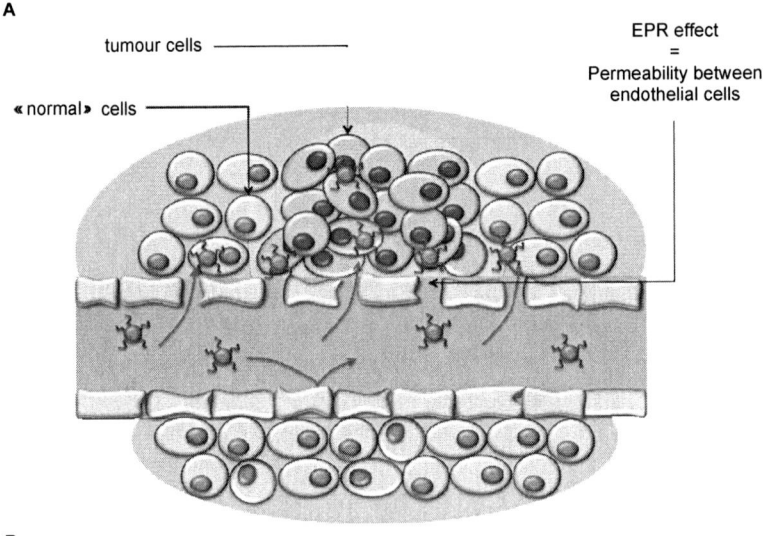

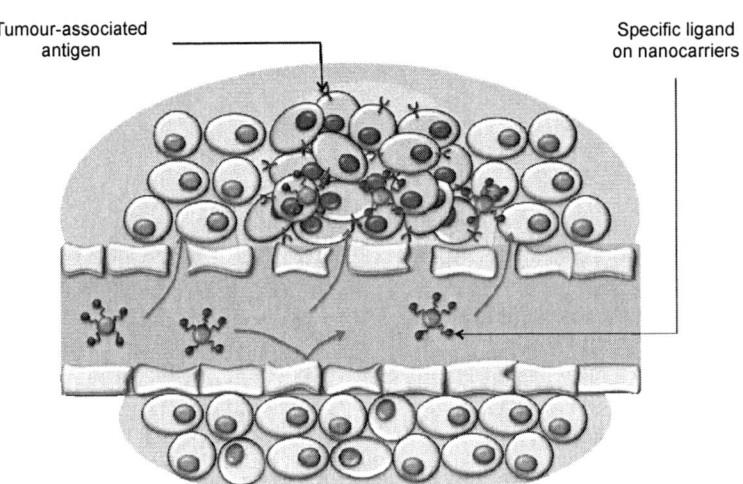

Fig. 6. Passive and active targeting of nanocarriers in brain tumour. (A) Passive targeting is based on the longevity of the nanocarrier in the blood and its accumulation in the tumour site *via* the EPR effect. (B) Active targeting is based on the functionalisation of the nanocarrier with ligands that bind to tumour-associated antigens over-expressed on tumour cells.

Color image of this figure appears in the color plate section at the end of the book.

which is over-expressed on the brain capillary endothelium and on the surface of proliferating brain tumour cells, has been tested (Fig. 7). The functionalisation of nanocarriers with an RGD sequence; peptides

Table 2. Systemic delivery of nanocarriers in experimental glioma models.

nanocarriers	loaded molecule	injection method	targeting	targeting molecule	glioma model	result	ref
liposomes	arsenic trioxide	iv	passive	/	rat C6	survival significantly prolonged	Zhao et al. 2008
polymer micelles	SN-38	iv	passive	/	human U87MG	survival significantly prolonged	Kuroda et al. 2010
PLGA NPs	doxorubicin	iv	passive	/	rat 101/8	survival significantly prolonged	Gelperina et al. 2010
CBSA NPs	aclaturin	iv	passive	/	rat C6	survival significantly prolonged	Lu et al. 2007
polysorbate NPs	doxorubicin	iv	semi-active	polysorbate-80	rat 101/8	survival significantly prolonged	Steiniger et al. 2004
polysorbate NPs	doxorubicin	iv	semi-active	polysorbate-80	rat 101/8	survival significantly prolonged and antiangiogenic effect	Hekmatara et al. 2009
PBCA NPs	doxorubicin	iv	semi-active	polysorbate-80	rat 101/8	survival significantly prolonged	Ambruosi et al. 2006
PBCA NPs	gemcitabin	iv	semi-active	polysorbate-80	rat C6	survival significantly prolonged	Wang et al. 2009
nanocapsules	indomethacin	ip	semi-active	polysorbate-80	rat C6	survival significantly prolonged	Bernardi et al. 2009
Superparamagnetic NPs	paclitaxel	iv	semi-active	magnetism	rat C6	survival significantly prolonged	Zhao et al. 2010
Liposomes	epirubucin	iv	active	TF	rat C6	survival significantly prolonged	Tian et al. 2010
PEG-DPSE micelles	paclitaxel	iv	active	NGR	rat C6	inhibition of tumoral growth	Zhao et al. 2010
PEG-DPSE micelles	paclitaxel	iv	active	RGD	human U87MG	survival significantly prolonged	Zhan et al. 2010
liposomes (PEG)	daunorubicin	iv	active	MAN and TF	rat C6	survival significantly prolonged	Ying et al. 2009
liposomes (PEG)	doxorubicin	ip	active	IL-13	human U87MG	survival significantly prolonged compare to doxorubicin liposome without IL-13	Madhankumar et al. 2009

SN-38:activedrugofirinotecan;PLGA:Polylactide-polyglycolide,CBSA:cationicbovineserumalbumin;NPs:Nanoparticles;PBCA:Polybutylcyanoacrylate; LNCs: Lipid nanocapsules; TF: Transferrin, PEG: polyethylene glycol; PEG-DSPE: 1,2-distearoyl-sn-glycero-3-phosphoethanolamine-N-[amino(polyethylene glycol)-2000]; NGR: Asn-Gly-Arg; RGD: Arg-Gly-Asp; MAN: p-aminophenyl-α-D-manopyroside; IL-13: Interleukin-13.

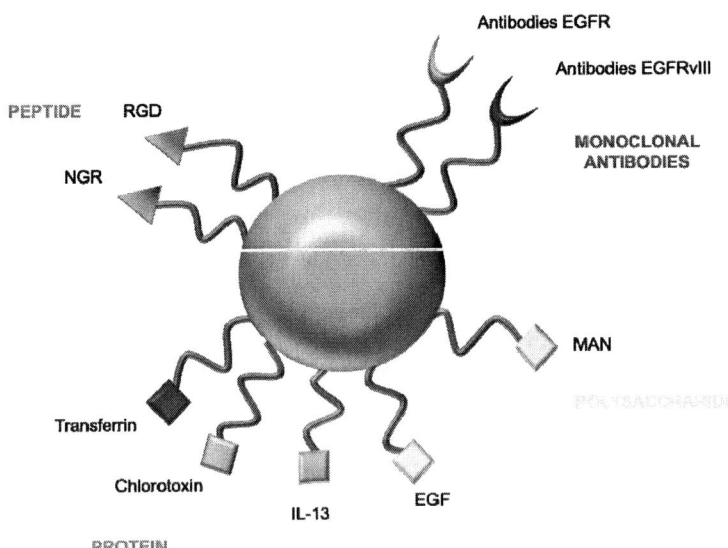

Fig. 7. Representation of ligands used to functionalise nanocarriers to target specifically brain tumour cells. Four types of ligands are mainly used in brain tumour therapy: monoclonal antobodies [anti-EGFR (epidermal growth factor receptor) and anti-EGFRvIII (EGFR variant III)], proteins [transferrin, chlorotoxin, EGF and interleukin-13 (IL-13)], peptides [RGD (Arg-Gly-Asp) and NGR (Asn-Gly-Arg)] and polysaccharide [Man (p-aminophenyl-α-D-manopyroside)].

containing the Asn-Gly-Arg (NGR) motif, p-aminophenyl-a-D-mannopyranoside (MAN) or interleukin-13 (IL-13) has also been performed (Fig. 7). RGD sequences facilitate interaction between drug delivery nanocarriers and some integrins such as integrin αvβ3, widely overexpressed on tumoural neovasculature and glioma cells. The motif NGR links to aminopeptidase N (CD13), a membrane-spanning molecule, overexpressed on tumour cells and most tumour endothelial cells. MAN has a specific affinity to the glucose transporter GLUT1 which is expressed mainly in the luminal surface of the brain capillaries and the choroid plexus.

These strategies have shown promising results in experimental glioma models; however, the major difficulty remains in knowing cell surface markers able to specifically target the BBB and tumour cells. At present, no unique glioma-specific antigen has been discovered due to the intratumoural heterogeneity and the inter-patient variability. However, despite these difficulties, the intravenous route remains a vital challenge for non-invasive administration and particularly for repeated administration for cancerous diseases in the brain.

FUTURE STRATEGIES OF TREATMENT (TABLE 3)

Now, after almost 40 years of research, nanoparticles have become a real platform for chemotherapy. However, chemotherapy may not be sufficient to induce glioma remission. Many studies have been working on combining various treatments (chemotherapy, radiotherapy and/or immunotherapy). For example, Allard et al. (2010) recently studied the association of radiotherapy with chemotherapy encapsulated in lipid nanocapsules, showing a synergistic effect on survival.

Over the last decade, gene therapy has undergone considerable expansion with improved knowledge of the genome and also with the discovery of siRNA. Gene therapy involves inserting a fragment of RNA or DNA into the cell that can directly/indirectly modify the expression of a target gene. Strategies consist of increasing the expression of a gene with the insertion of a new copy, or by inhibiting the expression of a gene with the insertion of siRNA. Due to renal filtration, nuclease degradation, and a poor cellular uptake, the encapsulation of these hydrophilic molecules is necessary to consider their use *in vivo*. As most nanocarriers have a predominantly hydrophobic solid or liquid core, the combination of nucleic acids with hydrophobic forms has necessitated the development of appropriate formulations. A possible association within liposomes has been largely developed: indeed, cationic lipids used for liposome formulation can react with the negative charge of nucleic acids and form stable complexes called lipoplexes. In addition, they have a good cellular uptake. Gupta et al. (2007) demonstrated this transfection ability in nude mice that received U87MG human glioma cells. An alternative form was developed by Morille et al. (2010a) and consisted of the encapsulation of lipoplexes within lipid nanocapsules. They showed long-circulating properties and accumulation by EPR effect in sub-cutaneous U87MG glioma model after intravenous injection. Moreover, examples of active targeting were realised by grafting TF or chlorotoxin. This latter agent is commonly used for labelling property, but in this case, chlorotoxin modified nanoparticles showed an improved cellular uptake and high gene expression.

Several studies worked on diminution of the expression of target gene, using nanocarriers delivering EGFR siRNA and EGFR as well as transforming growth factor beta (TGFβ) antisense oligonucleotides, and showed promising results in experimental animal glioma models. In fact, EGFR gene is often over-expressed in cancer cells like glioblastoma, and consequently conducts to high proliferative capacity and aggressive form. TGFβ is an anti-inflammatory factor that causes an immunosuppression and promotes tumoural development. In parallel, some studies worked on increasing the expression of target gene by using nanocarriers delivering

Table 3. Preclinical studies on gene, siRNA or antisens oligonucleotide delivery with nanocarriers in brain tumour therapy.

nanocarriers	loaded molecule	injection method	targeting	targeting molecule	Glioma model	result	ref
liposomes	GFP (ADN)	stereotaxy bolus	active	TAT peptide	human U87MG	good transfection *in vivo*	Gupta et al. 2007
LNCs	DNA	iv	passive	/	human U87MG	tumoural accumulation *in vivo* and expression	Morille et al. 2010b
MNPs	DNA (GFP)	iv	active	chlorotoxin	rat C6	DNA delivery in tumor and highly expression	Kievit et al. 2010
dendriworms	siRNA (EGFR)	CED	passive	/	human primary GBM	inhibition of EGFR by siRNA	Agrawal et al. 2009
PBCA NPs	antisens oligo-nucleotid (TGFβ)	ip	passive	/	rat F98	survival significantly prolonged	Schneider et al. 2008
folated PAMAM dendrimers	antisens oligo-nucleotide (EGFR)	stereotaxy bolus	passive	/	rat C6	survival significantly prolonged	Kang et al. 2010
CBSA-PEG NPs	plasmid pORF-hTRAIL	iv	passive	/	rat C6	inhibition of tumor growth	Lu et al. 2006
polypropylenimine dendrimers	TNFα	iv	active	TF	human T98G et A431	transfection *in vivo* in A431 and antiproliferative capacity in T98G	Koppu et al. 2010

GFP: Green fluorescence protein; TAT peptide: Transactivator of transcription peptide; LNCs: Lipid nanocapsules; TF: Transferrin; MNPs: Magnetic nanoparticles; EGFR: Epidermal growth factor receptor; CED: Convection enhanced delivery; PBCA: poly(α-butyl-cyanoacrylate); TGFβ: Transforming growth factor β; GBM: Glioblastomas; PAMAM: poly(amidoamine); CBSA: Cationic bovine serum, PEG: Polyethylene glycol, TNFα: Tumour necrosis factor α; TRAIL: TNF related apoptosis inducing ligand.

therapeutic DNA encoding tumour necrosis factor alpha (TNFα) or TNF-related apoptosis-inducing-ligand (TRAIL). TRAIL activates the extrinsic pathway of apoptosis by binding to TRAIL-receptors. TNFα is a pleiotropic inflammatory cytokine that promotes the immune response.

In conclusion, gene therapy for cancer represents a new therapeutic possibility in the context of brain tumours, and is gaining more and more interest due to recent successes achieved *in vivo*.

Summary Points

- Glioblastomas, more frequent primary brain tumours, still have a poor prognostic as the median survival rarely exceeds 1 year.
- The ineffectiveness of chemotherapy in glioblastomas is not linked to drug potency but mainly to the presence of BBB and the poor penetration of drug into the tumour. Drug-loaded nanocarriers such as dendrimers, micelles, liposomes and nanoparticles are able to protect drugs and improve their biodistribution and therapeutic index and represent a new hope for brain treatment.
- Local delivery strategies of drug-loaded nanocarriers to brain tumour (stereotaxy bolus and CED) showed promising results because they have the advantage to bypass the BBB and deliver high concentrations of drugs. However, they are not adaptable for repeated injections and for personalisation of patient treatment.
- Systemic delivery of drug-loaded nanocarriers offers the advantage to be non invasive and allow repeated injections over short periods. The development of stealth and functionalized nanocarriers to allow passive and active targeting in brain tumour respectively holds great promise for systemic delivery approaches.
- The gene therapy of brain tumour is more recent but first *in vivo* results with gene-loaded nanocarriers are promising and demonstrate the therapeutic potential of this strategy for brain cancer.

Key Terms–Definition

- *Blood-brain barrier (BBB)* is a specific structure observed in the brain; it regulates exchanges with the bloodstream and homeostasis.
- *Systemic strategy* consists in delivering therapeutic agents from the bloodstream after intravenous or intraperitoneal injection.
- *Local strategy* consists in injecting directly therapeutic agents into the tumour area. In the case of gliomas, this injection is executed into the brain by stereotaxy.
- *Convection-enhanced delivery (CED)* is a local technique allowing the display, *via* a constant pressure gradient, of greater volumes of distribution than by stereotaxy bolus.

- *Enhanced Permeability and Retention (EPR) effect* is a destabilisation of the vessel endothelium that causes the decrease of cell junction and the increase of the vessel permeability. This phenomenon is observed on vessels that are close to a tumour site.
- *Passive targeting* is a concept of "natural" accumulation in tumour area based on the long circulating property of nanocarriers and on the EPR effect.
- *Active targeting* is a concept that consists in functionalising nanocarriers with ligands to specifically target tumour cells or to guide carriers to the tumour site.

Abreviations

G	:	Generation
CBTRUS	:	Central brain tumours registry of the United States
BBB	:	Blood-brain barrier
MNPs	:	Magnetic nanoparticles
SPIONs	:	Superparamagnetic iron oxide nanoparticles
PLA	:	Polylactide
PLGA	:	Polylactide-co-glycolide
SLNs	:	Solid lipid nanoparticles
LNCs	:	Lipid nanocapsules
PEG	:	Polyethylene glycol
CED	:	Convection-enhanced delivery
TAA	:	Tumour associated antigen
EGFR	:	Epidermal growth factor receptor
EPR	:	Enhanced permeability retention effect
TF	:	Transferrin
MAN	:	P-aminophenyl-α-D-manopyroside
NGR	:	Asn-Gly-Arg
RGD	:	Arg-Gly-Asp
IL-13	:	Interleukin 13
GLUT1	:	Glucose transporter 1
TGFβ	:	Transforming growth factor β
TNFα	:	Tumour necrosis factor α
TRAIL	:	TNF-related apoptosis-inducing-ligand

References

Agrawal, A., D.H. Min, N. Singh, H. Zhu, A. Birjiniuk, G. von Maltzahn, T.J. Harris, D. Xing, S.D. Woolfenden, P.A. Sharp, A. Charest and S. Bhatia. 2009. Functional delivery of siRNA in mice using dendriworms. ACS Nano 3(9): 2495–504.

Allard, E., N.T. Huynh, A. Vessières, P. Pigeon, G. Jaouen, J.P. Benoit and C. Passirani. 2009. Dose effect activity of ferrocifen-loaded lipid nanocapsules on a 9L-glioma model. Int. J. Pharm. 379(2): 317–23.

Allard, E., D. Jarnet, A. Vessières, S. Vinchon-Petit, G. Jaouen, J.P. Benoit and C. Passirani. 2010. Local delivery of ferrociphenol lipid nanocapsules followed by external radiotherapy as a synergistic treatment against intracranial 9L glioma xenograft. Pharm. Res. 27(1): 56–64.

Ambruosi, A., S. Gelperina, A. Khalansky, S. Tanski, A. Theisen and J. Kreuter. 2006. Influence of surfactants, polymer and doxorubicin loading on the anti-tumour effect of poly(butyl cyanoacrylate) nanoparticles in a rat glioma model. J. Microencapsul. 23(5): 582–92.

Bernardi, A., E. Braganhol, E. Jäger, F. Figueiró, M.I. Edelweiss, A.R. Pohlmann, S.S. Guterres and A.M. Battastini. 2009. Indomethacin-loaded nanocapsules treatment reduces in vivo glioblastoma growth in a rat glioma model. Cancer Lett. 281(1): 53–63.

Brioschi, A.M., S. Calderoni, G.P. Zara, L. Priano, M.R. Gasco and A. Mauro. 2009. Chapter 11—Solid lipid nanoparticles for brain tumours therapy: State of the art and novel challenge. Prog. Brain Res. 180: 193–223.

Deeken, J.F. and W. Löscher. 2007. The blood-brain barrier and cancer: transporters, treatment and Trojan horses. Clin. Cancer Res. 13(6): 1663–74.

Gelperina, S., O. Maksimenko, A. Khalansky, L. Vanchugova, E. Shipulo, K. Abbasova, R. Berdiev, S. Wohlfart, N. Chepurnova and J. Kreuter. 2010. Drug delivery to the brain using surfactant-coated poly(lactide-co-glycolide) nanoparticles: influence of the formulation parameters. Eur. J. Pharm. Biopharm. 74(2): 157–63.

Grahn, A.Y., K.S. Bankiewicz, M. Dugich-Djordjevic, J.R. Bringas, P. Hadaczek, G.A. Johnson, S. Eastman and M. Luz. 2009. Non-PEGylated liposomes for convection-enhanced delivery of topotecan and gadodiamide in malignant glioma: initial experience. J. Neurooncol. 95(2): 185–97.

Gupta, B., T.S. Levchenko and V.P. Torchilin. 2007. TAT peptide-modified liposomes provide enhanced gene delivery to intracranial human brain tumour xenografts in nude mice. Oncol. Res. 16(8): 351–9.

Hadjipanayis, C.G., R. Machaidze, M. Kaluzova, L. Wang, A.J. Schuette, H. Chen, X. Wu and H. Mao. 2010. EGFRvIII antibody-conjugated iron oxide nanoparticles for magnetic resonance imaging-guided convection-enhanced delivery and targeted therapy of glioblastoma. Cancer Res. 70(15): 6303–12.

Harisinghani, M.G., J. Barents, P.F. Hahn, W.M. Desern, S.T. Abatabaei, C.H. vande Kaa, J. de la Rosette and R. Weissleder. 2003. Noninvasive detection of clinically occult lymph-node metastases in prostate cancer. N. Engl. J. Med. 348. 2491–U5.

Hekmatara, T., C. Bernreuther, A.S. Khalansky, A. Theisen, J. Weissenberger, J. Matschke, S. Gelperina, J. Kreuter and M. Glatzel. 2009. Efficient systemic therapy of rat glioblastoma by nanoparticle-bound doxorubicin is due to antiangiogenic effects. Clin. Neuropathol. 28(3): 153–64.

Huynh, N.T., C. Passirani, P. Saulnier and J.P. Benoît. 2009. Lipid nanocapsules : a new platform for nanomedicine. Int. J. Pharm. 379(2): 201–209.

Inoue, T., Y. Yamashita, M. Nishihara, S. Sugiyama, Y. Sonoda, T. Kumabe, M. Yokoyama and T. Tominaga. 2009. Therapeutic efficacy of a polymeric micellar doxorubicin infused by convection-enhanced delivery against intracranial 9L brain tumour models. Neuro Oncol. 11(2): 151–7.

Jain, R.A. 2000. The manufacturing techniques of various drug loaded biodegradable poly(lactide-co glycolide) (PLGA) devices. Biomaterials 21 (23): 2475e90.

Kang, C., X. Yuan, F. Li, P. Pu, S. Yu, C. Shen, Z. Zhang and Y. Zhang. 2010. Evaluation of folate-PAMAM for the delivery of antisense oligonucleotides to rat C6 glioma cells in vitro and in vivo. J. Biomed. Mater Res. A. 93(2): 585–94.

Kang, C., X. Yuan, Y. Zhong, P. Pu, Y. Guo, A. Albadany, S. Yu, Z. Zhang, Y. Li, J. Chang and J. Sheng. 2009. Growth inhibition against intracranial C6 glioma cells by stereotactic delivery of BCNU by controlled release from poly(D,L-lactic acid) nanoparticles. Technol Cancer Res. Treat. 8(1): 61–70.

Kievit, F.M., O. Veiseh, C. Fang, N. Bhattarai, D. Lee, R.G Ellenbogen and M. Zhang. 2010. Chlorotoxin labeled magnetic nanovectors for targeted gene delivery to glioma. ACS Nano. 4(8): 4587–94.

Kikuchi, T., R. Saito, S. Sugiyama, Y. Yamashita, T. Kumabe, M. Krauze, K. Bankiewicz and T. Tominaga. 2008. Convection-enhanced delivery of polyethylene glycol-coated liposomal doxorubicin: characterization and efficacy in rat intracranial glioma models. J. Neurosurg. 109(5): 867–73.

Koppu, S., Y.J. Oh, R. Edrada-Ebel, D.R. Blatchford, L. Tetley, R.J. Tate and C. Dufès. 2010. Tumour regression after systemic administration of a novel tumour-targeted gene delivery system carrying a therapeutic plasmid DNA. J. Control Release 143(2): 215–21.

Kuroda, J., J. Kuratsu, M. Yasunaga, Y. Koga, Y. Saito and Y. Matsumura. 2010. Potent antitumour effect of SN-38-incorporating polymeric micelle, NK012, against malignant glioma. Int. J. Cancer 124(11): 2505–11.

Langer, R. 1997. Tissue engineering: a new field and its challenges. Pharm. Res. 14(7): 840–1.

López, T., F. Figueras, J. Manjarrez, J. Bustos, M. Alvarez, J. Silvestre-Albero, F. Rodríguez-Reinoso, A. Martínez-Ferre and E. Martínez. 2010. Catalytic nanomedicine: a new field in antitumour treatment using supported platinum nanoparticles. *In vitro* DNA degradation and *in vivo* tests with C6 animal model on Wistar rats. Eur. J. Med. Chem. 45(5): 1982–90.

Lu, W., Q. Sun, J. Wan, Z. She and X.G. Jiang. 2006. Cationic albumin-conjugated pegylated nanoparticles allow gene delivery into brain tumours via intravenous administration. Cancer Res. 66(24): 11878.

Lu, W., J. Wan, Z. She and X. Jiang. 2007. Brain delivery property and accelerated blood clearance of cationic albumin conjugated pegylated nanoparticle. J. Control Release. 118(1): 38–53.

Madhankumar, A.B., B. Slagle-Webb, X. Wang, Q.X. Yang, D.A. Antonetti, P.A. Miller, J.M. Sheehan and J.R. Connor. 2009. Efficacy of interleukin-13 receptor-targeted liposomal doxorubicin in the intracranial brain tumour model. Mol. Cancer Ther. 8(3): 648–54.

Maeda, H., J. Wu, T. Sawa, Y. Matsumura and K. Hori. 2000. Tumour vascular permeability and the EPR effect in macromolecular therapeutics: a review. J. Control Release 65: 271–284.

Morille, M., T. Montier, P. Legras, N. Carmoy, P. Brodin, B. Pitard, J.P. Benoît and C. Passirani. 2010a. Long-circulating DNA lipid nanocapsules as new vector for passive tumour targeting. Biomaterials 31(2): 321–9.

Morille, M., C. Passirani, S. Dufort, G. Bastiat, B. Pitard, J.L. Coll and J.P. Benoît. 2011 Mar. *In vivo* gene expression in tumour after the systemic injection of PEGylated DNA lipid nanocapsules. Biomaterials 32(9): 2327–33.

Muldoon, L.L., C. Soussain, K. Jahnke, C. Johanson, T. Siegal, Q.R. Smith, W.A. Hall, K. Hynynen, P.D. Senter, D.M. Peereboom and E.A. Neuwelt. 2007. Chemotherapy delivery issues in central nervous system malignancy: a reality check. J. Clin. Oncol. 25(16): 2295–305.

Nanjwade, B.K., H.M. Bechra, G.K. Derkar, F.V. Manvi and V.K. Nanjwade. 2009. Dendrimers: emerging polymers for drug-delivery systems. Eur. J. Pharm. Sci. 38(3): 185–96.

Nicholson, C. and E. Sykova. 1998. Extracellular space structure revealed by diffusion analysis. Trends Neurosci. 21(5): 207e15.

Nishiyama, N. and K. Kataoka. 2006. Current state, achievements and future prospects of polymeric micelles as nanocarriers for drug and gene delivery. Pharmacol. Ther. 112(3): 630–48.

Prabha, S. and V. Labhasetwar. 2004. Critical determinants in PLGA/PLA nanoparticle-mediated gene expression. Pharm. Res. 21(2): 354e64.

Roger, M., A. Clavreul, M.C. Venier-Julienne, C. Passirani, L. Sindji, P. Schiller, C. Montero-Menei and P. Menei. 2010. Mesenchymal stem cells as cellular vehicles for delivery of nanoparticles to brain tumours. Biomaterials 31(32): 8393–8401.

Schneider, T., A. Becker, K. Ringe, A. Reinhold, R. Firsching and B.A. Sabel. 2008. Brain tumour therapy by combined vaccination and antisense oligonucleotide delivery with nanoparticles. J. Neuroimmunol. 195(1-2): 21–7.

Steiniger, S.C., J. Kreuter, A.S. Khalansky, I.N. Skidan, A.I. Bobruskin, Z.S. Smirnova, S.E. Severin, R. Uhl, M. Kock, K.D. Geiger and S.E. Gelperina. 2004. Chemotherapy of glioblastoma in rats using doxorubicin-loaded nanoparticles. Int. J. Cancer 109(5): 759–67.

Stupp, R., M.E. Hegi, W.P. Mason, M.J. van den Bent, M.J. Taphoorn, R.C. Janzer, S.K. Ludwin, A. Allgeier, B. Fisher, K. Belanger, P. Hau, A.A. Brandes, J. Gijtenbeek, C. Marosi, C.J. Vecht, K. Mokhtari, P. Wesseling, S. Villa, E. Eisenhauer, T. Gorlia, M. Weller, D. Lacombe, J.G. Cairncross and R.O. Mirimanoff. 2009. Effects of radiotherapy with concomitant and adjuvant temozolomide versus radiotherapy alone on survival in glioblastoma in a randomised phase III study: 5-year analysis of the EORTC-NCIC trial. Lancet Oncol. 10(5): 459–66.

Tian, W., X. Ying, J. Du, J. Guo, Y. Men, Y. Zhang, R.J. Li, H.J. Yao, J.N. Lou, L.R. Zhang and W.L. Lu. 2010. Enhanced efficacy of functionalized epirubicin liposomes in treating brain glioma-bearing rats. Eur. J. Pharm. Sci. 41(2): 232–43.

Ulrich, A.S. 2002. Biophysical Aspects of Using Liposomes as Delivery Vehicles. Bioscience Reports, Vol. 22, No. 2.

Vinchon-Petit, S., D. Jarnet, A. Paillard, J.P. Benoit, E. Garcion and P. Menei. 2010. In vivo evaluation of intracellular drug-nanocarriers infused into intracranial tumours by convection-enhanced delivery: distribution and radiosensitisation efficacy. J. Neurooncol. 97(2): 195–205.

Vonarbourg, A., C. Passirani, P. Saulnier and J.P. Benoit. 2006. Parameters influencing the stealthiness of colloidal drug delivery systems. Biomaterials. 27, 4356–4373.

Wang, C.X., L.S. Huang, L.B. Hou, L. Jiang, Z.T. Yan, Y.L. Wang and Z.L Chen. 2009. Antitumour effects of polysorbate-80 coated gemcitabine polybutylcyanoacrylate nanoparticles *in vitro* and its pharmacodynamics in vivo on C6 glioma cells of a brain tumour model. Brain Res. 1261: 91–9.

Wu, G., R.F. Barth, W. Yang, S. Kawabata, L. Zhang and K. Green-Church. 2006. Targeted delivery of methotrexate to epidermal growth factor receptor-positive brain tumours by means of cetuximab (IMC-C225) dendrimer bioconjugates. Mol. Cancer Ther. 5(1): 52–9.

Yang, W., R.F. Barth, G. Wu, T. Huo, W. Tjarks, M. Ciesielski, R.A. Fenstermaker, B.D. Ross, C.J. Wikstrand, K.J. Riley and P.J. Binns. 2009a. Convection enhanced delivery of boronated EGF as a molecular targeting agent for neutron capture therapy of brain tumours. J. Neurooncol. 95(3): 355–65.

Yang, W., R.F. Barth, G. Wu, W. Tjarks, P. Binns and K. Riley. 2009b. Boron neutron capture therapy of EGFR or EGFRvIII positive gliomas using either boronated monoclonal antibodies or epidermal growth factor as molecular targeting agents. Appl. Radiat. Isot. 67(7–8 Suppl): S328–31.

Ying, X., H. Wen, W.L. Lu, J. Du, J. Guo, W. Tian, Y. Men, Y. Zhang, R.J. Li, T.Y. Yang, D.W. Shang, J.N. Lou, L.R. Zhang and Q. Zhang. 2010. Dual-targeting daunorubicin liposomes improve the therapeutic efficacy of brain glioma in animals. J. Control Release 141(2): 183–92.

Yokosawa, M., Y. Sonoda, S. Sugiyama, R. Saito, Y. Yamashita, M. Nishihara, T. Satoh, T. Kumabe, M. Yokoyama and T. Tominaga. 2010. Convection-enhanced delivery of a synthetic retinoid Am80, loaded into polymeric micelles, prolongs the survival of rats bearing intracranial glioblastoma xenografts. Tohoku. J. Exp. Med. 221(4): 257–64.

Zhan, C., B. Gu, C. Xie, J. Li, Y. Liu and W. Lu. 2010. Cyclic RGD conjugated poly(ethylene glycol)-co-poly(lactic acid) micelle enhances paclitaxel anti-glioblastoma effect. J. Control Release 143(1): 136–42.

Zhao, B.J., X.Y. Ke, Y. Huang, X.M. Chen, X. Zhao, B.X. Zhao, W.L Lu, J.N. Lou, X. Zhang and Q. Zhang. 2010. The antiangiogenic efficacy of NGR-modified PEG-DSPE micelles containing paclitaxel (NGR-M-PTX) for the treatment of glioma in rats. J. Drug Target.

Zhao, M., C. Liang, A. Li, J. Chang, H. Wang, R. Yan, J. Zhang and J. Tai. 2010. Magnetic paclitaxel nanoparticles inhibit glioma growth and improve the survival of rats bearing glioma xenografts. Anticancer Res. 30(6): 2217–23.

Zhao, S., X. Zhang, J. Zhang, J. Zhang, H. Zou, Y. Liu, X. Dong and X. Sun. 2008. Intravenous administration of arsenic trioxide encapsulated in liposomes inhibits the growth of C6 gliomas in rat brains. J. Chemother. 20(2): 253–62.

16

Electrospinning of Nanofibres for Repair of the Injured Peripheral Nervous System

Dorothee Hodde,[1,a,]* José Luis Gerardo-Nava,[1,b] Ronald Deumens,[1,2] Jörg Mey[3] and Gary Anthony Brook[1,c]

ABSTRACT

Functional deficits caused by traumatic peripheral nerve injury remain a significant challenge in the clinic. The current gold standard treatment for bridging large injury-induced gaps in peripheral nerves is the transplantation of an autologous nerve, but even this strategy presents its own risks and by no means guarantees a useful degree of functional recovery. Recent advances in polymer chemistry and tissue engineering have

---

[1]Institute for Neuropathology, Medical Faculty, RWTH Aachen University, Pauwelsstrasse 30, D-52074 Aachen, Germany.
[a]E-mail: dhodde@ukaachen.de
[b]E-mail: jgerardonava@ukaachen.de
[c]E-mail: gbrook@ukaachen.de
[2]Department of Anesthesiology, Maastricht University Medical Center, P. Debyelaan 25, 6229 HX, Maastricht, The Netherlands; E-mail: r.deumens@maastrichtuniversity.nl
[3]Hospital Nacional de Paraplėjicos, SESCAM; E-mail: mey@bio2.rwth-aachen.de
*Corrresponding author

List of abbreviations after the text.

resulted in the design of increasingly sophisticated scaffolds that may act as biomimetic alternatives to the autograft. The adoption and modification of a technique that was first developed in the 1930s, i.e. the electrospinning of nanofibres, provides an innovative nanotechnological approach to peripheral nerve repair and presents particular advantages, such as a high substrate surface area-to-volume ratio, as well as the ability to control the orientation of the substrate guidance cues. Although a surprising degree of axon regeneration and functional tissue repair has been demonstrated with non-biodegradable, non-functionalized nanofibre scaffolds, current approaches to modify the surface properties of bioengineered nano-structured polymers herald the prospect of novel implantable devices for use in regenerative medicine that could, in the future, replace the autografted nerve.

INTRODUCTION

Peripheral nerve injury (PNI) interferes with the quality of life of individuals by leading to life-long disturbances of function, and in some instances to the development of neuropathic pain. The functional consequences of PNI depend on the severity and type of the injury (for a recent review see Deumens et al. 2010). Nerve crush injuries have a generally good prognosis because the architecture of the peripheral nerve extracellular matrix (ECM), including the basal lamina tubes, remains largely intact. The Schwann cell-filled basal lamina tubes (or endoneurial tubes) of the Wallerian-degenerated distal nerve stump provide the orientational cues responsible for guiding regenerating axons to their original targets, allowing re-innervation and functional recovery (Fig. 1). The consequences of peripheral nerve transection are somewhat more complicated, depending on the severity and extent of ECM disruption. As with crush-type injuries, successful tissue repair demands that regenerating axons enter appropriate (or the original) endoneurial tubes to re-establish functional connectivity. Simple transection-type injuries may be repaired surgically by end-to-end tension-free sutures which bring the proximal and distal nerve stumps together and maintain the continuity of the original fascicular pattern within the damaged nerve. Transection-type injuries that result in larger gaps, however, require the bridging of the defect by an axon-growth promoting implant. The current gold-standard for such intervention is the nerve autograft: the harvesting of sensory nerves from the patients themselves, to provide a bridging material with optimal cellular and molecular constituents (i.e., growth factors, cell adhesion molecules and ECM molecules) for regeneration and

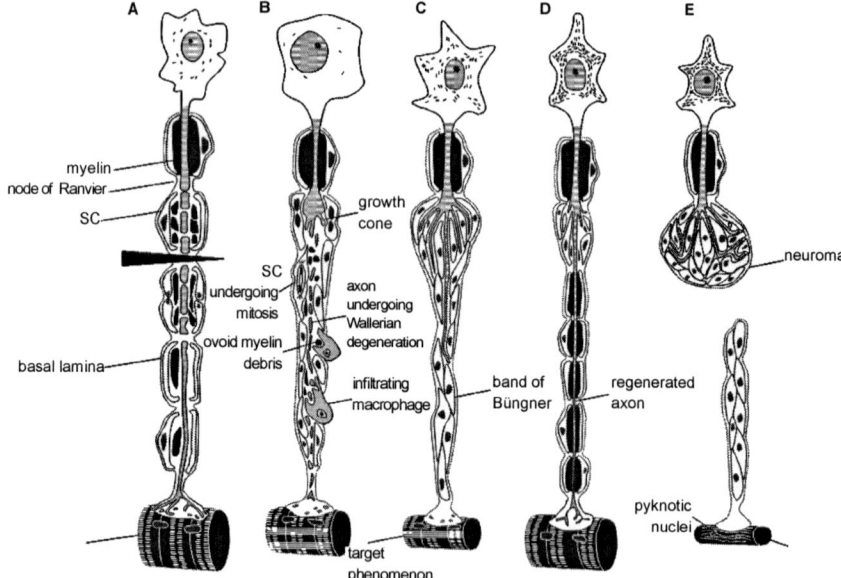

Fig. 1. **Schematic representation of the degenerative and regenerative events associated with peripheral nerve injury.** (A) During the early phase (first few days) after axonal injury (arrowhead), local degenerative events are accompanied by both retrograde and anterograde degeneration of axons and myelin. (B) During the intermediate phase (a few days to weeks), the anterograde pattern of Wallerian degeneration proceeds to completion with infiltrating macrophages contributing to the removal of tissue debris, and Schwann cells undergoing mitosis. The axotomised neuronal cell body undergoes reactive, chromatolytic changes and the severed proximal end of the axon develops regenerative axonal sprouts. (C) Of the numerous axonal sprouts that successfully traverse the injury site (during the first few weeks to months), some re-enter appropriate endoneurial tubes and continue to extend through the distal nerve stump. The target organ/tissue (in this case skeletal muscle) undergoes disuse atrophy. (D) Successful axon regeneration and the re-establishment of neurotransmission at the neuromuscular junction results in the retraction or dying-back of unsuccessful axon sprouts, the reversal of muscle fibre atrophy, of neuronal cell body chromatolysis and the establishment of maturing Schwann cell–axon interactions (including reduced internodal spacing). (E) Failure of regenerating axonal sprouts to cross the injury site (possibly due to the formation of a physical scarring barrier or the loss of a large segment of nerve) results in neuroma formation. The permanently denervated muscle fibres demonstrate severe atrophy, loss of their characteristic striations and pyknotic nuclei. Adapted with permission from (Deumens 2010). Copyright (2010) Elsevier. (Illustration designer Peter R. Schneider; Universitätszeichner, Bern, Switzerland).

repair. In such situations, even if regenerating axons are able to bridge the gap and penetrate the distal nerve stump, other confounding issues, such as inappropriate target innervation and polyneuronal innervation of motor end-plates contribute to a reduced efficiency of repair (Fig. 1). Thus only 40–50% of patients receiving autografts show functional benefits (for review, see Deumens et al. 2010). Furthermore, the autograft strategy presents numerous drawbacks:

(i) need for additional surgery to harvest the donor nerve,
(ii) limited availability of transplant material,
(iii) loss of function at the donor site,
(iv) risk of painful neuroma formation at the donor site (Fig. 1).

Enormous effort has been devoted to the development of alternative intervention strategies which may supplement or even replace the autograft in PNI. At present, only simple conduits have been approved by the US Food and Drug Administration (FDA) and the Conformité Européenne (CE) for use in human applications. These conduits are made from bovine type I collagen, poly(glycolic acid) (PGA) or poly(DL-lactide-ε-caprolactone) (Schlosshauer et al. 2006; Deumens et al. 2010). As described below, the scope for developing more sophisticated and efficacious micro- or nano-structured devices is, therefore, clearly evident.

IN VITRO EFFECTS OF SUBSTRATE TOPOGRAPHY AND ORIENTATION

The ability of any scaffold to efficiently direct axon regeneration across a physical gap is dictated by the presence of orientational cues that follow the longitudinally axis of the nerve guide. Throughout the development of conduits aimed at promoting peripheral nervous system (PNS) repair, such orientational cues have been provided by the walls of simple hollow conduits, the presence of longitudinally orientated fibres or pores, or even the presence of orientated ECM with or without accompanying axon growth promoting glia (e.g., Schmidt and Leach 2003). Indications that biomaterial microstructure may be modified to mimic, to a greater or lesser extent, the longitudinal geometry of peripheral nerves clearly demonstrated that the micron scale topography of fibres or grooves exerts profound effects on the guidance and orientation of regenerating cells and their processes (for reviews see Schmidt and Leach 2003; Deumens et al. 2010). It was reported that the greater surface curvature of small diameter fibres was able to promote an increasing tendency for longitudinally orientated cell process and neurite outgrowth in comparison to larger diameter fibres (i.e., diameter range studied: 5–500µm diameter) (Smeal et al. 2005; Smeal and Tresco 2008). The ability of fibres with diameters in the sub-micron range to exert a similarly powerful influence on the direction of cell migration and process growth by primary neural cell populations of the PNS was recently demonstrated (Yang et al. 2005; Schnell et al. 2007; Corey et al. 2007; Chow et al. 2007). Process extension and cell migration by mammalian Schwann cells, olfactory ensheathing cells and fibroblasts were all strongly influenced by the presence of orientated poly(ε-caprolactone)

(PCL) nanofibres, as was the direction of axonal growth from dorsal root ganglion (DRG) neurons (Schnell et al. 2007). Furthermore, the simple approach of bulk functionalizing PCL nanofibres by blending with 25% type-I collagen (C/PCL) prior to electrospinning resulted in significantly improved cell-substrate interactions (Schnell et al. 2007). The profound effect of nanofibre orientation on DRG neurite alignment and extent of growth as well as Schwann cell migration was confirmed and quantified using video microscopy and Fourier analysis (Corey et al. 2007; Chow et al. 2007; Klinkhammer et al. 2009; Bockelmann et al. 2010). An excellent example of this effect was provided by Bellamkonda and colleagues who demonstrated axonal and Schwann cell outgrowth from early postnatal rat pup DRG that had been explanted onto films of orientated or non-orientated poly(acrylonitrile-co-methylacrylate) (PAN-MA) nanofibres (Kim et al. 2008; Fig. 2). The influence of substrate orientation on glial cell form and function was further extended by demonstrating that human Schwann cell association with PCL nanofibres was related to the down-regulation of neurotrophin and neurotrophin receptor mRNA, as well as the up-regulation of myelin-related mRNAs for myelin associated glycoprotein (MAG) and P0, while the vast majority of genes, including myelin basic protein (MBP) and peripheral myelin protein 22 (PMP22), remained unaffected. In their publication, which is one of the few studies addressing the effects of electrospun nanofibres on gene expression, the up-regulation of some myelin-related genes was interpreted as an indication of nanofibre-induced Schwann cell differentiation (Chew et al. 2008).

THE DEVELOPMENT OF BIOMATERIALS FOR USE IN A NANOTECHNOLOGICAL APPROACH TO PNS REPAIR

The natural architecture of the wide range of tissues found in the body is usually of substantial relevance to their particular function. In the PNS, the geometry or topographical architecture of its component cells, processes and ECM plays a pivotal role in optimizing its ability to regenerate after injury (Pettigrew et al. 2001; Deumens et al. 2010). A widely adopted approach for tissue engineering of biomimetic devices is to attempt to reproduce elements of the native ECM architecture within the scaffold design. Since the ECM has a major influence on adherent cell form and function through its complement of signalling molecules and its topographical organization, the incorporation of entire ECM molecules (or more specifically, their functional peptide sequences) into scaffolds with controlled geometry or topography is anticipated to create a suitable environment for axon regeneration and functional recovery (Deumens et

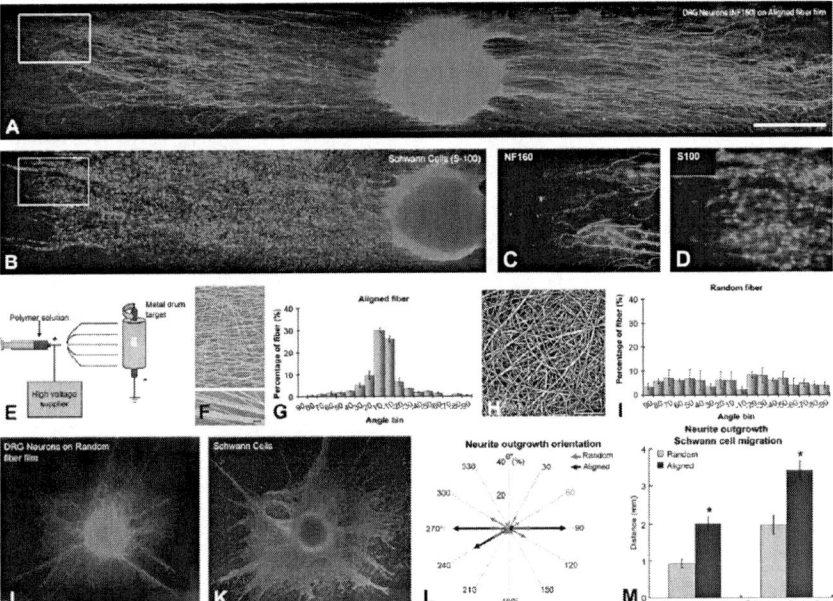

Fig. 2. Dorsal root ganglia (DRG) on aligned and randomly orientated electrospun nanofibres *in vitro*. (A) Extensive and orientated NF160-positive DRG neurons and axons on aligned nanofibres. (B) Extensive and orientated migration of S-100-positive Schwann cells on aligned nanofibres. (C) Higher power magnification of NF160-positive axonal profiles (from boxed area in A). (D) Higher power magnification of S-100-positive Schwann cell profiles (from boxed area in B). (E) Schematic diagram of aligned fibre fabrication by using a rotating drum. Random fibres were obtained using a flat target instead of on a rotating drum. (F) and (H) Scanning-electron microscopy of aligned and randomly orientated fibres, respectively. (G) and (I) Distribution of fibre alignment in aligned and randomly orientated samples, respectively. (J) Non-orientated, NF160-positive axon growth from DRG neurons on random nanofibres. (K) Non-orientated, S-100-positive Schwann cell migration on random nanofibres. (L) Quantitative comparison of neurite orientation and outgrowth on aligned and random fibres. Direction of arrows indicates the orientation of neurite outgrowth, and length of arrows indicates the rate of occurrence (percentage). (M) Quantitative comparison of the extent of neurite outgrowth and Schwann cell migration on the nanofibres. * $p < 0.05$. Error bars represent s.e.m.; scale bars: (F) 1 μm, (H) 30 μm, (J) and (K) 500 μm. Adapted and modified with permission from (Kim 2008). Copyright (2008) Elsevier.

al. 2010). The use of nanofibres in such an approach generates a substrate that, to some extent, mimics the sub-micron scale of the ECM and presents the advantage of an extremely high surface area-to-volume ratio which may optimize the delivery or presentation of bioactive molecules to the cell types expected to interact with the scaffold. Of the numerous approaches adopted for the generation of nanofibres, the present chapter will focus on the electrospinning method.

316 Nanomedicine and the Nervous System

ELECTROSPINNING

The process of electrospinning has been shown to be efficient for the fabrication of polymer fibres ranging in diameter from micrometres down to nanometres. In the electrospinning process, a polymer solution is pumped at a flow rate (usually below 1ml/h) through a syringe (spinneret), thereby forming a drop at the tip of the needle. Using a high voltage connected to the syringe, a strong electric field is created between the spinneret and the grounded collectors (target). When the electrical potential is increased beyond a critical value, electric repulsion overcomes the surface tension of the polymer drop, and a jet of polymer is ejected. The movement of charged molecules in the electromagnetic field results in a spiralling trajectory towards the collectors, whereby the solvent evaporates and thin, stretched polymer fibres are formed on the target. The basic collector, a single metal plate, renders non orientated, mesh like fibres (Fig 3A). However, the collection of fibres can be manipulated to achieve orientation (Fig. 3B-H). A routine method for orientated fibre collection is

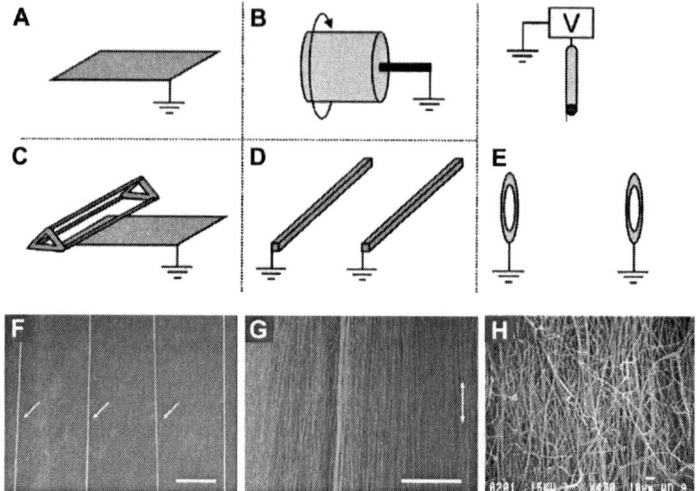

Fig. 3. Electrospinning of orientated and non-orientated nanofibres. (A-E) Schematic representation of various electrospinning collection systems showing (A) single plate configuration, (B) rotating drum, (C) triangular frame placed near single plate, (D) parallel dual plate and (E) the dual-grounded ring configuration. (F) Scanning-electron microscopy of low and (G) high density, orientated collagen poly(ε-caprolactone) nanofibres (e.g., arrows in F) that are obtained by the use of parallel dual plates. (H) Scanning-electron microscopy of orientated collagen nanofibres electrospun onto a rotating drum. Scale bars: (F) 30 µm and (G) 200 µm (H) 10 µm. (A-E) Adapted with permission from (Dalton 2005). Copyright (2004) Elsevier. (F and G) Adapted with permission from (Gearardo-Nava 2009) Copyright (2009) Future Medicine. (H) Adapted with permission from (Matthews 2002). Copyright (2002) American Chemical Society.

to electrospin onto a rotating drum collector (Matthews et al. 2002; Kim et al. 2008; Fig. 3B, H). An alternative approach is to generate suspended mats of orientated fibres between two parallel collectors, known as the gap method of alignment (Dalton et al. 2005; Gerarado-Nava et al. 2009; Fig. 3 D, E, F and G). Others have constructed hollow conduits from micro- and nanofibres by electrospinning onto a rotating copper wire which was removed at the end of the spinning process (Panseri et al. 2008).

THE CHOICE OF THE POLYMER

As mentioned earlier, the choice of polymer to generate the nanofibres will play a significant role in the ultimate degree of success of the scaffold in supporting axon regeneration and functional tissue repair. Certain material characteristics have already been identified as being ideal for optimal scaffold-host integration (see key facts of an optimal scaffold).

To date, a number of polymers have been used to produce fibres by means of the electrospinning technique. These can be classified as natural, synthetic and functionalized synthetic polymers.

NATURAL POLYMERS

Natural polymers possess an array of binding and signalling sequences that are important for promoting cell interactions and functional tissue repair. Collagen is a major ECM component that is readily manipulated for both hard- and soft tissue engineering and can form fibres following extrusion as well as electrospinning (e.g., Fig. 3H). However, relatively few natural polymers have been electrospun as substrates for nervous tissue repair. These include laminin, chitosan and gelatin (denatured collagen) (Cao et al. 2009; Cunha et al. 2010). The high cost of production, possible immunogenicity, stability and batch-to-batch variations in purity all limit their potential in nerve repair applications. In contrast to a number of synthetic polymers, personal experience has indicated that the electrospinning of high quality nanofibres from ECM proteins may prove difficult.

SYNTHETIC POLYMERS

In contrast to the above mentioned disadvantages associated with natural polymers, the purity of synthetic polymers is well defined and the composition can be designed to minimize immunological and inflammatory responses. Both biodegradable and non-biodegradable polymers have been considered for the development of nanofibre-based nerve regeneration strategies, but the degradable polymers would present the long term advantage of avoiding or minimizing foreign body reactions. Nanofibres from poly(lactic-co-glycolic acid) (PLGA), poly(L-lactic acid) (PLLA) and PCL are examples of biodegradable poly(α-hydroxy acids) that have been electrospun for nervous tissue repair. These polymers can be degraded by hydrolysis of their ester linkage resulting in degradation products with minimal toxicity to the host (for a review Cao et al. 2009). Examples of non-biodegradable polymers that have been electrospun and demonstrated to support an impressive degree of neural cell interactions *in vitro* (Fig. 2) as well as *in vivo* (Fig. 5) include PAN-MA (Kim et al. 2008; Clements et al. 2009). Nanofibre substrates from other synthetic polymers that have been investigated for nervous tissue repair include polyamide, poly(ε-caprolactone-co-ethyl ethylene phosphate) (PCLEEP), polydioxanone and a copolymer of methyl methacrylate and acrylic acid (PMMAAA) (Cao et al. 2009; Cunha et al. 2010).

A substantial disadvantage of synthetic polymers is their lack of specific cell recognition- and binding sites. The cellular interactions that such polymers support are largely due to electrostatic influences and the unspecific adsorption of serum- or medium containing proteins (e.g., fibronectin) to the polymer surface (Klinkhammer et al. 2010). Therefore, more recent strategies have attempted to tailor the functionalization of electrospun synthetic polymer nanofibres by coupling ECM molecules, integrin-activating peptide sequences or growth factors to the fibres themselves.

FUNCTIONALIZATION OF SYNTHETIC POLYMERS

The combination of both synthetic and natural-occurring materials offers the most promising approach to obtain scaffolds with the appropriate compliment of desired properties for PNS repair, including a defined and controllable degradation rate, the presence of cell recognition- and binding sites, or the controlled presentation and release of growth promoting molecules. A range of techniques for the functionalization of PLLA nanofibres with the ECM molecule, laminin, has been assessed. Comparison of polymer blending, covalent binding and simple adsorption revealed that blending was the simplest and most efficient method of fibre functionalization. Furthermore, PLLA-laminin blended nanofibres supported greater neurite outgrowth by PC12 neuroblasts than that observed with the other types of functionalization. This improved performance of the PLLA-laminin fibres, however, did not enhance cell viability (Koh et al. 2008). An additional advantage of the blending method over any modification applied after spinning is that the biological activity thus remains confined to the fibres and is absent on the surrounding two- or three-dimensional substrate.

Blending of 25% type-I collagen into PCL (C/PCL) nanofibres has also been demonstrated to enhance cell-substrate interactions with a range of PNS neural and non-neural cells types. Schwann cell-, olfactory nerve ensheathing cell- and fibroblast migration and process extension were all significantly improved on the blended C/PCL nanofibres. Furthermore, neurite orientation and outgrowth were also greater on C/PCL than on pure PCL nanofibres (Schnell et al. 2007). As an alternative to the incorporation of large, complex proteins to the surface of nanofibres, recent interest has developed in the use of short peptide sequences. This presents the advantages of using relatively inexpensive peptides of defined purity and sterility, as well as allowing specific cell binding sequences to be chosen for tailored functionalization.

The most commonly used peptide for surface modification is argine-glycine-aspartate (RGD) and its extended variant glycine-argine-glycine-aspartate-serine (GRGDS), both representing one of the many signalling domains derived from fibronectin, laminin and other ECM molecules (e.g., Klinkhammer et al. 2010). Recently, the modification of PCL fibres with star-shaped NCO-poly(ethylene glycol)-stat-poly(propylene glycol) (sPEG) as a covalent linker to incorporate the peptide GRGDS has been demonstrated (Klinkhammer et al. 2010). Comparison of pure PCL-fibres, PCL/sPEG-fibres and PCL/sPEG/GRGDS-fibres showed that functionalized fibres supported faster Schwann cell migration and greater axonal growth than non-functionalized fibres (Bockelmann et al. 2010). Others have demonstrated that covalently binding of sequences of fibronectin type III domains from the human tenascin-C molecule (i.e., VFDNFVLKIRDTKKQ and its extended form ADEGVFDNFVLKIRDTKKQ) to the surface of polyamide nanofibres resulted in enhanced neuronal attachment and neurite outgrowth compared to non-functionalized nanofibres (Ahmed et al. 2006).

In addition to ECM molecules and adhesion-related peptide sequences, nanofibres have been functionalized for the incorporation and delivery of certain growth factors. For example, PLLA nanofibres have been functionalized by covalent binding of heparin to the poly(ethylene glycol)-coated surface, followed by the addition of heparin-binding proteins, basic fibroblast growth factor (bFGF) and laminin (Fig. 4A and B). The effects of orientated nanofibres presenting laminin and/or bFGF were demonstrated to induce enhanced axon growth and cell migration (Patel et al. 2007, Fig 4C). Similarly, the loading of nerve growth factor (NGF) to PCL nanofibres has been reported to be facilitated by the introduction of

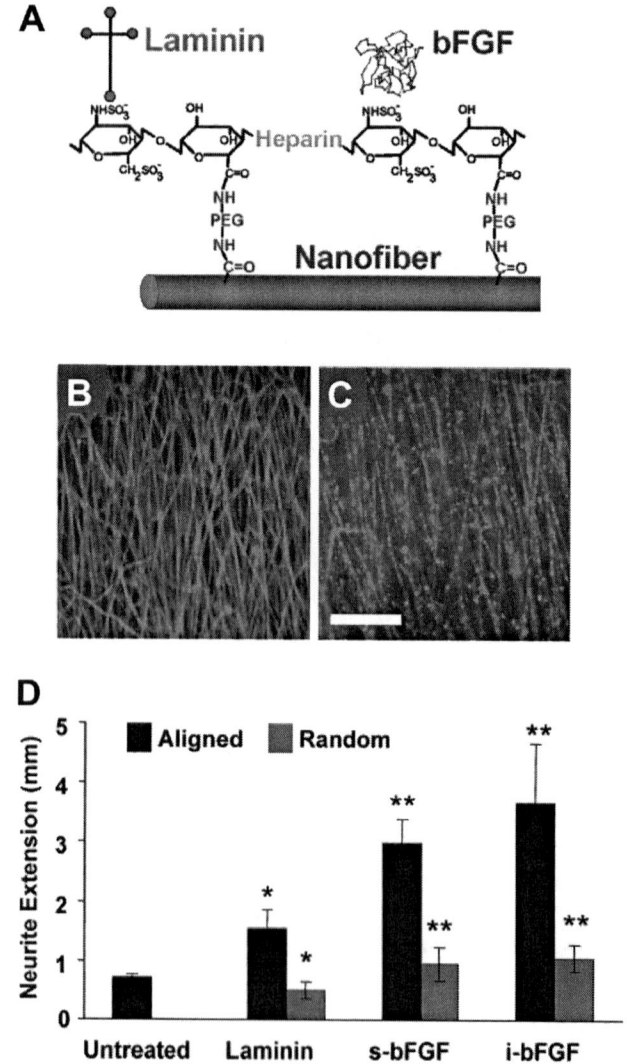

Fig. 4. Funcionalization of electrospun nanofibres with laminin and/or basic fibroblast growth factor (bFGF). (A) Immobilization of the heparin binding proteins bFGF and laminin on poly(L-lactic acid) nanofibres using poly(ethylene glycol) and heparin as linkers. (B) and (C) Immunocytochemistry for the detection of immobilized laminin and bFGF, respectively. (D) Quantitative assessment of the neurite outgrowth promoting effects of functionalized aligned and randomly orientated nanofibres. The values for all aligned samples were significantly greater ($p < 0.05$) than their respective randomly orientated samples with same chemical functionalization. *Significant difference ($p < 0.05$) compared to respective (random or aligned) untreated samples. **Significant difference ($p < 0.05$) compared to respective (random or aligned) untreated and laminin samples. Scale bars: (B) and (C) 10 μm. Adapted with permission from (Patel 2007). Copyright (2007) American Chemical Society.

bovine serum albumin (BSA) in electrospun PCL nanofibres. The presence of BSA allowed the controlled release of NGF from the fibre surface over a period of 28 days (Valmikinathan et al. 2009).

The ability to tailor the surface functionalization of nanofibre substrates presents the major advantage in regenerative medicine of the PNS by enabling enhanced Schwann cell migration, resulting in the implant being seeded by migrating and proliferating host Schwann cells. These cells in turn, would provide the range of surface adhesion molecules, neurotrophic factors and ECM molecules known to be beneficial for supporting axon regeneration. Such a self-seeding capacity of the implanted scaffold would circumvent the time-consuming processes of isolation and proliferation of autologous Schwann cells required for *in vitro* seeding and subsequent implantation.

IN VIVO EFFECTS OF IMPLANTING NANOFIBRE-CONTAINING AXON GROWTH PROMOTING SCAFFOLDS OR CONDUITS

Although numerous *in vitro* studies have been performed on neural cell-nanofibre interactions, to date relatively few *in vivo* investigations have been performed to assess the extent of functional repair following the implantation of nanofibre-containing scaffolds/conduits into traumatically injured PNS.

The simplest application of electrospinning technology has been to prepare membranes from non-orientated nanofibres which could then be rolled to form the walls of hollow conduits (Chew et al. 2007). A modification of this approach was to electrospin nanofibres directly onto a single rotating wire collector which was adopted for the preparation of electrospun PLGA/PCL conduits that were used to bridge relatively small (i.e., 1cm) gaps of the adult rat sciatic nerve (Panseri et al. 2008). Axon growth through the conduit and remyelination of regenerated axons was observed over a four months period. However, some axonal growth was mis-routed through the wall of the nanofibre conduit and could be detected on its external surface. The conduits supported the progressive recovery of tactile sensory function (as indicated by the von Frey test) as well as a partial recovery of evoked plantar compound muscle action potentials. Although such conduits appeared biocompatible by the absence of any noticeable encapsulation, multi-nucleated giant cells could be detected within the conduit wall as an indication of a foreign body response (Panseri et al. 2008). Detailed *in vivo* studies on the influence of nanofibre topography on the foreign body response suggested that aligned nanofibres minimize the host response and promote increased graft-host integration (Cao et al. 2010). Hollow conduits with walls composed of aligned PCL/PCLEEP electrospun nanofibres (directed longitudinally or circumferentially) have been shown to support axon regeneration across 15mm gaps of the rat sciatic nerve over a period of three months. Surprisingly, both longitudinally- and circumferentially orientated nanofibre conduits supported the same degree of axon regeneration but the incorporation of the glial cell line-derived neurotrophic factor (GDNF) into longitudinally orientated PCLEEP fibres further promoted the degree of axon regeneration and functional recovery (Chew et al. 2007).

An alternative approach for constructing nanofibre-based conduits has been to electrospin fibres prepared from PAN-MA onto the surface of a spinning drum, effectively generating a thin film of densely packed and orientated nanofibres. The films have then been removed, trimmed to an appropriate dimension and 10–12 samples stacked on top of each other to form the inner, longitudinally orientated scaffolds within hollow polysulphone nerve conduits (Kim et al. 2008). The tissue regeneration supported by these crudely designed conduits over a four month period was determined across moderate sized lesion (17 mm) of the adult rat tibial nerve. A surprisingly good quality of tissue repair was supported by such non-biodegradable PAN-MA conduits in the absence of any prior biochemical functionalization or cell-seeding. Behavioural analysis (grid walk), electrophysiology and correlative morphometric analysis of lesioned animals receiving PAN-MA nanofibre-containing conduits demonstrated fewer foot-falls, recovery of motor and sensory compound action potentials (although signal latency was substantially delayed), and axon regeneration across the lesion which resulted in the re-establishment of neuromuscular junctions on the previously denervated skeletal muscles (Kim et al. 2008). The scaffolds containing aligned nanofibres performed significantly better than those containing non-orientated nanofibres and for several tests close to or the same as autografted controls. The surprisingly good tissue repair achieved with non-degradable PAN-MA conduits indicated a number of advantages of such conduits over conventional hollow conduits or even autografts: PAN-MA conduits (or better still, conduits made from biodegradable polymers) could easily be prepared in a ready-to-use (off the shelf) manner with unlimited availability, they could be sterilized and stored, and would be easy to handle. Also, it was suggested that it would be possible to pre-customize the conduits for particular types of nerve injury, for example by pre-setting the conduit diameter or length (Kim et al. 2008). Despite the rather surprising observations reported by Kim and colleagues with the non-functionalized polymer conduit, it has now been widely accepted that the most efficacious designs of conduits or scaffolds for PNS repair should incorporate extra components such as growth factors, ECM molecules or axon growth promoting cells. Each of these aspects of conduit/scaffold improvement has been the subject of more detailed review articles (cited in Deumens et al. 2010). Refinement of the electrospinning approach allowed the positioning of extremely fine films of aligned, electrospun PAN-MA into the hollow polysulphone conduits. This more sophisticated approach demonstrated that superior axon growth and conduction velocity of regenerating axons was supported by single PAN-MA films rather than multiple films (Clements et al. 2009, Fig. 5). This data supports the principle already highlighted by others (Ngo et al. 2003) that minimal levels of appropriately positioned topographical cues are capable of supporting substantial axon regeneration and functional tissue repair.

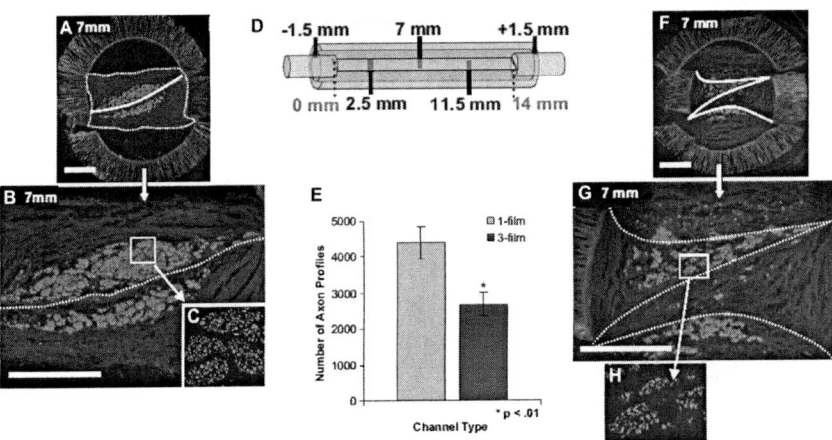

Fig. 5. Cross-section of polysulphone conduits containing either one or three films of aligned, electrospun nanofibres, taken at the midpoint of the conduit (a distance of 7 mm) 13 weeks after implantation into the lesioned tibial nerve. (A) The brightly fluorescent core of NF160-positive axons is consolidated around the single aligned thin-film (marked by a solid white line). (B) and (C) Magnified images of NF160-positive axons shown in (A). Near the channel midpoint, axons were aggregated into isolated groups. (D) Schematic representation of the nerve guidance conduit that was grafted into a 14 mm gap after transection of the tibial nerve of adult rats. (E) Axon profiles were quantified at the midpoint of each channel (7 mm) at the 13 week time point. Despite containing smaller regeneration cables, the 1-film channels supported significantly higher numbers of regenerating axonal profiles. (F) The axonal core was fragmented asymmetrically around the three thin films, the approximate locations of which are marked with white dotted lines. (G) and (H): Magnified images of (F). Axons showed a tendency to aggregate in tight groups near the channel midpoint. Error bars represent s.e.m.; scale bars: (A), (B), (F) and (G) 400μm. Adapted and modified with permission from (Clements 2009). Copyright (2009) Elsevier.

Applications to other Areas of Health and Disease

As mentioned earlier, electrospinning can be used to generate aligned as well as randomly orientated substrates from a range of natural and synthetic polymers. This versatility, in particular for spinning randomly orientated nanofibres into sheets with a high, but controllable degree of porosity and functionalization has resulted in the technology being adopted for a wide range of health (e.g., filtration, barrier fabrics, wipes, personal care) and disease related applications (e.g., for regenerative medicine related to either soft- or hard tissues including skin-, vascular-, cardiac -, muscular-, skeletal and CNS systems; Sell et al. 2009). Nanofibre scaffolds, for example, have been used for dressing wounds where they reduce fluid loss, inhibit invasion by micro-organisms and support rapid wound healing. Electrospinning of tubular structures has been developed for vascular repair strategies, and meshes of aligned nanofibres have been used as a substrate for the generation of longitudinally organised myotubes. The application of electrospinning to diseases and disorders of the many body systems is largely in its infancy, for example the influences of nanofibre-based substrates on CNS glial- and neuronal cell form and function although substantial, remains somewhat undefined. Nonetheless, it is anticipated that the advantages presented by electrospinning will be further developed and exploited in regenerative medicine.

Key Facts for Optimal Scaffold Design

- The material should be biocompatible to prevent inflammation and immune responses after implantation.
- Biodegradable materials are preferable because the scaffold should eventually be replaced by host cells and their related ECMs. In this regard the degradation rate may play an important role in graft-host integration but the ideal rate has yet to be determined. Scaffolds from non-degradable materials have the risk of encapsulation and eventual constriction.
- The products of scaffold degradation should be non-toxic.
- The mechanical properties of the scaffold should be similar to those of the surrounding host tissues to avoid local irritation/tissue damage at the graft-host interface and prevent scaffold collapse during movement-associated muscular activity.
- The scaffold should be sufficiently porous to allow the exchange of nutrients and metabolites to maintain cell viability and function, and should avoid infiltration by cells that interfere with functional recovery.
- Orientational cues should ideally be present in the scaffold to promote directed axon regeneration across the lesion site.

Key Facts for Electrospinning of Nanofibres

- The first patent describing the process of developing polymeric fibres via electrospinning was obtained in 1934 by A. Formhals.
- Parameters known to influence the quality of electrospun nanofibres include: the nature and molecular weight of the polymer, the viscosity and conductivity of the polymer solution, the applied voltage, distance between the tip of the needle and the collectors, as well as ambient parameters (temperature, humidity and air velocity).
- Electrospun nanofibres possess the advantage of a high surface area-to volume ratio.
- Orientated as well as non-orientated fibres can be generated using different collection techniques.
- Functionalization of the electrospun nanofibres is possible with a range of bioactive molecules including molecules or peptides derived from the ECM and growth factors.

Definitions

Atrophy: A decrease in size of a body organ, tissue, or cell due to disease, injury or disuse (e.g., disuse and shrinkage of a muscle after peripheral nerve injury).

Autograft: Tissue transplanted from one part of the body to another in the same individual.

Chromatolysis: The dissolution of protein-building elements (Nissl substance) from the cytoplasm of injured neurons.

Dorsal root ganglia: Accumulation of sensory neuronal cell bodies forming well defined and encapsulated nodules on sensory nerves adjacent to the spinal cord.

Extracellular matrix (ECM): A group of molecules produced by cells and secreted into the extracellular space that serves as a scaffold for cell attachment and growth.

Extrusion: The process of forming a polymer into a defined shape or configuration by forcing it through a suitably shaped mould or die.

Foreign body response: An immunological/inflammatory reaction evoked by the detection of foreign material within the body (e.g., tissue-encapsulation of an implant is a means by which the host will try to isolate a foreign element from the rest of the body).

Glia (or glial cell): Glia are non-excitable cells that surround and provide support for neurons (e.g., nutrition, maintaining homeostatic balance of the ionic components of the extracellular fluid, forming myelin, assisting in the clearance of debris after cell damage or death).

Grid walk: a behavioural test that quantifies the number of stepping mistakes made by an experimental animal as it crosses a wire grid-like walkway (of defined spacing) providing a quantifiable measure of motor and sensory function.

Neuroma (traumatic): An overgrowth of nerve fibres and their supporting cells at the distal end of a lesioned peripheral nerve, resulting in the formation of a bulbous swelling.

Pyknotic nucleus: The nucleus of a dying cell that decreases in size and becomes darker in histological stains due to the condensation of its nuclear material.

Schwann cell: A glial or support cell that assists neuronal function by ensheathing or wrapping around axons in peripheral nerves. The degree of insulation for electrical conduction is dependent on the amount of fatty myelin sheath that they produce. Schwann cells provide a range of surface adhesion molecules, neurotrophic factors and ECM molecules known to be beneficial for supporting axon regeneration.

von Frey test: A test that uses monofilaments of varying thickness to determine the threshold force (usually applied to the underside of the paw) which provokes a withdrawal reflex.

Wallerian degeneration: Degeneration of a nerve fibre and its myelin sheath distal to the point of injury following the lesion-induced separation of the axon from its cell body.

Summary Points

- The complications associated with the harvesting of autologous nerve grafts and their moderate success in promoting functional tissue repair highlight the need for the development of bioengineered scaffolds, the functional and mechanical properties of which can be tailored to promote optimal repair.
- *In vitro* experiments have revealed the importance of topographical cues, namely fibre diameter in the sub-micron range and orientation, in supporting directional axon regeneration, cell growth and migration.
- Electrospinning provides a simple and versatile technique for generating nanofibres, and the range collection methods available allows for control of fibre orientation and density.
- The advantages presented by both synthetic and natural polymers can be combined in the generation of orientated, functionalized nanofibres presenting both physical and chemical cues capable of mimicking the ECM. Such nanofibres represent excellent substrates for inclusion in peripheral nerve repair strategies.

- The few *in vivo* studies that have been performed to date have demonstrated the potential of non-functionalized nanofibre-based constructs in peripheral nerve repair. However, the introduction of bioactive molecules (i.e., whole ECM or growth factor molecules, or peptide sequences that code for biological activity) promise to enhance the degree of graft-host integration and functional recovery.

Abbreviations

bFGF	:	Basic fibroblast growth factor
BSA	:	Bovine serum albumin
CE	:	Conformité Européenne
C/PCL	:	Collagen/poly(ε-caprolactone)
DRG	:	Dorsal root ganglion
ECM	:	Extracellular matrix
FDA	:	Food and drug administration
GDNF	:	Glial cell line-derived neurotrophic factor
GRGDS	:	Glycine-argine-glycine-aspartate-serine
MAG	:	Myelin associated glycoprotein
MBP	:	Myelin basic protein
NGF	:	Nerve growth factor
PAN-MA	:	Poly(acrylonitrile-co-methylacrylate)
PCL	:	Poly(ε-caprolactone)
PCLEEP	:	PCL- ethyl ethylene phosphate
PGA	:	Poly(glycolic acid)
PMMAAA	:	Methyl methacrylate and acrylic acid
PMP22	:	Peripheral myelin protein 22
PLA	:	Poly(lactic acid)
PLGA	:	Poly(lactic-co-glycolic acid)
PLLA	:	Poly(L-lactic acid)
PNI	:	Peripheral nerve injury
PNS	:	Peripheral nervous system
RGD	:	Argine-glycine-aspartate
sPEG	:	Star-shaped NCO-poly(ethylene glycol)-stat-poly(propylene glycol)

References

Ahmed, I., H. Liu, P.C. Mamiya, A.S. Ponery, A.N. Babu, T. Weik, M. Schindler and S. Meiners. 2006. Three-dimensional nanofibrillar surfaces covalently modified with tenascin-c-derived peptides enhance neuronal growth *in vitro*. J. Biomed. Mater. Res. A. 76: 851–860.

Bockelmann, J., K. Klinkhammer, A. von Holst, N. Seiler, A. Faissner, G.A. Brook, D. Klee and J. Mey. 2010. Functionalization of electrospun poly(ε-caprolactone) fibers with the extracellular matrix-derived peptide grgds improves guidance of schwann cell migration and axonal growth. Tissue Eng Part A. (online ahead of print).

Cao, H., T. Liu and S.Y. Chew. 2009. The application of nanofibrous scaffolds in neural tissue engineering. Adv. Drug Deliv. Rev. 61: 1055–1064.

Cao, H., K. McHugh, S.Y. Chew and J.M. Anderson. 2010. The topographical effect of electrospun nanofibrous scaffolds on the *in vivo* and *in vitro* foreign body reaction. J. Biomed. Mater. Res. A. 93: 1151–1159.

Chew, S.Y., R. Mi, A. Hoke and K.W. Leong. 2007. Aligned protein-polymer composite fibers enhance nerve regeneration: A potential tissue-engineering platform. Adv. Funct. Mater 17: 1288–1296.

Chew, S.Y., R. Mi, A. Hoke and K.W. Leong. 2008. The effect of the alignment of electrospun fibrous scaffolds on schwann cell maturation. Biomaterials 29: 653–661.

Chow, W.N., D.G. Simpson, J.W. Bigbee and R.J. Colello. 2007. Evaluating neuronal and glial growth on electrospun polarized matrices: bridging the gap in percussive spinal cord injuries. Neuron Glia. Biol. 3: 119–126.

Clements, I.P., Y.T. Kim, A.W. English, X. Lu, A. Chung and R.V. Bellamkonda. 2009. Thin-film enhanced nerve guidance channels for peripheral nerve repair. Biomaterials 30: 3834–3846.

Corey J.M., D.Y. Lin, K.B. Mycek, Q. Chen, S. Samuel, E.L. Feldman and D.C. Martin. 2007. Aligned electrospun nanofibers specify the direction of dorsal root ganglia neurite growth. J. Biomed. Mater. Res. A. 83: 636–645.

Cunha, C., S. Panseri and S. Antonini. 2010. Emerging nanotechnology approaches in tissue engineering for peripheral nerve regeneration. Nanomedicine. (online publication ahead of print).

Dalton, P.D., D. Klee and M. Möller. 2005. Electrospinning with dual collection rings. Polymer 46: 611–614.

Deumens, R., A. Bozkurt, M.F. Meek, M.A.E. Marcus, E.A.J. Joosten, J. Weis and G.A. Brook. 2010. Repairing injured peripheral nerves: Bridging the gap. Prog. Neurobiol. 92: 245–76.

Gerardo-Nava, J., T. Führmann, K. Klinkhammer, N. Seiler, J. Mey, D. Klee, M. Möller, P.D. Dalton and G.A. Brook. 2009. Human neural cell interactions with orientated electrospun nanofibers *in vitro*. Nanomedicine 4: 11–30.

Kim,Y.T., V.K. Haftel, S. Kumar and R.V. Bellamkonda. 2008. The role of aligned polymer fiber-based constructs in the bridging of long peripheral nerve gaps. Biomaterials 29: 3117–3127.

Klinkhammer, K., J. Bockelmann, C. Simitzis, G.A. Brook, D. Grafahrend, J. Groll, M. Möller, J. Mey and D. Klee. 2010. Functionalization of electrospun fibers of poly(epsilon-caprolactone) with star shaped nco-poly(ethylene glycol)-stat-poly(propylene glycol) for neuronal cell guidance. J. Mater. Sci. Mater. Med. 21: 2637–2651.

Klinkhammer, K., N. Seiler, D. Grafahrend, J. Gerardo-Nava, J. Mey, G.A. Brook, M. Möller, P.D. Dalton and D. Klee. 2009. Deposition of electrospun fibers on reactive substrates for *in vitro* investigations. Tissue Eng. Part C. Methods 15: 77–85.

Koh, H.S., T. Yong, C.K. Chan and S. Ramakrishna. 2008. Enhancement of neurite outgrowth using nano-structured scaffolds coupled with laminin. Biomaterials 29: 3574–3582.
Matthews, J.A., G.E. Wnek, D.G. Simpson and G.L. Bowlin. 2002. Electrospinning of collagen nanofibers. Biomacromolecules 3: 232–238.
Ngo, T.T.B., P.J. Waggoner, A.A. Romero, K.D. Nelson, R.C. Eberhart and G.M. Smith. 2003. Poly(l-lactide) microfilaments enhance peripheral nerve regeneration across extended nerve lesions. J. Neurosci. Res. 72: 227–238.
Panseri, S., C. Cunha, J. Lowery, U. Del Carro, F. Taraballi, S. Amadio, A. Vescovi and F. Gelain. 2008. Electrospun micro- and nanofiber tubes for functional nervous regeneration in sciatic nerve transections. BMC Biotechnol. 8.
Patel, S., K. Kurpinski, R. Quigley, H. Gao, B.S. Hsiao, M.M. Poo and S. Li. 2007. Bioactive nanofibers: synergistic effects of nanotopography and chemical signaling on cell guidance. Nano Lett. 7: 2122–2128.
Pettigrew, D.B., K.P. Shockley and K.A. Crutcher. 2001. Disruption of spinal cord white matter and sciatic nerve geometry inhibits axonal growth *in vitro* in the absence of glial scarring. BMC Neurosci. 2.
Schlosshauer, B., L. Dreesmann, H.E. Schaller and N. Sinis. 2006. Synthetic nerve guide implants in humans: a comprehensive survey. Neurosurgery 59: 740–7; discussion 747–8.
Schmidt, C.E. and J.B. Leach. 2003. Neural tissue engineering: strategies for repair and regeneration. Annu. Rev. Biomed. Eng. 5: 293–347.
Schnell, E., K. Klinkhammer, S. Balzer, G. Brook, D. Klee, P. Dalton and J. Mey. 2007. Guidance of glial cell migration and axonal growth on electrospun nanofibers of poly-epsilon-caprolactone and a collagen/poly-epsilon-caprolactone blend. Biomaterials 28: 3012–3025.
Sell, S.A., M.J. McClure, K. Garg, P.S. Wolfe and G.L. Bowlin. 2009. Electrospinning of collagen/biopolymers for regenerative medicine and cardiovascular tissue engineering. Adv. Drug Deliv. Rev. 61: 1007–1019.
Smeal, R.M., R. Rabbitt, R. Biran and P.A. Tresco. 2005. Substrate curvature influences the direction of nerve outgrowth. Ann. Biomed. Eng. 33: 376–382.
Smeal, R.M. and P.A. Tresco. 2008. The influence of substrate curvature on neurite outgrowth is cell type dependent. Exp. Neurol. 213: 281–292.
Valmikinathan, C.M., S. Defroda and X. Yu. 2009. Polycaprolactone and bovine serum albumin based nanofibers for controlled release of nerve growth factor. Biomacromolecules 10: 1084–1089.
Yang, F., R. Murugan, S. Wang and S. Ramakrishna. 2005. Electrospinning of nano/micro scale poly(l-lactic acid) aligned fibers and their potential in neural tissue engineering. Biomaterials 26: 2603–2610.

17

Bridging the Gap for Regeneration after CNS Injury by Nanofiber Scaffolds: use of SAPNS for Traumatic Brain and Spinal Cord Injury

Jiasong Guo,[1,a,5] *Gilberto K. K. Leung*[6] *and Wutian Wu*[1,b,2,3,4,]*

ABSTRACT

Central nervous system (CNS) injury is a group of serious conditions for which only a limited number of treatment strategies are currently available. Within the injured spinal cord or brain,

---

[1]Department of Anatomy, Li Ka Shing Faculty of Medicine, The University of Hong Kong, 21 Sassoon Road, Hong Kong SAR, China.
[a]E-mail: jiasongguo@yahoo.com.cn
[b]E-mail: wtwu@hkucc.hku.hk
[2]State Key Laboratory of Brain and Cognitive Sciences.
[3]Research Center of Reproduction, Development and Growth, Li Ka Shing Faculty of Medicine, The University of Hong Kong, Pokfulam, Hong Kong SAR, China.
[4]Joint Laboratory for Brain Function and Health (BFAH), Jinan University and The University of Hong Kong, Guangzhou, China.
[5]Department of Histology and Embryology, Southern Medical University, Guangzhou, China.
[6]Department of Surgery, Li Ka Shing Faculty of Medicine, The University of Hong Kong, Pokfulam, Hong Kong SAR, China; E-mail: gilberto@hkucc.hku.hk
*Corresponding author

List of abbreviations after the text.

mechanical trauma may result in tissue loss, demyelination and axonal degeneration, as well as a series of cellular and molecular responses. Recent researches indicated that tissue engineering with biomaterial scaffolds is a potential therapeutic strategy for CNS regeneration after trauma. A common approach in tissue engineering is to create three-dimensional structures that mimic the naturally occurring extracellular matrix, which contain fibrous proteins of 50 to 500 nm in diameter. Conventional artificial biomaterials processing techniques are unable to produce fibers smaller than 10 μm in diameter and are therefore not suitable for producing scaffolds for neuro-regenerative purposes. Nanofiber scaffolds are novel biomaterials which consist of individual nano-scale fibers. Nanofiber scaffolds have the advantages of mimicking the extracellular matrix, allowing cellular infiltration, and promoting tissue regeneration. In this chapter, the authors will focus on the properties of existing nanofiber scaffolds and discuss their potential roles in CNS regeneration.

Keywords: Central nervous system; traumatic brain injury; spinal cord injury; tissue engineering; nanofiber scaffolds; regeneration.

INTRODUCTION

Central Nervous System Injury

Traumatic brain injury (TBI) and spinal cord injury (SCI) are serious conditions for which only limited treatment strategies are currently available. In addition to the high mortality rates associated with these conditions, the societal costs as well as the personal loss sustained by victims and their families are considerable.

Traumatic brain injury

TBI is a leading cause of morbidity and mortality in individuals under the age of 45 years. TBI commonly results in significant sensorimotor deficits (e.g., loss of ambulation and fine motor skills) as well as psychological and cognitive impairment (e.g., depression, memory loss) (Werner and Engelhard 2007). In the United States (U.S.), TBI affects over 1.5 million people every year. It results in long-term disabilities and mortalities in 124,000 and 50,000 patients, respectively (Selassie et al. 2008).

Spinal cord injury

Spinal cord injury (SCI) may result from direct penetrating injuries or extrinsic compression from displaced bone fragments and herniated discs after vertebral column trauma. Traumatic SCI incurs a significant economic burden to society and has a significant, negative impact on quality of life in survivors. To date, there are approximately 2.5 million sufferers of SCI worldwide. Within the European Union alone, at least 330,000 people are living with SCI sequels and around 11,000 new cases are being diagnosed every year. In the U.S., the corresponding figures are 300,000, and 12,000, respectively (Straley et al. 2010) .

Gap Formation as a Major Obstacle in CNS Regeneration

Within the injured spinal cord or brain, mechanical trauma initiates a series of cellular and molecular responses. Significant cell death may occur within 12 hours after injury, followed by a cascade of secondary events, including the opening of the blood–brain or blood–spinal cord barrier, inflammation, and the onset of edema, ischemia, excitotoxicity, free radical release, and altered cell signaling and gene expression. These secondary events may result in further cell death, demyelination and axonal degeneration. The subsequent formation of glial scar gives rise to an in hospital microenvironment for neuronal regeneration (Fawcett 2006), especially when associated with the presence of a tissue 'gap' that may extend well beyond the area of the initial insult (McDonald et al. 2002).

Within minutes to hours after CNS injury, a large number of activated microglia migrates to the area of injury, together with blood-derived macrophages if the vasculature is also damaged. These cells serve the important functions of destroying invading micro-organisms and removing potentially deleterious debris (Ladeby et al. 2005). These activities may be associated with significant tissue loss with the formation of fluid-filled cysts, which are typically lined by reactive gliosis. By the 3rd to 5th day post-injury, oligodendrocyte precursors are recruited from the surrounding tissues, and meningeal cells may also migrate into the CNS through penetrating injuries. These cells interact with astrocytes, leading to the formation of the astrocytic glial limitans (Li and David 1996). The final structure of the glial scar is predominantly made up of tightly interwoven astrocytes surrounded by extracellular matrix (ECM) (Fawcett and Asher 1999). Because of the physical properties of the glial scar and the presence of inhibitory molecules, growth cones from regenerating axons are unable to extend across the tissue gap (Fawcett and Asher 1999). Anatomical repair and functional recovery is often unsatisfactory.

Regeneration after CNS Injury

Most tissues of the human body are capable of regeneration to varied extents. For a long time, however, the CNS was considered to be non-regenerative. This opinion was widely accepted since 1550 BC (Mitchell 1872). In early 20th century, Ramon y Cajal declared, "Once the development has ended, the founts of growth and regeneration of the axons and dendrites dried up irrevocably. In adult centers the nerve paths are something fixed, ended, immutable" (Cajal 1959). However, recent researches have demonstrated the potential capacity of CNS neurons for re-growth (Benfey and Aguayo 1982). Today, it is generally held that the injured CNS may regenerate if provided with the appropriate treatment.

The Role of Tissue Engineering in CNS Repair

For regeneration to occur after traumatic CNS injury, axonal growth is required to cross the glial scars and lesion cavity towards the appropriate target tissue. This process may be potentially facilitated by the applications of modern tissue engineering techniques. Tissue engineering with biomaterial scaffolds is an emerging interdisciplinary field that applies the principles of engineering and life sciences to the development of biological substitutes which can restore, maintain or promote tissue functions. Biomaterial scaffolds may play a pivotal role in neuro-regeneration by serving as a matrix for cellular infiltration, proliferation, differentiation and three-dimensional regrowth. The use of biomaterial scaffolds may also reduce glial scar formation. They can be designed to contain molecular motifs for cell adhesion to facilitate axonal regrowth further. Some novel biomaterials may also serve as bridges to guide the regenerating axons in restoring connections with target organs.

A biomaterial scaffold that is suitable for clinical use must have several important properties. Firstly, the material must be biocompatible and be able to function without interrupting normal physiological processes. Secondly, the scaffold must not promote or initiate adverse tissue reactions. Thirdly, production of the biomaterial must be simple yet versatile enough to produce a wide array of configurations appropriate for use within the CNS. Following implantation, the material must be biodegradable or can be readily incorporated through innate remodeling mechanisms. Biomaterial scaffolds may consist of either natural or synthetic materials.

Natural materials

Natural materials are similar to ECM in terms of their properties for cell adhesion and migration. They are biodegradable. To date, numerous experimental studies have reported promising results from the use of natural material in promoting axonal regeneration or tissue reconstruction. Natural materials that can be used to construct tissue scaffolds include agarose, alginate, chitosan, methylcellulose, nitrocellulose, collagen, fibrin, fibronectin and hyaluronic acid. The main disadvantage of these materials is that it is difficult to purify them; incomplete purification may result in immune system activation by the implant (Holmes 2002).

Artificial materials

Artificial or synthetic materials have many advantages over natural materials including the minimized risk of contamination from biological pathogens. Artificial polymers can be tailored to possess a wide range of mechanical properties which may be combined to meet specific needs. These include the ability for controlled drug release, tissue repair and tissue reconstruction. Synthetic biomaterials may be biodegradable or non-biodegradable.

Nondegradable materials

Several nondegradable materials have been used to fabricate nerve guidance channels, including silicone, polyacrylonitrile/polyvinylchloride, poly (tetrafluoro-ethylene), and poly (2-hydroxyethyl methacrylate). Their main disadvantages are their propensities to initiate inflammatory responses within the host tissue. Their persistent mass effect may also result in neural compression, and they may require a second surgery for their removal (Belkas et al. 2005). Although nondegradable channels such as those constructed with poly (2-hydroxyethyl methacrylate) have been shown to support axonal regeneration in the injured rat spinal cords (Tsai et al. 2006), they are generally considered not suitable for CNS reconstruction. As a result, nondegradable materials are not the focus of current research efforts.

Degradable materials

Degradable artificial materials are preferred over nondegradable materials since the former provoke less tissue reaction and do not require second-stage removal. Biodegradable materials that have been studied for use in CNS repair include mainly the polymer family of poly (α-hydroxyacids), poly (β-hydroxybutyrate) and nanofiber scaffolds.

The family of poly (α-hydroxyacids) include synthetic polymers and copolymers such as poly (glycolic acid) (Wong et al. 2007), poly (lactic acid) (Cai et al. 2007), and poly (lactic acid-co-glycolic acid) (Wang et al. 2008). Several methods have been identified for controlling the degradation and mechanical properties of these polymers by manipulating the ratio of monomer units, the stereochemistry of the monomer units, and the molecular weight distribution of chains. The use of poly (α-hydroxyacids) in neural repair is, however, limited by the fact that these polymers tend to degrade *in vivo* by hydrolysis and produce acidic degradation products that are potentially neurotoxic (Straley et al. 2010).

Poly (β-hydroxybutyrate) exists in microorganisms as energy storage granules and can be obtained from bacterial cultures. After implantation, Poly (β-hydroxybutyrate) is slowly degraded to produce β-hydroxybutyric acid, which is a common, non-toxic metabolite in mammals. Its slow degradation also limits the accumulation of acidic degradation products when compared with poly-α-hydroxyacids. Poly (β-hydroxybutyrate) has been shown to promote regeneration in the injured spinal cord and because of its good tensile strength and elasticity, the risk of causing compression on growing neurons is minimized (Novikova et al. 2008).

Nanofibers have shown great promises in neuro-regenerative studies and will be the focus of the following discussion.

NANOFIBER SCAFFOLDS FOR CNS REGENERATION

A common approach in tissue engineering is to create three-dimensional (3-D) structures which mimic the naturally occurring ECM. There are two main components within the ECM, namely, polysaccharides and fibrous proteins. In normal biological tissues, the diameters of fibrous proteins in the ECM range from 50 to 500 nm. Conventional artificial biomaterials processing techniques are, however, unable to produce fibers smaller than 10 μm in diameter. Tissue scaffolds created by these relatively large fibers do not permit infiltration by host or grafted cells. To create a suitable 3-D microenvironment for tissue repair, scaffolds need to contain fibers that are significantly smaller than cells (i.e., nanofibers) so that cells can become imbedded, similar to the situation within the normal ECM. To date, there are three fiber-fabrication techniques which can consistently produce nanofibers to this standard, namely, self-assembly, electrospinning, and phase separation (Barnes et al. 2007; Jayaraman et al. 2004).

Existing Nanofiber Scaffolds

Self-assembling peptide nanofiber scaffolds

Self-assembling peptide nanofiber scaffolds (SAPNS) contain peptides molecules characterized by periodic repeats of alternating ionic hydrophilic and hydrophobic amino acids. These give rise to a typical β-sheet structure with distinct polar and nonpolar surfaces. Under physiological conditions and at the appropriate pH, ionic strength, temperature and concentration, these β-sheets would spontaneously self-assemble to form 3-D peptide scaffolds (Koutsopoulos et al. 2009; Zhang 2008; Zhang et al. 1993). Zhang and coworkers are leading authorities on the chemical and structural principles that dictate this self-assembly process. The first member of SAPNSs discovered was EAK16-II (AEAEAKAKAEAEAKAK), which originated from a segment in a yeast protein, Zuotin (Zhang et al. 1992). The most widely used SAPNS designed by Zhang's group is RADA16-I (RADARADARADARADA), in which lysine and glutamate residues are replaced by arginine and aspartate substitutes. RADA16-I was designed and characterized for salt-facilitated nanofiber scaffold formation. In aqueous solutions, the alanine residues would form overlapping hydrophobic interactions, whilst on the charged polar surface, both positive and negative charges are packed together through intermolecular ionic interactions. The resultant β-sheet structures are stable across a broad range of temperatures, pH and in the presence of high concentrations of denaturing agents such as urea and guanidium hydrochloride. Individual nanofibers created by SAPNS have diameters of around 10 nm, and nanopores that range from a few nanometers to a few hundred nanometers in size. The nanofiber density correlates with the concentration of peptide solution and retains extremely high hydration of greater than 99% in water (5–10 mg/ml, w/v) (Zhang 2008).

Other related self-assembling peptides have also been developed, including the MAX1 and MAX8 β-hairpin peptide systems (Kretsinger et al. 2005), the helical coiled-coils (Moutevelis and Woolfson 2009), and triple-helical supramolecules (Yamazaki et al. 2008). Recently, isolucine-lysine-valine-alanine-valine (IKVAV)-containing peptide amphiphile molecules have been developed for neuro-regeneration studies (Silva et al. 2004). The molecular design of this scaffold incorporates the pentapeptide epitope IKVAV found in laminin and the material has been shown to promote neurite sprouting and direct neurite growth (Wei et al. 2007; Yamazaki et al. 2008). IKVAV-containing nanofiber scaffolds has also been found to induce selective differentiation of neural progenitor cells into neurons while suppressing astrocytic differentiation (Silva et al. 2004).

SAPNS have several advantages over other natural and artificial biomaterials. SAPNS provide a 3-D environment for cell growth and migration (Guo et al. 2007), similar to that of native ECM, with nanofibers of around 10 nm and nanopores of 5 and 200 nm in diameter (Holmes 2002; Koutsopoulos et al. 2009; Zhang et al. 1993). SAPNS have excellent physiological compatibility as well as minimal cytotoxicity. Their degradation products are natural amino acids, which can potentially be re-used by nearby cells (Davis et al. 2006). SAPNS do not provoke immune response or tissue inflammation when introduced into animals (Holmes 2002). In the form of injectable liquid, SAPNS conform readily to the shape of lesion cavities (Ellis-Behnke et al. 2006a; Guo et al. 2007). They have haemostatic properties (Ellis-Behnke et al. 2006b), and may act as drug-delivery systems (Li et al. 2009). These properties render SAPNS highly promising candidates for use in neuro-regenerative studies.

Eletrospun nanofiber scaffolds

Electrospinning is a well-established fabrication process capable of producing ultra-fine fibers by electrically charging a suspended droplet of polymer melt or solution (Barnes et al. 2007). For instance, polyethylene, polypropylene, and polyester may be electrospun into nano-fibers from melted polymer solution in vacuum (Cao et al. 2009). Electrospinning may also be used to create fibers bearing various features and chemical compositions using materials such as synthetic polymers [e.g., poly (lactide-co-glycolide), polycaprolactone, poly(lactic acid), Poly(glycolic acid), and polydioxanone], as well as natural polymers such as collagen, hyaluronic acid, gelatin, elastin, fibrinogen, hemoglobin, and myoglobin. Moreover, novel nanofibers may be produced by combining solutions of both synthetic and natural polymers (Badami et al. 2006; Barnes et al. 2007; Cao et al. 2009; Ji et al. 2006).

The characteristics of electrospun nanofiber scaffolds can be readily controlled by adjusting the electrospinning parameters. For instance, fiber diameter may be controlled by altering the concentration/viscosity of the polymer solution. The size and shape of the electrospun scaffold may be tailored to the grounded target geometry, and rotation of the latter may further alter fiber alignment. Scaffold thickness can also be controlled by the volume of polymer used (Barnes et al. 2007).

Electrospun nanofibers have been shown to be suitable for use in cultures of dorsal root ganglion neurons, Schwann cells, and cortical neurons (Corey et al. 2007; Schnell et al. 2007). Ramakrishna and coworkers also demonstrated that electrospun nanofibers supported neuron growth and differentiation both *in vivo* and *in vitro* (Yang et al. 2005).

Phase separation

Phase separation technique was originally used for the production of porous polymer membranes. Recently, this technique has been applied in the development of 3-D nanofiber scaffolds. It permits considerable control over the characteristics of nanofiber scaffolds during production from a variety of biodegradable aliphatic polyesters. The diameters of fibers range from 50 to 500 nm. Phase separation involves a thermodynamic separation of a polymer solution into a polymer-rich component and a polymer-poor/solvent-rich component. Briefly, a polymer is dissolved in solution, followed by phase separation induced either thermally (most common method) or through the addition of a nonsolvent to the polymer solution to create a gel. Water is then used to extract the solvent from the gel (Ma et al. 2005). Compared with self-assembly, phase separation is a simpler technique and does not require specialized equipment. The mechanical properties and architecture of the scaffold can be controlled by varying the polymer/porogen concentrations. However, this technique is limited only to a selected number of polymers and remains largely a laboratory-scale method (Barnes et al. 2007). There is as yet no reported study on the use of nanofibers produced by phase separation in neural tissue engineering.

Effects of Nanofiber Scaffolds on the Injured CNS

The potential applications of nanofibers in tissue engineering and regenerative medicine have attracted considerable enthusiasm and related experimental studies on CNS injuries have generated promising and interesting findings.

Traumatic brain injury

The pioneering work of using nanofiber scaffold in the treatment of TBI was conducted by Ellis-Behnke et al. (Ellis-Behnke et al. 2006a). With an animal model of optic tract transection, the effects of RADA16-I, a type I SAPNS, on the wounded brain was studied. RADA16-I was found to promote knitting of the wounded tissues in both young and adult animals. To verify anatomical evidence of axonal regeneration across the lesion site, cholera-toxin subunit B fragment, which could be taken up in retinal ganglion cells and transported to axon terminals, was injected into the vitreous humor of the animal eye. Labeled regenerated axons were found to grow across the optic tract lesion and terminate in the superior colliculus in over 90% of the SAPNS-treated animals. Behavioral testing also revealed significant functional recovery of vision at 90-day post-treatment.

In another model designed to mimic tissue loss in acute TBI, Guo et al. studied the use of SAPNS on a surgically created lesion cavity in rodents (Guo et al. 2009). Treatment with SAPNS was found to result in significantly better healing of the lesion cavity. When compared with saline treatment, SAPNS-treated brain tissue had smaller lesion cavities, and less astrocytic and microglial infiltration. The SAPNS graft integrated well with the host tissue, and many host cells, including astrocytes, microglia and macrophages, were found to migrate into and survive within the grafted SAPNS. Unlike other solid or gelatiniform biomaterials, SAPNS, in the form of an injectable liquid, was found to conform readily to any lesion cavities in the brain. Taken together, findings from these early studies suggested that SAPNS may potentially mitigate secondary injury after TBI by minimizing glial reaction and inflammatory responses within and around injured brain tissues (Guo et al. 2009).

Spinal cord injury

In 2007, Meiners et al. reported the use of electrospun nanofibers for the treatment of experimental SCI (Meiners et al. 2007). The polymeric nanofiber mat was cross-linked in the presence of an acid catalyst, resulting in the formation of a network of filaments of around 180 nm and nanopores of 100–800 nm in diameter. The scaffold was then implanted into a lesion created by over-hemisection of the thoracic spinal cord. The scaffold did not provoke any significant tissue response in the spinal cord and was associated with lesser degree of scarring when compared with the control group. The nanofiber was found to integrate well with the host tissue with no evidence of cavitation. Axonal regeneration was detected across the lesion site. However, the regenerating axons were restricted to grow along the surface contours of the tissue scaffold due to the random orientation of the nanofibers. It was only when the folds of the nanofibers fell parallel by chance to the spinal cord axis that axons would extend with a parallel configuration.

In contrast, treatment with SAPNS produced more satisfactory outcomes. In the pioneering studies by Tysseling-Mattiace et al., compressed spinal cords in mice were intraspinally microinjected with IKVAV-SAPNS (Tysseling-Mattiace et al. 2008). The treatment resulted in greater oligodendrocyte cell number as well as reduced apoptotic cell death and glial scarring at the lesion site. Moreover, IKVAV treatment was found to promote the regeneration of injured motor and sensory axons, with partial recovery of locomotor function. These results were later confirmed in another study in which IKVAV-SAPNS treatment was found to also increase the number of serotonin-containing fibers caudal to the lesion. The presence of these fibers was thought to be at least partially responsible for the observed improvement in behavioral function. (Tysseling et al. 2010).

The potential use of SAPNS as a cell-carrier has also been explored (Guo et al. 2007). Guo et al. studied the effects of SAPNS pre-cultured with adult rat Schwann cells (SCs) or embryonic neural progenitor cells (NPCs) in a rodent spinal cord dorsal column transection model (Guo et al. 2007). Six-weeks after implantation, the SAPNS graft was found to bridge the lesion area and integrate with the host tissue with no obvious cavititation. Inflammatory response and ED1-positive cells around the lesion site were reduced significantly. Moreover, a large amount of host cells were found to have migrated into the implants, together with extensive blood vessel formation and axonal regeneration. In animals treated with SAPNS pre-cultured with SCs or NPCs, the transplanted cells were found to have survived within the SAPNS graft and migrated into the host tissue (Guo et al. 2007). Some of the SCs were also found to have matured with tube-like morphologies, and some of the NPCs had differentiated into neurons, astrocytes and oligodendrocytes. When compared with treatment with SAPNS alone, SAPNS combined with SCs or NPCs was associated with more axonal regrowth (Guo et al. 2007).

Our unpublished data also showed that SAPNS is a good material for controlled drug release. When the injured spinal cord was treated with the mixture of SAPNS and Chondroitinase ABC, which is an enzyme for digesting chondroitin sulfate proteoglycans (CSPGs), the CSPGs concentration in the host tissue was decreased significantly and was associated with more axonal regeneration and functional recovery (Fig. 1).

Applications to Areas of Health and Disease

Nanofiber scaffolds are potentially suitable for clinical use. Synthetic biomaterials such as SAPNS are non-toxic, non-immunogenic, and biodegradable. SAPNS can be applied as an injectable liquid that requires only very basic storage facilities. Surgical procedures performed for the treatment of CNS diseases often results in the formation of large cavities and raw surfaces. The abilities of SAPNS to stop bleeding rapidly and to fill up and conform to these surgical cavities make it a potential surgical haemostat that is also conducive to regeneration. For instance, SAPNS may be applied to the operative site after the removal of a brain tumor or the evacuation of an intracerebral haematoma. Nanofiber scaffolds may also act as potential drug and/or cell carriers for regenerative therapy. One of the major obstacles in regenerative treatment is to maintain transplanted cells or exogenous growth factors at the appropriate state and concentration within the treatment site. Nanofiber scaffolds may be exploited as a transport medium that is capable of maintaining controlled drug release as well as the survival, proliferation, differentiation and integration of transplanted cells. An example is transplantation therapy for patients with Parkinson's' disease in which SAPNS incubated with dopaminergic cells is injected into the substantial nigra. In daily neurosurgical practice, nanofiber scaffolds carrying neurotrophic factors may also be used to promote axonal regeneration across the sites of cranial or peripheral nerve transection. Other potential clinical applications include the use of nanofiber scaffolds as carriers of radioactive isotopes, chemotherapeutic agents and antimicrobial drugs.

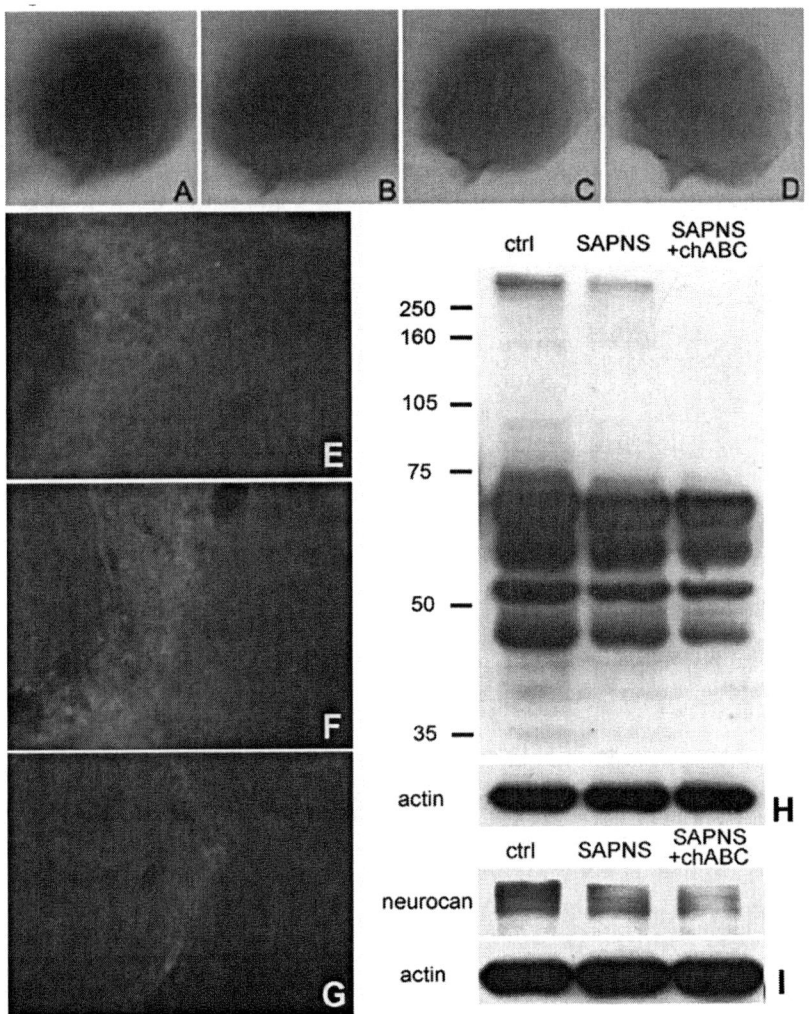

Fig. 1. SAPNS is a good material for chemical control release (unpublished data). (A-D) the mixture of SAPNS and typan blue was merged in saline for 1h, 1d, 1w, and 4w respectively, showed the dye was released slowly. When the injured spinal cord was treated with the mixture of SAPNS and Chondroitinase ABC (chABC), which is an enzyme for digesting chondroitin sulfate proteoglycans (CSPGs), the CSPGs concentration in the host tissue was decreased significantly. (E-G) Immunohistochemistry showed the CSPG expression in control (E), SAPNS alone treated (F) and SAPNS with chABC treated (G). (H, I) Western blotting showed the total CSPGs (H) and neurocan (I), which is a component of CSPGs, in different treated injured spinal cord.

Color image of this figure appears in the color plate section at the end of the book.

Key Facts

- Traumatic brain injury is a leading cause of morbidity and mortality in individuals under the age of 45 years.
- Traumatic brain injury affects over 1.5 million people every year.
- To date, there are approximately 2.5 million sufferers of spinal cord injury worldwide.
- Gap formation is a major obstacle in central nervous system regeneration.
- The injured brain or spinal cord has the potential capacity to regenerate if provided with the appropriate treatment.
- Nanofiber scaffolds have the advantages of mimicking the naturally occurring ECM and promoting tissue regeneration.

Summary Points

- Traumatic brain injury and spinal cord injury are devastating conditions which are commonly associated with the formation of tissue gap that is in hospital to regeneration.
- The injured brain or spinal cord has the potential capacity for regeneration if given the suitable environment, and biomaterial scaffolds may provide three-dimensional frameworks for tissue regrowth and integration.
- There are different types of biomaterial scaffolds consisting of naturally occurring or synthetic materials. The latter may be biodegradable or non-biodegradable. Nanofiber scaffolds are biodegradable synthetic materials particularly suitable for clinical use.
- Self-assembling peptide nanofiber scaffolds consist of peptides molecules which would self-assemble to form three-dimensional nano-structures that resemble the naturally occurring ECM. They are highly porous and may facilitate the infiltration, survival and integration of cells. They have been shown to be able to stop bleeding and promote the healing of injured CNS tissues.
- Self-assembling peptide nanofiber scaffolds can act as drug- and/or cell-carriers, and are therefore potentially useful for regenerative therapies for a large variety of CNS conditions such as traumatic injuries, degenerative and neoplastic diseases.

Definitions

Traumatic brain injury (TBI): TBI occurs when an external force injures the brain traumatically. Neuronal loss secondary to primary and secondary brain insults, commonly result in significant sensorimotor deficits (e.g., loss of ambulation, and fine motor skills) as well as psychological and cognitive impairment (e.g., depression, memory loss).

Spinal cord injury (SCI): SCI are damages to the spinal cord that may result in the loss of sensory or motor functions. SCI may be caused by direct penetrating injuries or extrinsic compression after vertebral column trauma. Injury in the spinal cord in the thoracic, lumbar, or sacral segments may cause paraplegia. Injury to the spinal cord in the cervical region may cause quadriplegia.

Tissue engineering: Tissue engineering is an emerging interdisciplinary technology and methodology aimed to assist and accelerate the regeneration and repairing of defective and damaged tissues. Biomaterial scaffolds design is the most important aspect of tissue engineering. Biomaterial scaffolds may be designed to contain bimolecular motifs and/or (stem) cell to further facilitate tissue regeneration.

Nanofiber scaffolds: The optimal biomaterials for tissue engineering should create three-dimensional structures which mimic the naturally occurring ECM. Conventional artificial biomaterials processing techniques are, however, unable to produce fibers smaller than 10 µm in diameter which do not permit infiltration by host or grafted cells. Recently, three fiber-fabrication techniques, such as self-assembly, electrospinning, and phase separation, have been used to produce scaffolds with nano-scale fibers. The nanofiber scaffolds have the advantages of mimicking the ECM and promoting tissue regeneration.

Self-assembling peptide nanofiber scaffolds (SAPNS): SAPNSs are newly emerged nanobiomaterials, with peptides molecules that are characterized by periodic repeats of alternating ionic hydrophilic and hydrophobic amino acids. Under physiological or other conditions, the peptide can spontaneously self-assemble to form nanofibers with diameters of around 10 nm. SAPNS have several major advantages when compared with other biomaterials: they provide a truly 3-D environment similar to that of native ECM; have excellent physiological compatibility as well as minimal cytotoxicity; do not provoke immune response or tissue inflammation; conform readily to the shape of lesion cavities; have haemostatic properties, and may act as cell and/or drug-delivery systems.

Abbreviations

CNS	:	Central nervous system
TBI	:	Traumatic brain injury
SCI	:	Spinal cord injury
ECM	:	Extracellular matrix
3-D	:	Three dimensional

SAPNS : Self-assembling peptide nanofiber scaffolds
SCs : Schwann cells
NPCs : Neural progenitor cells
ChABC : Chondroitinase ABC
CSPGs : Chondroitin sulfate proteoglycans

References

Badami, A.S., M.R. Kreke, M.S. Thompson, J.S. Riffle and A.S. Goldstein. 2006. Effect of fiber diameter on spreading, proliferation, and differentiation of osteoblastic cells on electrospun poly(lactic acid) substrates. Biomaterials 27: 596–606.

Barnes, C., S. Sell, E. Boland, D. Simpson and G. Bowlin. 2007. Nanofiber technology: Designing the next generation of tissue engineering scaffolds[?]. Advanced Drug Delivery Reviews 59: 1413–1433.

Belkas, J.S., C.A. Munro, C.A. Shoichet, M.S.M. Johnston and R. Midha. 2005. Long-term *in vivo* biomechanical properties and biocompatibility of poly(2-hydroxyethyl methacrylate-co-methyl methacrylate) nerve conduits. Biomaterials 26: 1741–1749.

Benfey, M. and A.J. Aguayo. 1982. Extensive elongation of axons from rat brain into peripheral nerve grafts. Nature 296: 150–152.

Cai, J., K.S. Ziemba, G.M. Smith and Y. Jin. 2007. Evaluation of cellular organization and axonal regeneration through linear PLA foam implants in acute and chronic spinal cord injury. J. Biomed. Mater. Res. A 83: 512–520.

Cajal, R. 1959. Degeneration and regeneration in the nervous system. New York: Haffner.

Cao, H., T. Liu and S.Y. Chew. 2009. The application of nanofibrous scaffolds in neural tissue engineering. Adv. Drug Deliv. Rev. 61: 1055–1064.

Corey, J.M., D.Y. Lin, K.B. Mycek, Q. Chen, S. Samuel, E.L. Feldman and D.C. Martin. 2007. Aligned electrospun nanofibers specify the direction of dorsal root ganglia neurite growth. J. Biomed. Mater. Res. A 83: 636–645.

Davis, M.E., P.C. Hsieh, T. Takahashi, Q. Song, S. Zhang, R.D. Kamm, A.J. Grodzinsky, P. Anversa and R.T. Lee 2006. Local myocardial insulin-like growth factor 1 (IGF-1) delivery with biotinylated peptide nanofibers improves cell therapy for myocardial infarction. Proc. Natl. Acad. Sci. USA 103: 8155–8160.

Ellis-Behnke, R.G., Y.X. Liang, S.W. You, D.K. Tay, S. Zhang, K.F. So and G.E. Schneider. 2006a. Nano neuro knitting: peptide nanofiber scaffold for brain repair and axon regeneration with functional return of vision. Proc. Natl. Acad. Sci. USA 103: 5054–5059.

Ellis-Behnke, R.G., Y.X. Liang, D.K. Tay, P.W. Kau, G.E. Schneider, S. Zhang, W. Wu and K.F. So. 2006b. Nano hemostat solution: immediate hemostasis at the nanoscale. Nanomedicine 2: 207–215.

Fawcett, J.W. 2006. Overcoming inhibition in the damaged spinal cord. J. Neurotrauma 23: 371–383.

Fawcett, J.W. and R.A. Asher. 1999. The glial scar and central nervous system repair. Brain Res. Bull. 49: 377–391.

Guo, J., H. Su, Y. Zeng, Y.X. Liang, W.M. Wong, R.G. Ellis-Behnke, K.F. So and W. Wu. 2007. Reknitting the injured spinal cord by self-assembling peptide nanofiber scaffold. Nanomedicine 3: 311–321.
Guo, J., K.K.G. Leung, H. Su, Q. Yuan, L. Wang, T.H. Chu, W. Zhang, J.K.S. Pu, G.K.P. Ng, W.M. Wong, X. Dai and W. Wu. 2009. Self-assembling peptide nanofiber scaffold promotes the reconstruction of acutely injured brain. Nanomedicine 5: 345–351.
Holmes, T.C. 2002. Novel peptide-based biomaterial scaffolds for tissue engineering. Trends Biotechnol. 20: 16–21.
Jayaraman, K., M. Kotaki, Y. Zhang, X. Mo and S. Ramakrishna. 2004. Recent advances in polymer nanofibers. J. Nanosci. Nanotechnol. 4: 52–65.
Ji, Y., K. Ghosh, X.Z. Shu, B. Li, J.C. Sokolov, G.D. Prestwich, R.A. Clark and M.H. Rafailovich. 2006. Electrospun three-dimensional hyaluronic acid nanofibrous scaffolds. Biomaterials 27: 3782–3792.
Koutsopoulos, S., L.D. Unsworth, Y. Nagai and S. Zhang. 2009. Controlled release of functional proteins through designer self-assembling peptide nanofiber hydrogel scaffold. Proc. Natl. Acad. Sci. USA 106: 4623–4628.
Kretsinger, J.K., L.A. Haines, B. Ozbas, D.J. Pochan and J.P. Schneider. 2005. Cytocompatibility of self-assembled beta-hairpin peptide hydrogel surfaces. Biomaterials 26: 5177–5186.
Ladeby, R., M. Wirenfeldt, D. Garcia-Ovejero, C. Fenger, L. Dissing-Olesen, I. Dalmau and B. Finsen. 2005. Microglial cell population dynamics in the injured adult central nervous system. Brain Res. Brain Res. Rev. 48: 196–206.
Li, F., J. Wang, F. Tang, J. Lin, Y. Zhang, E.Y. Zhang, C. Wei, Y.K. Shi and X. Zhao. 2009. Fluorescence studies on a designed self-assembling peptide of RAD16-II as a potential carrier for hydrophobic drug. J. Nanosci. Nanotechnol. 9: 1611–1614.
Li, M.S. and S. David. 1996. Topical glucocorticoids modulate the lesion interface after cerebral cortical stab wounds in adult rats. Glia 18: 306–318.
Ma, Z., M. Kotaki, R. Inai and S. Ramakrishna. 2005. Potential of nanofiber matrix as tissue-engineering scaffolds. Tissue Eng. 11: 101–109.
McDonald, J.W., D. Becker, C.L. Sadowsky, J.A. Jane Sr., T.E. Conturo and L.M. Schultz. 2002. Late recovery following spinal cord injury. Case report and review of the literature. J. Neurosurg. 97: 252–265.
Meiners, S., I. Ahmed, A.S. Ponery, N. Amor, S.L. Harris, V. Ayres, Y. Fan, Q. Chen, R. Delgado-Rivera and A.N. Babu. 2007. Engineering electrospun nanofibrillar surfaces for spinal cord repair: a discussion. Polym. Int. 56: 1340–1348.
Mitchell, S. 1872. Injuries of nerves. Philadelphia: J.B. Lippincott & Co.
Moutevelis, E. and D.N. Woolfson. 2009. A periodic table of coiled-coil protein structures. J. Mol. Biol. 385: 726–732.
Novikova, L.N., J. Pettersson, M. Brohlin, M. Wiberg and L.N. Novikov. 2008. Biodegradable poly-beta-hydroxybutyrate scaffold seeded with Schwann cells to promote spinal cord repair. Biomaterials 29: 1198–1206.
Schnell, E., K. Klinkhammer, S. Balzer, G. Brook, D. Klee, P. Dalton and J. Mey. 2007. Guidance of glial cell migration and axonal growth on electrospun nanofibers of poly-epsilon-caprolactone and a collagen/poly-epsilon-caprolactone blend. Biomaterials 28: 3012–3025.

Selassie, A.W., E. Zaloshnja, J.A. Langlois, T. Miller, P. Jones and C. Steiner. 2008. Incidence of long-term disability following traumatic brain injury hospitalization, United States, 2003. J. Head Trauma Rehabil. 23: 123–131.

Silva, G.A., C. Czeisler, K.L. Niece, E. Beniash, D.A. Harrington, J.A. Kessler and S.I. Stupp. 2004. Selective differentiation of neural progenitor cells by high-epitope density nanofibers. Science 303: 1352–1355.

Straley, K.S., C.W. Foo and S.C. Heilshorn. 2010. Biomaterial design strategies for the treatment of spinal cord injuries. J. Neurotrauma 27: 1–19.

Tsai, E.C., P.D. Dalton, M.S. Shoichet and C.H. Tator. 2006. Matrix inclusion within synthetic hydrogel guidance channels improves specific supraspinal and local axonal regeneration after complete spinal cord transection. Biomaterials 27: 519–533.

Tysseling-Mattiace, V.M., V. Sahni, K.L. Niece, D. Birch, C. Czeisler, M.G. Fehlings, S.I. Stupp and J.A. Kessler. 2008. Self-assembling nanofibers inhibit glial scar formation and promote axon elongation after spinal cord injury. J. Neurosci. 28: 3814–3823.

Tysseling, V.M., V. Sahni, E.T. Pashuck, D. Birch, A. Hebert, C. Czeisler, S.I. Stupp and J.A. Kessler. 2010. Self-assembling peptide amphiphile promotes plasticity of serotonergic fibers following spinal cord injury. J. Neurosci. Res.

Wang, Y.C., Y.T. Wu, H.Y. Huang, H.I. Lin, L.W. Lo, S.F. Tzeng and C.S. Yang. 2008. Sustained intraspinal delivery of neurotrophic factor encapsulated in biodegradable nanoparticles following contusive spinal cord injury. Biomaterials 29: 4546–4553.

Wei, Y.T., W.M. Tian, X. Yu, F.Z. Cui, S.P. Hou, Q.Y. Xu and I.S. Lee. 2007. Hyaluronic acid hydrogels with IKVAV peptides for tissue repair and axonal regeneration in an injured rat brain. Biomed. Mater 2: S142–146.

Werner, C. and K. Engelhard. 2007. Pathophysiology of traumatic brain injury. Br. J. Anaesth 99: 4–9.

Wong, D.Y., S.J. Hollister, P.H. Krebsbach and C. Nosrat. 2007. Poly(epsilon-caprolactone) and poly (L-lactic-co-glycolic acid) degradable polymer sponges attenuate astrocyte response and lesion growth in acute traumatic brain injury. Tissue Eng. 13: 2515–2523.

Yamazaki, C.M., S. Asada, K. Kitagawa and T. Koide. 2008. Artificial collagen gels via self-assembly of de novo designed peptides. Biopolymers 90: 816–823.

Yang, F., R. Murugan, S. Wang and S. Ramakrishna. 2005. Electrospinning of nano/micro scale poly(L-lactic acid) aligned fibers and their potential in neural tissue engineering. Biomaterials 26: 2603–2610.

Zhang, S. 2008. Designer self-assembling Peptide nanofiber scaffolds for study of 3-d cell biology and beyond. Adv. Cancer Res. 99: 335–362.

Zhang, S., T. Holmes, C. Lockshin and A. Rich. 1993. Spontaneous assembly of a self-complementary oligopeptide to form a stable macroscopic membrane. Proc. Natl. Acad. Sci. USA 90: 3334–3338.

Zhang, S., C. Lockshin, A. Herbert, E. Winter and A. Rich. 1992. Zuotin, a putative Z-DNA binding protein in Saccharomyces cerevisiae. EMBO J. 11: 3787–3796.

18

Nanoparticle-iron Chelator Conjugates as Multifunctional Disease-modifying Drugs for Prevention and Treatment of Alzheimer's Disease

Gang Liu,[1,a,] Ping Men,[1,b] George Perry[2] and Mark A. Smith[3]*

ABSTRACT

Accumulating studies suggest that chelation therapy has clinical benefit in patients with Alzheimer's disease (AD). Additionally, experimental investigations show that chelating agents may have the ability to target multiple etiologies in AD by preventing reactive oxidative species formation and a detrimental cycle of

---

[1]Department of Radiology, University of Utah, Salt Lake City, Utah 84108, USA.
[a]E-mail: gang.liu@hsc.utah.edu
[b]E-mail: ping.men@hsc.utah.edu
[2]College of Sciences, University of Texas at San Antonio, San Antonio, Texas 78249, USA; Department of Pathology, Case Western Reserve University, Cleveland, Ohio 44106, USA; E-mail: George.Perry@utsa.edu
[3]Department of Pathology, Case Western Reserve University, Cleveland, Ohio 44106, USA; E-mail: mark.smith@case.edu
*Corresponding author

List of abbreviations after the text.

oxidative stress, and by mitigating amyloid-β (Aβ) aggregation and neurodegeneration. Although chelating agents may provide an attractive therapeutic approach as multifunctional disease-modifying drugs for prevention and treatment of AD, the use of currently available chelating drugs has been hampered since many pose poor blood brain barrier penetration capability due to their large size and hydrophilicity, and many have toxicity at therapeutically effective concentrations. To overcome these obstacles, a new approach of nanoparticle-chelator conjugation has been developed utilizing the nanoparticles as transport vehicles to both deliver the chelators into the brain and bring the metal-chelator complexes out of the brain. This approach can also reduce the chelator toxicity while preserving the chelator metal binding ability. Importantly, the nanoparticle-iron chelator conjugates show the capability to complex iron ion, to inhibit Aβ aggregation, to mitigate Aβ associated neurotoxicity and to prevent neurons from oxidative damage *in vitro*. Further development of these conjugates will likely provide advantageous chelation therapeutics as multifunctional disease-modifying drugs for prevention and treatment of AD as well as other neurodegenerative diseases.

INTRODUCTION

Alzheimer's disease (AD) is an age-related, non-reversible, neurodegenerative brain disorder characterized by progressive damage to thought, memory, language and eventually even the ability to carry out the simplest tasks (Smith 1998a). This disease is the fourth leading cause of death among people aged 65 and older and the most common form of dementia, from which an estimated 5.1 million people suffer currently in the United State alone (Minino et al. 2007; Yaari and Corey-Bloom 2007). Unfortunately, not only are the causes of the AD process still unknown, but also there is no cure for the disease. At present, two drug classes dominate AD treatments including (1) cholinesterase inhibitors and, more recently, (2) an NMDA (N-methyl-D-Aspartate) receptor antagonist (Melnikova 2007). These available US Food and Drug Administration (FDA) approved drugs, however, provide only partial benefit to select patients and fall tremendously short as adequate means of therapeutic management (Liu et al. 2005; Marlatt et al. 2005). Therefore, there is an urgent need for improved treatments and, in particular, for disease-modifying drugs (DMDs) for AD (Klafki et al. 2006). To this end, iron chelating agents may provide a valuable approach as multifunctional DMDs because evidence shows that redox active iron plays a key role in AD initiation and progress by generating reactive oxidative species, inducing a detrimental cycle of oxidative stress, amyloid-β (Aβ) aggregation, and neurodegeneration (Liu et al. 2009a; Duce et al. 2010; Avramovich-Tirosh et al. 2007; Storr et al. 2009). However, available chelating drugs are limited in their potential with regard to their bioavailability, such as blood brain barrier (BBB) penetration, and their possible toxicity (Liu et al. 2005). To overcome these obstacles, we review the development of iron chelators conjugated to nanoparticles that are able to deliver target chelators to the brain with an absence of neurotoxicity, and to bring the chelator-iron complexes out of the brain. Such an approach will likely provide safer, more effective iron chelating therapeutics with a highly advantageous mode of attack on multiple etiologies in AD.

CHELATING AGENTS AS MULTIFUNCTIONAL DMDs FOR AD

Evidence

Because it is believed that a number of disease pathologies are involved in AD development, a novel strategy with multifunctional DMDs that aim simultaneously at several therapeutic targets may provide more effective approaches to slow or stop the initiation and progression of AD. As evidence, the combination therapy of cholinesterase inhibitors and NMDA receptor antagonists, which act differently on the chemistry of the brain, shows a statistically significant benefit as compared with mono-therapy with regard to measures of cognitive function, activities of daily living, behavior and clinical global status (Tariot et al. 2004). A multifunctional DMD may also be safer than a combination of several mono-functional drugs for AD treatment so as to avoid drug interactions and limit the side-effects of each drug.

In recent years, new chelators as AD therapeutic agents have been studied because they may have the capability as multifunctional DMDs to target multiple etiologies in AD (Avramovich-Tirosh et al. 2007; Storr et al. 2009; Liu et al. 2009b). Studies have shown that chelation agents can provide at least a "three pronged" mode of action to prevent and treat AD. Firstly, since iron ions, as redox-active centers, lead to free radical generation and oxidative stress (Castellani et al. 2007; Smith et al. 1997), which contribute to the initiation and promotion of neurodegeneration (Casadesus et al. 2004; Markesbery 1997), chelation agents that selectively bind to, remove, and/or "redox silence" iron have been considered an attractive therapeutic target for AD (Liu et al. 2005). Some chelation agents also have the ability to scavenge free radicals, thus mitigating free radical caused damage (Storr et al. 2009; Morel et al. 1992). Secondly, since iron ions are suggested to play an important role in the self-assembly and neurotoxicity of Aβ (Exley 2006), not surprisingly, Aβ toxicity is markedly attenuated by such chelating agents (Rottkamp et al. 2001; Schubert and Chevion 1995) and blocking Aβ aggregation by chelation may provide a valuable therapeutic approach (Cohen and Kelly 2003). Thirdly, since oxidative stress, some of which is consequent to metal-mediated processes (Sayre et al. 2000), is associated with increased Aβ—a consequence of the coordinated upregulation of amyloid-β protein precursor (AβPP) (Yan et al. 1995) and β- and γ-secretases (Tamagno et al. 2008; Yang et al. 2003)—it is also not surprising that treatment of AβPP-overexpressing transgenic mice, a model of AD that displays significant Aβ deposition and oxidative stress (Duce et al. 2010; Pappolla et al. 1998; Smith et al. 1998b), with chelating agents results in less Aβ deposition (Adlard et al. 2008; Cherny et al. 2001). Indeed, metal chelating compounds, such as desferrioxamine (DFO), ethylenediaminetetraacetic acid (EDTA) and iodochlorhydroxyquin (clioquinol), have been studied to treat AD patients and provided significant clinical improvement (Crapper-McLachlan et al. 1991; Regland et al. 2001; Ritchie et al. 2003; Casdorph 2001).

Impediments

However, some obstacles concerning chelator bioavailability, such as BBB penetration and toxic side-effects, have hindered further investigation, thus limiting both the understanding of the pathologic role of iron dysregulation in AD and the evaluation of the efficacy and safety of chelation therapy. For instance, DFO, an iron chelator approved by the FDA for the treatment of iron overload, has been used to treat AD patients, and slows progression of AD (Crapper-McLachlan et al. 1991), but it has serious side effects including neurotoxicity and neurological changes (Blake et al. 1985) and cannot penetrate the BBB due to its large molecule size and hydrophilic nature (Lynch et al. 2000). EDTA is also a hydrophilic chelator and shows the same lack of BBB penetration (Abbott et al. 1985). While lipophilic chelators with small molecular weight, like 1,2-dimethyl-3-hydroxyl-4-pyridinone (L1, a bi-dentate iron chelating drug), have the ability to penetrate the BBB, they have considerable neurotoxicity (Hider et al. 1994a) and do not remove iron from the brain despite effectively binding iron, which can result in additional toxicity in the brain (Crowe and Morgan 1994). Additionally, clioquinol is a lipophilic molecule and able to enter the brain, however its use has had to be stopped due to the consideration of its serious toxicity (Mao and Schimmer 2008). Thus, the use of chelators as multifunctional DMDs for further studies on AD treatment is currently limited.

NANOPARTICLE-CHELATOR BRAIN DELIVERY: IMPROVED BIOAVAILABILITY AND LOWERED TOXICITY

Although a large body of studies supports chelating agents as a potential multifunctional therapeutic approach to prevent and treat AD, the difficulties presented by bi-directional crossing of the BBB and the potential toxicity of the currently available chelators hamper further studies. In fact, however, these obstacles may be overcome by utilizing nanoparticle delivery technology, which has shown promise in brain targeting, improved drug efficacy and reduced drug toxicity (Kreuter 2001; Kreuter et al. 2002). Although the mechanisms by which the nanoparticles penetrate BBB are not fully known, there are several possibilities that likely work in combination to achieve the desired effect (Bonda et al. 2011). For example, (1) increasing retention of nanoparticles in blood-brain capillaries combined with absorption into capillary walls to create a higher concentration gradient, enhances nanoparticle transport across endothelial cell layers into the brain; (2) a surfactant effect that would lead to membrane fluidization and/or to opening tight junctions between endothelial cell, thus enhancing drug permeability; (3) an inhibition of the efflux system (i.e., P-glycoprotein) via polysorbate-80 (a nanoparticle coating proven to yield the most effective delivery of drugs across the BBB (Alyautdin et al. 1998; Kreuter et al. 2002; Schroeder et al. 1998)); (4) endocytosis and/or transcytosis through the BBB. While each of these mechanisms is quite possible, the most probable one seems to be the receptor-mediated endocytosis of the nanoparticles.

Specifically, a particular class of nanoparticles with apolipoprotein E (ApoE) preferentially absorbed on the surface is able to cross the BBB by mimicking low-density lipoprotein (LDL), enabling them to interact with the LDL receptor, and thereby resulting in their uptake by brain endothelial cells (Kreuter 2001; Kreuter et al. 2002). Several studies confirm endocytosis of polysorbate-80 coated nanoparticles (Alyautdin et al. 2001; Borchard et al. 1994; Kreuter et al. 1995; Ramge et al. 2000). In addition, a recent study using polysorbate-80 coated nanoparticles to deliver rivastigmine, a drug for the treatment of AD, has shown increased brain concentration of the drug in an animal model (Wilson et al. 2008). Interestingly, absorption of ApoE on the surface of polysorbate-80 coated nanoparticles coincides with the latter's BBB penetration (Kreuter et al. 2002), providing key insights into the mechanism of action of the receptor-mediated endocytosis.

Indeed, our studies have shown that nanoparticles covalently conjugated to chelators have the potential to deliver the chelators into the brain without altering iron chelating capability while limiting the chelator toxicity. Additionally, this nanoparticle-iron chelator conjugate offers distinct advantages over many other chelators (Bonda et al. 2011). For instance: (1) the chelator need not be lipophilic to enter the BBB, as the nanoparticle can carry across any compound to which it is covalently bound (Liu et al. 2010). This directly eliminates much of the toxicity associated with the small, lipophilic chelators. (2) As noted above, nanoparticle delivery allows the use of hexadentate iron chelators that were previously too large and hydrophilic to cross the BBB (Alyautdin et al. 1998; Schroeder et al. 1998). (3) Iron-complexed chelators (i.e., those that have complexed with the transition metal within the brain) may be effectively removed from the brain via nanoparticle transportation, thus completing the treatment regimen. As to the latter, it is equally necessary for a metal chelator to be able to exit the brain through the BBB as it is to enter it; a complexed chelator must therefore leave the brain after retaining its substrate. With lipophilic chelators, their removal from the brain becomes problematic once they have complexed with a transition metal, due to their resulting altered lipophilicity (Liu et al. 2010; Porter et al. 1988); they can no longer exit the BBB and themselves become progenitors of metal-associated oxidative damage (Liu et al. 2005; Liu et al. 2006).

A nanoparticle delivery system, however, theoretically provides an adequate escape route for the chelator (Liu et al. 2010). That is, the chelator may be removed from the brain via the system of the apolipoprotein which the nanoparticle conjugates most likely mimic (Kreuter 2001; Liu et al. 2006). Consequently, the nanoparticle-chelator system would enable the full, desired effect of a metal chelator without any toxic side effect (Bonda et al. 2011). Although some new chelating agents for AD have been developed with the ability to enter the brain, very few studies have considered the development of chelating agents with the capability of leaving the place after complexing excess metals (Youdim et al. 2004). In addition, the unchanged iron binding capability of the chelating agents after nanoparticle conjugation make it possible to target the multiple etiologies in AD associated with imbalanced iron in the brain. Thus, our development of the nanoparticle-iron chelator conjugation could have profound impacts on discovery of safer, more effective multifunctional DMDs for AD prevention and treatment, which are supported by our key studies as follows.

Iron Binding of the Nanoparticle-iron Chelator Conjugates

The nanoparticle-iron chelator conjugates (NICCs) can be generated via covalently linking nanoparticles and chelators, both of which contain functional groups for the conjugation using several standard methods (Liu et al. 2009a,c, 2010). It is very important that the conjugated chelators with nanoparticles should still retain their iron binding ability in order to be used for chelation. To demonstrate this capability, two types of NICCs as prototypes were used; one was a L1 derivative, a bidentate iron chelator, conjugated with nanoparticles (Nano-N2PY) and the other was DFO, a hexadentate iron chelator, conjugated with nanoparticles (Nano-DFO1). After reaction with ferric iron (Liu et al. 2009a,c), the results showed that about two bidentate chelator molecules of Nano-N2PY complexed with one iron atom (Fig. 1A). This indicated that the surface linkages of nanoparticles served as backbones and the nitrogen atoms in amide groups might be involved in iron complexing, thereby converting two of these bidentate chelators into one hexadentate chelator, which improved the iron binding stability of the conjugates. The hexadentate iron chelators also had advantages that include kinetic stability, concentration independence of iron affinity, and low toxicity (Hider et al. 1994b). Results also showed that DFO of Nano-DFO1 still retained the 1:1 complexing with iron (Fig. 1B), indicating the conjugation with nanoparticles did not affect the iron binding ability of the DFO.

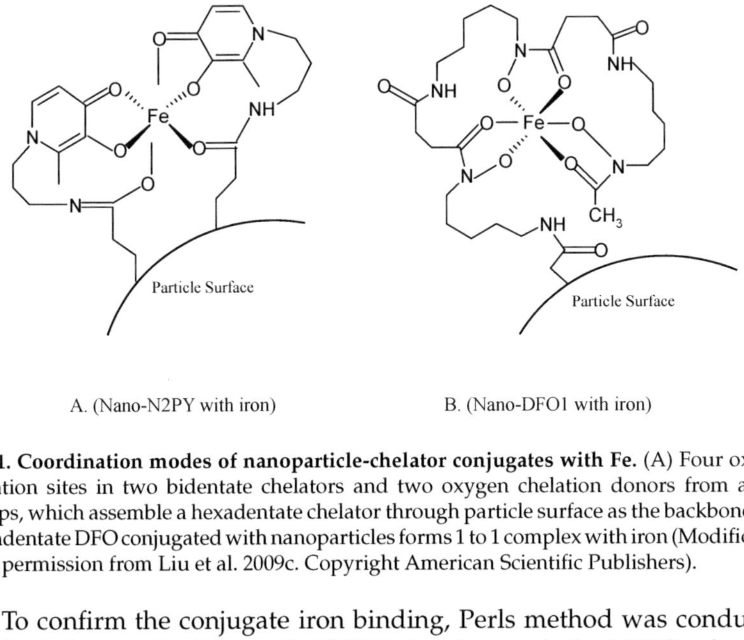

Fig. 1. Coordination modes of nanoparticle-chelator conjugates with Fe. (A) Four oxygen chelation sites in two bidentate chelators and two oxygen chelation donors from amido groups, which assemble a hexadentate chelator through particle surface as the backbones. (B) Hexadentate DFO conjugated with nanoparticles forms 1 to 1 complex with iron (Modification with permission from Liu et al. 2009c. Copyright American Scientific Publishers).

To confirm the conjugate iron binding, Perls method was conducted for ferric iron stain (Liu et al. 2009b). As observed, the NICCs changed their color from white to blue, implying the presence of iron. For further examination, the samples were examined using transmission electron microscopy (TEM), which showed ferric ferrocyanide granules on the nanoparticle surface (Fig. 2A), and indicated that iron was chelated with the conjugates.

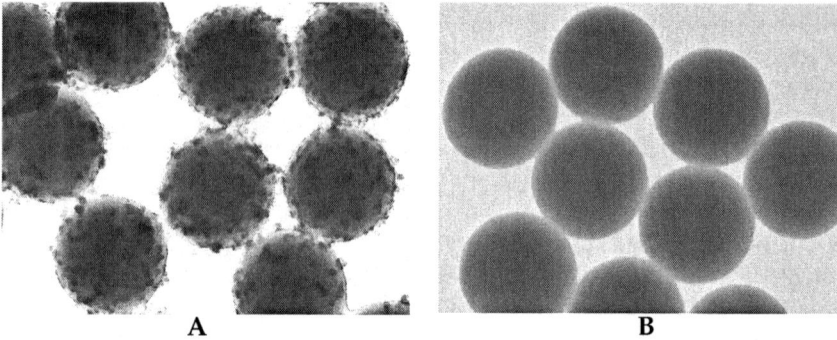

Fig. 2. Nanoparticle TEM images. (A) Nanoparticles with chelator conjugation and iron binding, and (B) Nanoparticles without both reactions. (Modification with permission from Liu et al. 2009b. Copyright Elsevier Limited).

Inhibiting Effects of NICCs on Aβ Aggregation

A large body of studies suggests that neurotoxicity of Aβ may result from the formation of protease-resistant oligomeric and fibrillar forms of Aβ (Selkoe 1991) and blocking Aβ aggregation may provide a valuable therapeutic approach (Cohen and Kelly 2003). Nanoparticles of C60 fullerence have been reported as an inhibitor for Aβ fibrillation in *in vitro* studies (Kim and Lee 2003; Podolski et al. 2007). In contrast, TiO_2 nanoparticles have reportedly been able to promote Aβ fibrillar formation (Wu et al. 2008). It is possible that the surface properties of nanoparticles play a key role in manipulating Aβ aggregation (Rocha et al. 2008). In our study, the ability of nanoparticle-iron chelator conjugates to prevent Aβ aggregation was examined, and results showed that the conjugates (Nano-N2PY as prototype) were able to prevent Aβ aggregate formation *in vitro* under physiological conditions (Fig. 3B) (Liu et al. 2009b).

Protective Effects of NICCs on Aβ Associated Neurocytotoxicity

Our study showed that the nanoparticle-iron chelator conjugates had the ability to block Aβ aggregate formation, and this formation was suggested as a key contributor to neurodegeneration in AD. It would be very interesting to know whether such ability of conjugates could lead to protection of neurons from Aβ associated toxicity. To answer this question, using Nano-N2PY and Nano-DFO1 as prototypes, the ability of the conjugate to inhibit Aβ cytotoxicity was examined *in vitro* with human cortical neuronal cells (HCN-1A) (Liu et al. 2009a). Results showed that Aβ induced toxicity to the neurons compared to controls, while the conjugates significantly mitigated the Aβ associated cytotoxicity to the cells (Fig. 4A). It is important that the

Fig. 3. Images of fluorescence microscopy. (A) Precipitates of Aβ aggregates readily form in PBS without nanoparticle iron chelator conjugates. (B) No Aβ precipitates form in PBS-containing nanoparticle iron chelator conjugates (Modification with permission from Liu et al. 2009b. Copyright Elsevier Limited).

Color image of this figure appears in the color plate section at the end of the book.

conjugates at the given concentration did not cause significant toxicity to the neurons. In addition, the neuronal cell proliferation tests found the conjugate protected the cells against Aβ-associated cytotoxicity, and again the conjugates alone did not show adverse effects on the neuronal cells (Fig. 4B).

The Ability of NICCs to Prevent free Radical Damage

Harmful free radicals can be generated by transition metals such as iron. These highly reactive species can cause all kinds of biological damage including DNA breakdown, protein oxidation and lipid-membrane peroxidation, which has been considered as a key contributor to AD (Halliwell and Gutteridge 1999). To demonstrate the capability of the nanoparticle-iron chelator conjugates to protect against free radical damage, Nano-N2PY was evaluated against hydrogen peroxide-induced oxidative damage to DNA using an immunocytochemistry assay of 8-hydroxy-2'-deoxyguanosine (8-oxo-dG). HCN-1A cells were first treated with hydrogen peroxide in PBS or the same amount of hydrogen peroxide with Nano-N2PY. Untreated controls were also established. Then, the cells were assessed with the 8-oxo-dG assay and analyzed by fluorescence microscopy. These results clearly show the ability of the NICCs to mitigate oxidative damage of DNAs (Fig. 5).

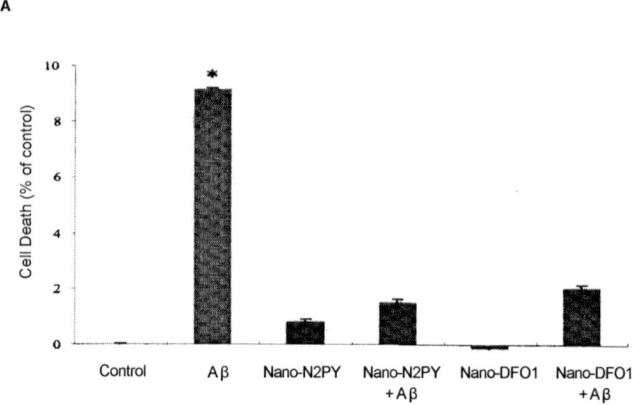

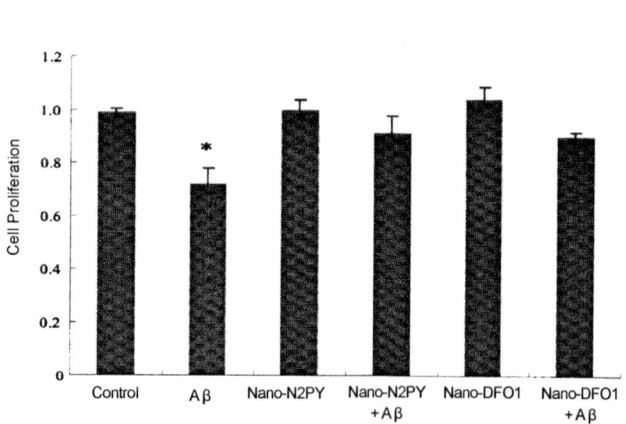

Fig. 4. **Evaluation of cytotoxicity and cell proliferation.** (A) The cytotoxicity of nanoparticle-iron chelator conjugates, Aβ only, and the conjugates with Aβ when incubated with neuronal cells as measured by Cytotoxicity LDH Detection assay. Values, as % of control, were represented as mean ± standard errors (n = 4–6; *significantly different from other groups at $P < 0.05$). (B) Effects of nanoparticle-iron chelator conjugates, Aβ only, and the conjugates with Aβ on cell proliferation of neuron cells as determined by WST-1 assay. Results were represented as mean ± standard errors (n = 3–5; *significantly different from other groups at $P < 0.05$). (With permission from Liu et al. 2009a. Copyright Elsevier Limited).

Evaluating the Possibility of NICCs to Enter and Leave the Brain *in vitro*

To examine the ability of NICCs to enter the brain and leave the brain after complexing metals, two-dimensional polyacrylamide gel electrophoresis

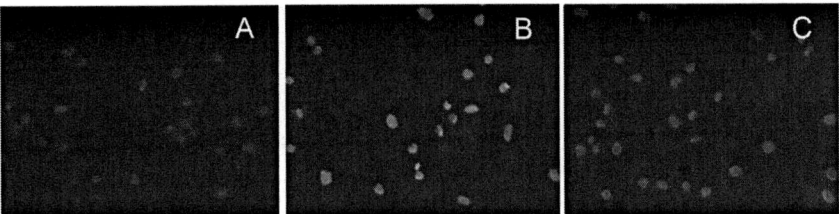

Fig. 5. **Immunocytochemistry assay of 8-oxo-dG.** A, B and C show the untreated neuronal cells, the cells treated with hydrogen peroxide and the cells treated with both hydrogen peroxide/Nano-N2PY, respectively. The cells of Fig. B are much brighter than that of Fig. A and C, which indicates that there is more oxidative damage of DNAs in the cells treated with hydrogen peroxide than the untreated cells and the cells treated with hydrogen peroxide/Nano-N2PY (Unpublished material).

Color image of this figure appears in the color plate section at the end of the book.

(2-D PAGE) analyses were performed to evaluate the protein absorption patterns on free NICCs and the conjugates with iron (Liu et al. 2006, 2009a). Our studies showed that the polysorbate 80 coated conjugates had the ability to preferentially absorb ApoE on their surface (Fig. 6B), while the conjugates complexed with iron preferred to absorb ApoA (Fig. 6C). These results were very interesting because the selected ApoE absorption of the free NICCs and the preferential ApoA absorption of the conjugates with iron might allow them to mimic the ApoE and ApoA nanoparticles, respectively. This feature could allow the conjugates both to enter the brain and then exit the brain using LDL transport mechanisms (Liu et al. 2006, 2009a; Davson and Segal 1996).

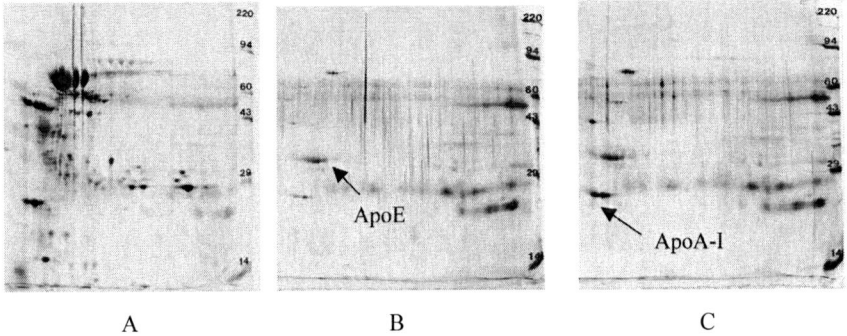

Fig. 6. **Images of plasma protein patterns examined by 2-D PAGE.** (A) Plasma; (B) Nanoparticle iron chelator conjugates coated with polysorbate 80; and (C) The conjugates complexed with iron (With permission from Liu et al. 2006. Copyright Elsevier Limited).

To further examine their ability to cross the BBB, the NICCs were investigated for their BBB penetration with a cell culture BBB model (Dehouck et al. 1996) and the nanoparticle iron chelator conjugates across the BBB were examined. Figure 7 shows the TEM imaging of the nanoparticle-iron chelator conjugates that had penetrated the BBB.

In vivo pharmacokinetic investigation of NICC ability to target and leave the brain has been conducted. Our preliminary results show that the conjugates have such ability, which paves the road to further develop these novel NICCs as multifunctional DMDs for prevention and treatment of AD.

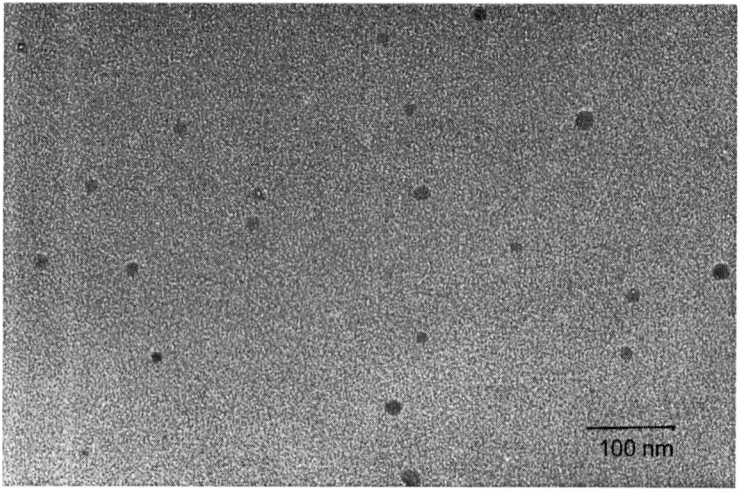

Fig. 7. Examination of nanoparticle-iron chelator conjugates with a cell culture BBB model. The BBB penetrated nanoparticle-iron chelator conjugates detected by TEM (Unpublished material).

APPLICATIONS TO OTHER AREAS OF HEALTH AND DISEASE

Accumulating studies have suggested that excess iron plays an important role not only in AD initiation and progression, but also in some other neurological diseases such as Parkinson's and Huntington's diseases, Friedrich ataxia and Hallervorden Spatz syndrome. Therefore, our development will likely provide safer and more effective chelation therapeutics for prevention and treatment of these aforementioned diseases. Additionally, our studies may also lead to the discovery of advantageous chelation pharmaceutical agents coupled with various radioactive/non-radioactive metal ions for neurological diseases' diagnosis and treatment.

Key Facts of Alzheimer's Disease and Novel Chelation Therapeutics

- Alzheimer's Disease (AD), named after Dr. Alois Alzheimer who described this disease in 1906, is the most common form of dementia and the fourth leading cause of death among people aged 65 and above (Minino et al. 2007; Yaari and Corey-Bloom 2007).
- AD is a progressive, degenerative, and irreversible brain disorder, two main features of which are many abnormal clumps (amyloid plaques) and tangled bundles of fibers (neurofibrillary tangles) in the brain (Smith 1998a).
- Currently, an estimated 24 million people suffer from dementia worldwide, and the affected population will exponentially increase to 42 million by year 2020 and to 81 million by year 2040 if no effective prevention and treatment become available (Ferri et al. 2005; Hebert et al. 2001; Scorer 2001).
- AD also puts a heavy economic burden on society, with over $100 billion annual cost of patient care in the United States alone (Minino et al. 2007; Yaari and Corey-Bloom 2007). Thus, this disease has presented a health problem with an enormous impact on individuals, families, the health care system, and society as a whole.
- Unfortunately, currently available drugs for AD treatment have shown only modest effects in modifying the clinical symptoms with relatively short periods, and none has demonstrated the ability to cure AD or to stop the disease progression (Melnikova 2007; Klafki et al. 2006). Therefore, there is an urgent need for discovering new drugs for AD treatments.
- Among other developing AD therapeutics, iron-chelating agents may serve as an alternative therapeutic approach because evidence shows that excess iron in the brain play a key role in AD development (Liu et al. 2009a; Duce et al. 2010; Avramovich-Tirosh et al. 2007; Storr et al. 2009).
- However, available chelating drugs still have serious drawbacks regarding their bioavailability and potential toxicity (Liu et al. 2005). Our development of new chelators coupled with nanoparticle delivery technology shows promise to overcome these obstacles and may lead to novel, safer, more effective therapeutics for prevention and treatment of AD.

Definitions and Explanations

Donepezil (Aricept), rivastigmine (Exelon), and galantamine (Reminyl)—the FDA-approved medications used to treat AD symptoms—are prescribed to treat mild to moderate AD symptoms. Donepezil has been recently approved to treat severe AD as well.

Memantine (Namenda), the newest FDA-approved AD medication, is prescribed to treat moderate to severe AD symptoms.

A chelating agent is a compound that can form several bonds to a single metal ion, resulting in a stable chelating agent-metal ion complex.

A nanoparticle is a microscopic particle whose size is measured in nanometers; the thickness of a sheet of paper is about 100,000 nanometers. Here, the nanoparticles can be composed of either natural or synthetic macromolecules as therapeutic adjuvants and/or drug carriers.

The nanoparticle-iron chelator conjugates (NICCs) can be generated via linking nanoparticles and chelators with covalent chemical bonds. The conjugated chelators with nanoparticles still retain their iron binding ability and can thus be used for chelation.

Blood brain barrier (BBB) serves as a separation of circulating blood and the brain extracellular fluid in the central nervous system, and strictly limits and specifically controls the passage of substances between the blood and the cerebral extracellular space.

Summary Points

- Nanoparticle-iron chelator conjugation shows promise in targeting the brain, delivering large, hydrophilic chelators across the BBB, reducing chelator toxicity.
- Importantly, the nanoparticle-iron chelator conjugation may also provide the metal-chelator complexes with a viable escape mechanism to leave the brain.
- The nanoparticle-iron chelator conjugates have very promising potential to be able to act as multifunctional DMDs by chelating redox active iron, inhibiting Aβ aggregation, mitigating the neurotoxicity associated with Aβ aggregates and preventing neurons against oxidative damage.
- Further development of nanoparticle-iron chelator conjugates could lead to discovery of novel, safer and more effective chelation therapeutics as multifunctional DMDs for prevention and treatment of AD.
- This development will also provide new therapeutics for other neurodegenerative diseases as well as new neuro-imaging agents.

Abbreviations

Aβ	:	amyloid-β
AD	:	Alzheimer's disease
ApoA	:	apolipoprotein A
ApoE	:	apolipoprotein E

AβPP	:	amyloid-β protein precursor
BBB	:	blood brain barrier
DFO	:	desferrioxamine
DMDs	:	disease-modifying drugs
2D-PAGE	:	two-dimensional polyacrylamide gel electrophoresis
EDTA	:	ethylenediaminetetraacetic acid
HCN-cells	:	human cortical neuronal cells
NICCs	:	nanoparticle-iron chelator conjugates
L1	:	1,2-dimethyl-3-hydroxyl-4-pyridinone
LDL	:	low-density lipoprotein
NMDA	:	N-methyl-D-Aspartate
8-Oxo-dG	:	8-hydroxy-2'-deoxyguanosine
TEM	:	transmission electron microscopy
FDA	:	US Food and Drug Administration

References

Abbott, N.J., M. Bundgaard and H.F. Cserr. 1985. Tightness of the Blood-brain barrier and evidence for brain interstitial fluid flow in the cuttlefish, *Sepia Officinalis*. J. Physiol. 368: 213–26.

Adlard, P.A., R.A. Cherny, D.I. Finkelstein, E. Gautier, E. Robb, M. Cortes, I. Volitakis, X. Liu, J.P. Smith, K. Perez, K. Laughton, Q.X. Li, S.A. Charman, J.A. Nicolazzo, S. Wilkins, K. Deleva, T. Lynch, G. Kok, C.W. Ritchie, R.E. Tanzi, R. Cappai, C.L. Masters, K.J. Barnham and A.I. Bush. 2008. Rapid restoration of cognition in Alzheimer's transgenic mice with 8-hydroxy quinoline analogs is associated with decreased interstitial Abeta. Neuron. 59: 43–55.

Alyautdin, R.N., E.B. Tezikov, P. Ramge, D.A. Kharkevich, D.J. Begley and J. Kreuter. 1998. Significant entry of tubocurarine into the brain of rats by adsorption to polysorbate 80-coated polybutylcyanoacrylate nanoparticles: an *in situ* brain perfusion study. J. Microencapsul. 15: 67–74.

Alyaudtin, R.N., A. Reichel, R. Lobenberg, P. Ramge, J. Kreuter and D.J. Begley. 2001. Interaction of poly(butylcyanoacrylate) nanoparticles with the blood-brain barrier *in vivo* and *in vitro*. J. Drug Target. 9: 209–221.

Avramovich-Tirosh ,Y., T. Amit, O. Bar-Am, H. Zheng, M. Fridkin and M.B.H. Youdim. 2007. Therapeutic targets and potential of the novel brain-permeable multifunctional iron chelator-monoamine oxidase inhibitor drug, M-30, for the treatment of Alzheimer's disease. J. Neurochem. 100: 490–502.

Blake, D.R., P. Winyard, J. Lunec, A. Williams, P.A. Good, S.J. Crewes, J.M. Gutteridge, D. Rowley, B. Halliwell, A. Cornish and R.C. Hider. 1985. Cerebral and ocular toxicity induced by desferrioxamine, Q. J. Med. 56: 345–355.

Bonda, D.J., G. Liu, P. Men, G. Perry and M.A. Smith. 2011. Nanoparticle delivery of transition-metal chelators to the brain: Oxidative stress will never see it coming! Prog. Brain Res. (in press).

Borchard, G., K.L. Audus, F. Shi and J. Kreuter. 1994. Uptake of surfactant-coated poly(methyl methacrylate)-nanoparticles by bovine brain microvessel endothelial cell monolayers. Int. J. Pharm. 110: 29–35.

Casadesus, G., M.A. Smith, X. Zhu, G. Aliev, A.D. Cash, K. Honda, R.B. Petersen and G. Perry. 2004. Alzheimer disease: evidence for a central pathogenic role of iron-mediated reactive oxygen species, J. Alzheimers Dis. 6: 165–169.

Casdorph, H.R. 2001. EDTA chelation therapy: efficacy in brain disorders. pp. 142–163. In: E.M. Cranton. [Ed.] A Textbook on EDTA Chelation Therapy. Hamtpon Roads Publishing Company, Inc., Charlottesville, USA.

Castellani, R.J., P.I. Moreira, G. Liu, J. Dobson, G. Perry, M.A. Smith and X. Zhu. 2007. Iron: the redox-active center of oxidative stress in Alzheimer disease. Neurochem. Res. 32: 1640–1645.

Cherny, R.A., C.S. Atwood, M.E. Xilinas, D.N. Gray, W.D. Jones, C.A. McLean, K.J. Barnham, I. Volitakis, F.W. Fraser, Y. Kim, X. Huang, L.E. Goldstein, R.D. Moir, J.T. Lim, K. Beyreuther, H. Zheng, R.E. Tanzi, C.L. Masters and A.I. Bush. 2001. Treatment with a copper-zinc chelator markedly and rapidly inhibits beta-amyloid accumulation in Alzheimer's disease transgenic mice, Neuron 30: 665–676.

Cohen, F.E. and J.W. Kelly. 2003. Therapeutic approaches to protein-misfolding diseases. Nature. 426: 905–909.

Crapper McLachlan, D.R., A.J. Dalton, T.P. Kruck, M.Y. Bell, W.L. Smith, W. Kalow and D.F. Andrews. 1991. Intramuscular desferrioxamine in patients with Alzheimer's disease, Lancet 337: 1304–1308.

Crowe, A. and E.H. Morgan. 1994. Effects of chelators on iron uptake and release by the brain in the rat, Neurochem. Res. 19: 71–76.

Davson, H. and M.B. Segal. 1996. Physiology of the CSF and Blood-brain barriers. CRC, Boca Raton. USA.

Dehouck, M.P., B. Dehouk, L. Fenart and R. Cecchelli. 1996. Blood-brain barrier *in vitro* Rapid evaluation of strategies for achieving drug targeting to the central Nerous system. pp. 143–146. In: P.O. Couraud and D. Scherman [Eds.] Biology and physiology of the blood-brain barrier, transport, cellular interactions, and brain pathologies.

Duce, J.A., A. Tsatsanis, M.A. Cater, S.A. James, E. Robb, K. Wikhe, S.L. Leong, K. Perez, T. Johanssen, M.A. Greenough, H.H. Cho, D. Galatis, R.D. Moir, C.L. Masters, C. McLean, R.E. Tanzi, R. Cappai, K.J. Barnham, G.D. Ciccotosto, J.T. Rogers and A.I. Bush. 2010. Iron-export ferroxidase activity of β-amyloid precursor protein is inhibited by zinc in Alzheimer's disease. Cell 142: 857–67.

Exley, C. 2006. Aluminium and iron, but neither copper nor zinc, are key to the precipitation of beta-sheets of Abeta (42) in senile plaque cores in Alzheimer's disease, J. Alzheimers Dis. 10: 173–177.

Ferri, C.P., M. Prince, C. Brayne, H. Brodaty, L. Fratiglioni, M. Ganguli, et al. 2005. Global prevalence of dementia: A Delphi consensus study. Lancet, 366(9503): 2112–2117.

Halliwell, B. and J.M.C. Gutteridge. 1999. Free Radicals in Biology and Medicine. Oxford University Press, New York. USA.

Hebert, L.E., L.A. Beckett, P.A. Scherr and D.A. Evans. 2001. Annual incidence of Alzheimer disease in the United States projected to the years 2000 through 2050. Alzheimer Disease and Associated Disorders. 15: 169–173.
Hider, R.C., O. Epemolu, S. Singh and J.B. Porter. 1994a. Iron chelator design, Adv. Exp. Med. Biol. 356: 343–349.
Hider, R.C., J.B. Porter and S. Singh. 1994b. The design of therapeutically useful iron chelators. pp. 353–371. *In*: R.J. Bergeron and G.M. Brittenham. [eds.] The Development of Iron Chelators for Clinical Use, CRC, Boca Raton. USA.
Kim, J.E. and M. Lee. 2003. Fullerene inhibits beta-amyloid peptide aggregation. Biochem. Biophys. Res. Commun. 303: 576–9.
Klafki, H.W., M. Staufenbiel, J. Kornhuber and J. Wiltfang. 2006. Therapeutic approaches to Alzheimer's disease. Brain 129: 2840–2855.
Kreuter, J. 2001. Nanoparticulate systems for brain delivery of drugs, Adv. Drug Deliv. Rev. 47: 65–81.
Kreuter, J., D. Shamenkov, V. Petrov, P. Ramge, K. Cychutek, C. Koch-Brandt and R. Alyautdin. 2002. Apolipoprotein-mediated transport of nanoparticle-bound drugs across the blood–brain barrier, J. Drug Target 10: 317–325.
Kreuter, J., R.N. Alyautdin, D.A. Kharkevich and A.A. Ivanov. 1995. Passage of peptides through the blood-brain barrier with colloidal polymer particles (nanoparticles). Brain Res. 674: 171–174.
Liu, G., M.R. Garrett, P. Men, X. Zhu, G. Perry and M.A. Smith. 2005. Nanoparticle and other metal chelation therapeutics in Alzheimer disease. Biochim. Biophys. Acta. 1741: 246–52.
Liu, G., P. Men, P.L. Harris, R.K. Rolston, G. Perry and M.A. Smith. 2006. Nanoparticle iron chelators: a new therapeutic approach in Alzheimer disease and other neurologic disorders associated with trace metal imbalance. Neurosci. Lett. 406: 189–193.
Liu, G., P. Men, G. Perry and M.A. Smith. 2009a. Chapter 5—Development of iron chelator-nanoparticle conjugates as potential therapeutic agents for Alzheimer disease. Prog. Brain Res. 180: 97–108.
Liu, G., P. Men, W. Kudo, G. Perry and M.A. Smith. 2009b. Nanoparticle-chelator conjugates as inhibitors of amyloid-beta aggregation and neurotoxicity: A novel therapeutic approach for Alzheimer disease. Neuroscience Letters 455: 187–190.
Liu, G., P. Men, G. Perry and M.A. Smith. 2009c. Metal chelators coupled with nanoparticles as potential therapeutic agents for Alzheimer's disease. Journal of Nanoneuroscience 1: 42–55.
Liu, G., P. Men, G. Perry and M.A. Smith. 2010. Nanoparticle and iron chelators as a potential novel Alzheimer therapy. Methods Mol. Biol. 610: 123–144.
Lynch, S.G., T. Fonseca and S.M. Levine. 2000. A multiple course trial of desferrioxamine in chronic progressive multiple sclerosis. Cell Mol. Biol. (Noisy-le-grand) 46: 865–869.
Markesbery, W.R. 1997. Oxidative stresshypothesis in Alzheimer's disease, Free Radic. Biol. Med. 23: 134–147.
Mao, X., A.D. Schimmer. 2008. The toxicology of Clioquinol. Toxicol. Lett. 182: 1–6.

Marlatt, M.W., K.M. Webber, P.I. Moreira, H.G. Lee, G. Casadesus, K. Honda, X. Zhu, G. Perry and M.A. Smith. 2005. Therapeutic opportunities in Alzheimer disease: one for all or all for one? Curr. Med. Chem. 12: 1137–47.

Melnikova, I. 2007. Therapies for Alzheimer's disease. Nat. Rev. Drug Discovery. 6: 341–2.

Minino, A.M., M.P. Heron, S.L. Murphy and K.D. Kochanek. 2007. Deaths: final data for 2004. Natl. Vital Stat. Rep. 55: 1–119.

Morel, I., J. Cillard, G. Lescoat, O. Sergent, N. Pasdeloup, A.Z. Ocaktan, M.A. Abdallah, P. Brissot and P. Cillard. 1992. Antioxidant and free radical scavenging activities of the iron chelators pyoverdin and hydroxypyrid-4-ones in iron-loaded hepatocyte cultures: comparison of their mechanism of protection with that of desferrioxamine. Free radical biology & medicine 13: 499–508.

Pappolla, M.A., Y.J. Chyan, R.A. Omar, K. Hsiao, G. Perry, M.A. Smith and P. Bozner. 1998. Evidence of oxidative stress and *in vivo* neurotoxicity of beta-amyloid in a transgenic mouse model of Alzheimer's disease: a chronic oxidative paradigm for testing antioxidant therapies *in vivo*, Am. J. Pathol. 152: 871–877.

Podolski, I.Y., Z.A. Podlubnaya, E.A. Kosenko, E.A. Mugantseva, E.G. Makarova, L.G. Marsagishvili, M.D. Shpagina, Y.G. Kaminsky, G.V. Andrievsky and V.K. Klochkov. 2007. Effects of hydrated forms of C60 fullerene on amyloid 1-peptide fibrillization *in vitro* and performance of the cognitive task. J. Nanosci. Nanotechnol. 7: 1479–85.

Porter, J.B., M. Gyparaki, L.C. Burke, E.R. Huehns, P. Sarpong, V. Saez and R.C. Hider. 1988. Iron mobilization from hepatocyte monolayer cultures by chelators: the importance of membrane permeability and the iron-binding constant. Blood 72: 1497–1503.

Ramge, P., R.E. Unger, J.B. Oltrogge, D. Zenker, D. Begley, J. Kreuter and H. Von Briesen. 2000. Polysorbate-80 coating enhances uptake of polybutylcyanoacrylate (PBCA)-nanoparticles by human and bovine primary brain capillary endothelial cells. Eur. J. Neurosci. 12: 1931–1940.

Regland, B., W. Lehmann, I. Abedini, K. Blennow, M. Jonsson, I. Karlsson, M. Sjogren, A. Wallin, M. Xilinas and C.G. Gottfries. 2001. Treatment of Alzheimer's disease with clioquinol. Dement. Geriatr. Cogn. Disord. 12: 408–414.

Ritchie, C.W., A.I. Bush, A. Mackinnon, S. Macfarlane, M. Mastwyk, L. MacGregor, L. Kiers, R. Cherny, Q.X. Li, A. Tammer, D. Carrington, C. Mavros, I. Volitakis, M. Xilinas, D. Ames, S. Davis, K. Beyreuther, R.E. Tanzi and C.L. Masters. 2003. Metal-protein attenuation with iodochlorhydroxyquin (clioquinol) targeting Abeta amyloid deposition and toxicity in Alzheimer disease: a pilot phase 2 clinical trial. Arch. Neurol. 60: 1685–1691.

Rocha, S., A.F. Thunemann, C. Pereira Mdo, M. Coelho, H. Mohwald and G. Brezesinski. 2008. Influence of fluorinated and hydrogenated nanoparticles on the structure and fibrillogenesis of amyloid beta-peptide. Biophys. Chem. 137: 35–42.

Rottkamp, C.A., A.K. Raina, X. Zhu, E. Gaier, A.I. Bush, C.S. Atwood, M. Chevion, G. Perry and M.A. Smith. 2001. Redox-active iron mediates amyloid-beta toxicity. Free Radic. Biol. Med. 30: 447–450.

Sayre, L.M., G. Perry, P.L. Harris, Y. Liu, K.A. Schubert and M.A. Smith. 2000. *In situ* oxidative catalysis by neurofibrillary tangles and senile plaques in Alzheimer's disease: a central role for bound transition metals. J. Neurochem. 74: 270–279.

Schroeder, U., P. Sommerfeld, S. Ulrich and B.A. Sabel. 1998. Nanoparticle technology for delivery of drugs across the blood-brain barrier. J. Pharm. Sci. 87: 1305–1307.

Schubert, D. and M. Chevion. 1995. The role of iron in beta amyloid toxicity. Biochem. Biophys. Res. Commun. 216: 702–707.

Scorer, C.A. 2001. Preclinical and clinical challenges in the development of disease-modifying therapies for Alzheimer's disease. Drug Discovery Today 6: 1207–1219.

Selkoe, D.J. 1991. The molecular pathology of Alzheimer's disease. Neuron 6: 487–98.

Smith, M.A. 1998a. Alzheimer's diseasse. International Review Neurobiology 42: 1–54.

Smith, M.A., K. Hirai, K. Hsiao, M.A. Pappolla, P.L. Harris, S.L. Siedlak, M. Tabaton and G. Perry. 1998b. Amyloid-beta deposition in Alzheimer transgenic mice is associated with oxidative stress. J. Neurochem. 70: 2212–5.

Smith, M.A., P.L. Harris, L.M. Sayre and G. Perry. 1997. Iron accumulation in Alzheimer disease is a source of redox-generated free radicals. Proc. Natl. Acad. Sci. USA 94: 9866–9868.

Storr, T., L.E. Scott, M.L. Bowen, D.E. Green, K.H. Thompson, H.J. Schugar and C. Orvig. 2009. Glycosylated tetrahydrosalens as multifunctional molecules for Alzheimer's therapy. Dalton Trans. 16: 3034–43.

Tamagno, E., M. Guglielmotto, M. Aragno, R. Borghi, R. Autelli, L. Giliberto, G. Muraca, O. Danni, X. Zhu, M.A. Smith, G. Perry, D.G. Jo, M.P. Mattson and M. Tabaton. 2008. Oxidative stress activates a positive feedback between the gamma- and betasecretase cleavages of the beta-amyloid precursor protein. J. Neurochem. 104: 683–695.

Tariot, P.N., M.R. Farlow, G.T. Grossberg, S.M. Graham, S. McDonald and I. Gergel. 2004. Memantine treatment in patients with moderate to severe Alzheimer disease already receiving donepezil: a randomized controlled trial. JAMA. 291: 317–24.

Wilson, B., M.K. Samanta, K. Santhi, K.P. Kumar, N. Paramakrishnan and B. Suresh. 2008. Poly(nbutylcyanoacrylate)nanoparticles coated with polysorbate 80 for the targeted delivery of rivastigmine into the brain to treat Alzheimer's disease. Brain Res. 1200: 159–68.

Wu, W.H., X. Sun, Y.P. Yu, J. Hu, L. Zhao, Q. Liu, Y.F. Zhao and Y.M. Li. 2008. TiO2 nanoparticles promote beta-amyloid fibrillation *in vitro*. Biochem. Biophys. Res. Commun. 373: 315–8.

Yaari, R. and J. Corey-Bloom. 2007. Alzheimer's disease. Semin. Neurol. 27: 32–41.

Yan, S.D., S.F. Yan, X. Chen, J. Fu, M. Chen, P. Kuppusamy, M.A. Smith, G. Perry, G.C. Godman, P. Nawroth, J.L. Sweier and D. Stern. 1995. Non-enzymatically glycated tau in Alzheimer's disease induces neuronal oxidant stress resulting in cytokine gene expression and release of amyloid beta-peptide, Nat. Med. 1: 693–699.

Yang, L.B., K. Lindholm, R. Yan, M. Citron, W. Xia, X.L. Yang, T. Beach, L. Sue, P. Wong, D. Price, R. Li and Y. Shen. 2003. Elevated beta-secretase expression and enzymatic activity detected in sporadic Alzheimer disease. Nat. Med. 9: 3–4.

Youdim, M.B., G. Stephenson and D. Ben Shachar. 2004. Ironing iron out in Parkinson's disease and other neurodegenerative diseases with iron chelators: a lesson from 6-hydroxydopamine and iron chelators, desferal and VK-28. Ann. N. Y. Acad. Sci. 1012: 306–25.

19

The Pharmacological and Toxicological Profiles of Carbon Nanotubes as Drug Carriers Toward Central Nervous System

Yingge Zhang[1,]* *and Changxiao Liu*[2]

ABSTACT

Owning to their structural robustness and synthetic versatility, carbon nanotubes (CNT), a kind of classical nanomaterials, have been utilized in multiple biomedical applications; especially their use as drug carriers has attracted great attention. Recently, CNT-based target drug delivery systems (TDDS) have been designed for the treatment of hard-to-cure diseases have attracted much attention, especially the TDDS for the treatment of those diseases that develop from the neurons of central nervous systems. While many authors reported the positive uses, there are also studies

---

[1]Institute of Pharmacology and Toxicology, Beijing Academy of Medical Sciences, 27 Taiping Road, Beijing 100850, PRC; E-mail: zhangygm@126.com
[2]New Drug Evaluation Center, Tianjin Institute of Pharmaceutical Research, 308 Western Anshan Road, Tianjin 300193, PRC; E-mail: zhangygm@126.com
*Corresponding author

List of abbreviations after the text.

showing unfavorable effects of CNT such as DNA damages, which demonstrate realistic dilemmas researchers can face while choosing carbon-based nanostructured materials in biomedicine. For the safe use of CNT, identification of pharmacological and toxicological profiles is of critical importance. Recently, this issue has been investigated in detail with SWCNT-based TDDS. It was demonstrated that lysosomes are the pharmacological target organelles and mitochondria are the toxicological organelles of SWCNT. Pharmacologically, the gastrointestinally absorbed SWCNT were lysosomotropic, which is important for SWCNT to release the drugs it carries. Toxicologically, SWCNT also entered mitochondria in large doses, which is the basis for its negative effects. III PI3K and LAMP-2A genes were involved in such an organelle preference and some chemicals have regulative effects on it. In lysosomes of the neurons in central nervous system, the drugs released from SWCNT to develop curative effects. In mitochondria, SWCNT resulted in collapse of mitochondrial membrane potentials, giving rise to overproduction of ROS, leading to damage of mitochondria, which was followed by lysosomal and cellular injury. There are differences in the dose-effects relations between the distributions of SWCNT in lysosomes and mitochondria. Based on this difference, SWCNT can be successfully used to deliver drugs into the brain for treatment of Alzheimer's disease with a safety range as high as 12. The reason for this high safety range is that it can be ensured, By controlling doses, SWCNT can only or mainly enter lysosomes, the pharmacological organelles, and no or less enter mitochondria, the toxicological organelles. Here we discuss this problem based on results of some recent studies.

INTRODUCTION

Carbon nanotubes (CNTs) are believed to be important classical nanomaterials although they have been found for no more than 20 years (Iijima 1991) because of their wide use in various fields (Chen et al. 2010; Yang et al. 2010; Bao et al. 2010). Recently their application in neuroscience has also attracted increasing attention of people (Malarkey and Parpura 2010). Structurally, carbon nanotubes (CNT) are composed of sheets of graphene that form into a cylinder, either with a single wall, single-walled carbon nanotube (SWCNT), or multiple wall, multi-walled carbon nanotubes (MWCNT). Another kind of CNT, double-walled CNTs (DWCNT) composed of two concentric graphene cylinders, represents an intermediate structure between MWCNTs and SWCNTs. The diameter of CNTs ranges typically from 0.4 to 2 nm for SWCNT and 2 to 100 nm for MWCNT while the lengths of these nano carbon structures can get to several hundred micrometers. The arrangement of the carbon atoms in the wall of CNTs can take several conformations, including armchair, chiral or zigzag ones. These conformations determine the physical and chemical properties of these carbon nanostructures; all armchair CNTs are conductive (metallic), while the zigzag and chiral CNTs can be either metallic or semiconducting. There are a variety of ways for the synthesis of CNT, including chemical vapor deposition, electric arc discharge and laser ablation. After manufacture, CNTs can be modified to enable them to perform new functions and improve their biocompatibility by attaching various chemical groups to them. Some functional groups, such as lipids, DNA and various peptides, can often be simply adsorbed to the wall of CNTs. If a more firm and durable attachment is desired and needed, compounds may be covalently linked to CNTs. This is most often done without too much difficulty by incubating CNTs with strong oxidizing agents, which add carboxyl groups to the ends of the tubes and any defect sites. Then, other groups can be added, usually converting the carboxyl group to acyl chloride, which can then be reacted with the compound of interest. For more information on the modifications of CNT structures see previous chapters (Bekyarova et al. 2005). Although CNTs have shown promise in many fields of biomedicine, here we discuss the use of CNTs in neuroscience, focusing mainly on recent developments with regard to their use as drug carriers in the treatment of some diseases in the central nervous systems, with a brief coverage of earlier studies. For more information on the earlier applications of CNTs in neurobiology see Malarkey and Parpura's article (2007).

CNTs AS DRUG CARRIERS TARGETING NEURONS IN CENTRAL NERVOUS SYSTEM

Early works have demonstrated that CNTs can translocate in cells. CNTs can interact with mammalian cells and enter cells via cytoplasmic translocation (Kostarelos et al. 2007); therefore deliver a range of therapeutic reagents inside the cell. For example, plasmid DNA has been internalized inside the cell and the expression of the plasmid-carried marker genes are enhanced (Cai et al. 2005; Gao et al. 2006). Other macromolecules including proteins (Liu et al. 2005), polymers (Kam and Dai 2005) and single-stranded DNA (Cherukuri et al. 2004) have also been internalized by coating onto CNTs and through the interaction of CNTs with mammalian cells.

Our recent work demonstrated SWCNTs can be used to deliver drugs into the neurons in brain (Yang et al. 2010). We first examined whether SWCNTs can be absorbed through gastrointestinal tract. SWCNTs were administered in the way of gastrogavage in mice, which was deliberately fed with fluent foods free from any particulate substances and any substances that may deposit in cells or organelles to form particles (particulate-free food, PFF). PFF feeding began 48 hours before the administration of SWCNTs through the end of the experiments. Raman spectrum showed the SWCNTS containing 99.455% carbon, 0.439% oxygen, 0.106% silicon atoms and no iron (Fig. 1). To obtain suspension for gastrogavage, SWCNT were sheared in dimethyl formamide by a hi-speed dispersator and were centrifuged. The precipitation was re-suspended in normal saline by sonication (Liu et al. 1998) and allowed to stand still for 10 min. The supernatant fluid was then collected and re-centrifuged, and shortened SWCNT was obtained, which was dried and re-suspended in normal saline

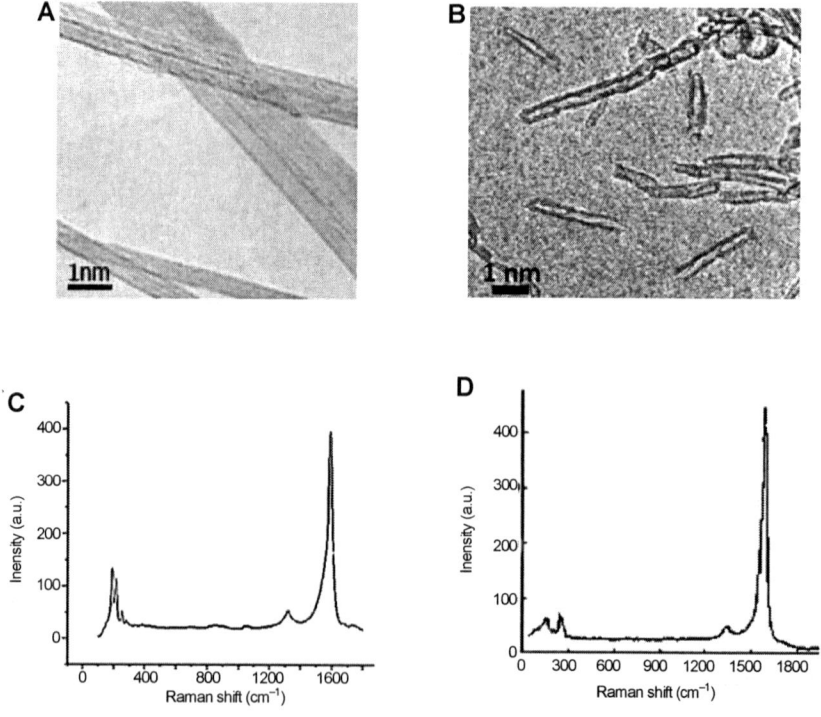

Fig. 1. The electron microscope image and Raman spectrum of SWCNT. A: TEM image of SWCNT; B: TEM image of the shearing-shortened SWCNT; C: the Raman spectrum of the SWCNT before gastrogavage; D: the Raman spectrum of the shearing-shortened SWCNT.

by sonication and incubated for 10 ~ 20 minutes before gastrogavage. The shortened SWCNT had a length of about 50 ~ 300 nm (Fig. 1C). Shearing and brief sonication did not change the chemical composition of CNT as shown by the similar Raman spectrum before and after the shearing (Fig. 1D) (Sato et al. 2005). At the 48th hour after the beginning of PFF feeding, gastrogavege of SWCNTs began, once a day for 10 days. Within 30 minutes after the last gastrogavege, the mice were sacrificed and the internal organs were taken out and observed under transmission electron microscope (TEM). Results demonstrated that SWCNTs can be absorbed through columnar cells of the intestinal membrane, which lays the foundation for their use as an oral carrier of drugs. The absorbed SWCNTs may enter into the neurons in the brain through neurite axoplasm transportation. Intracellularly, SWCNTs distributed in the lysosomes of in neurons (Fig. 2), which lays the foundations for SWCNTs to be used as an drug carrier targeting the neurons in the central nervous system.

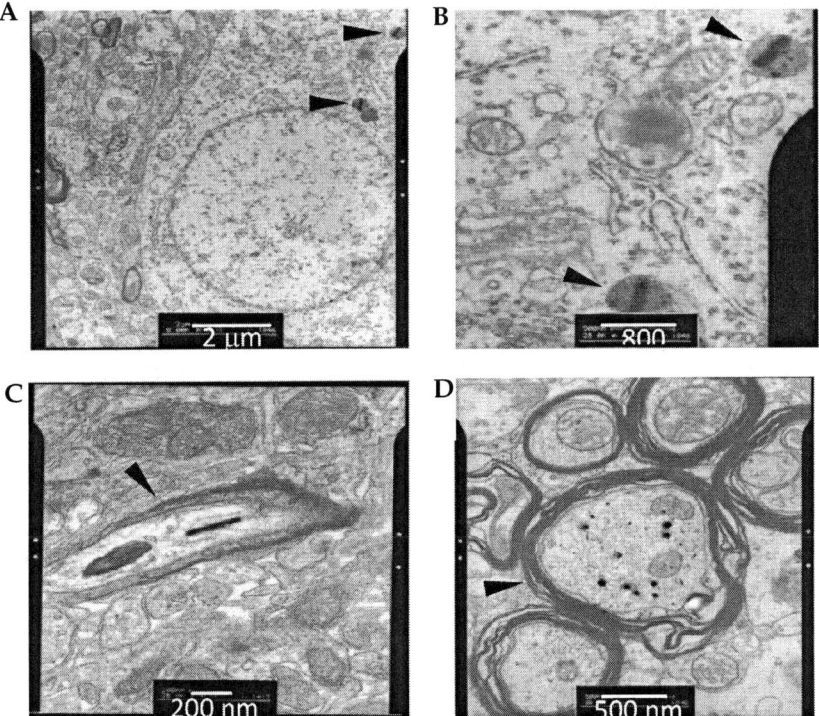

Fig. 2. The cell distribution of SWCNT in brain examined by TEM A: a neuron containing SWCNT (arrows); B: SWCNT in 2 lysosomes indicated by arrows in a; C:: section parallel to the longitudinal axe of one neurite, which was confirmed by the sheath (arrow). SWCNT is fiber-like; D: cross section of neurite, which was confirmed by the sheath (arrow). SWCNT were dot-like.

Acetylcholine (ACh) is a natural transmitter of the cholinergic nervous system, which plays the role of the signal transmission between neurons and was closely related to the functional activities of neurons, such as learning, memory and thinking. Alzheimer disease (AD) is a neurodegenerative disease originating from the lack of ACh because of the loss of the neuronal function to synthesize ACh. And therefore, ACh should be the most effective drug for the treatment of AD but ACh can't be used in the treatment of AD, since there is no way to deliver it into brain because of its strong positive charges, which inhibits its entry extremely into brain through blood-brain barrier (BBB). For the reasons mentioned above, ACh is believed to have only pharmacological significance while no clinical significance. Aimed at the solving of this problem to seek a novel drug for AD, we designed an experiment (Fig. 3). An SWCNT-based ACh delivery system (SWCNT-ACh) was prepared by incubating of SWCNTs with ACh. ACh can be adsorbed onto SWCNTs as shown by Raman spectrum although the mechanism remains to be made clear (Fig. 3). Through observing the effects of SWCNT-ACh on the experimental AD, we can know whether SWCNTs can carry ACh into brain (Yang et al. 2010).

An ideal drug carrier should be administered orally. To detect whether SWCNTs can carry ACh from the gastrointestinal tract through intestinal wall, blood circulation and BBB to the neurons in hippocampus of the brain, AD animal models were made by intraperitoneal single dose injection of 20 mg/kg kainic acid (KA) (Liang et al. 2007), which is a neurotoxic amino acid and can damage neurons to cause dementia and has been used as an agent in the preparation of animal models of AD (Liang et al. 2007). The learning and memory capabilities were tested before KA injection and re-tested 24 h after KA injection. AD mice would have lower merits in comparison with themselves before KA injection as well as the normal mice. The mice of therapeutic groups were gastrogavaged with SWCNT-ACh on the 20th hour after KA injection. If oral SWCNT-ACh has higher learning and memory merits, it would be confirmed that SWCNT can carry ACh into brain. The SWCNT-ACh was used as a therapeutic agent for the model AD in comparison with single SWCNT and free ACh. Results demonstrated that SWCNT-ACh can improve the learning and memory deficit of AD mice to the level similar to the normal mice while single SWCNT and free ACh cannot (Fig. 4). The excellent therapeutic effects of SWCNT-ACh on experimental AD indicated that SWCNTs delivered ACh into the neurons successfully, providing a novel way for the treatment of AD as promising oral carriers of ACh targeting the neurons in the central nervous system.

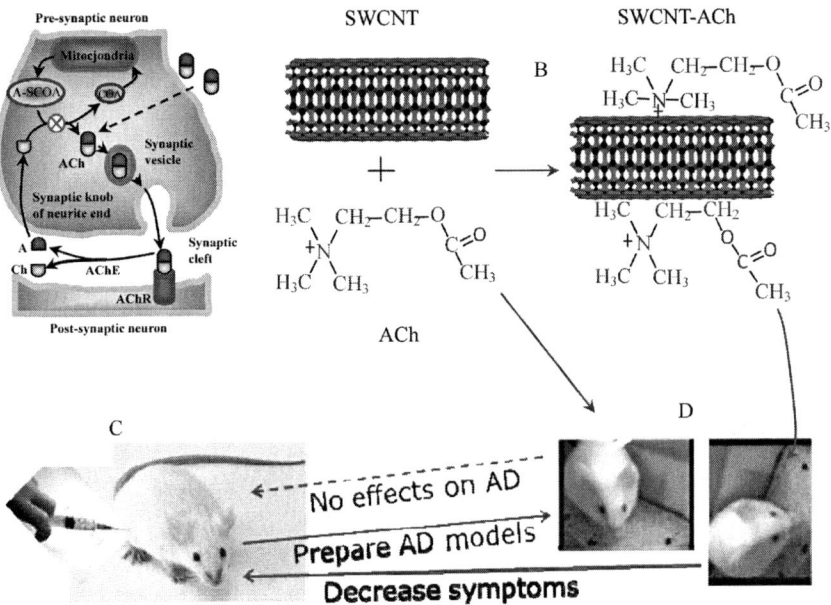

Fig. 3. The Mechanism of Alzheimer's Disease to develop and the protocol of the experiments to verify the capability of SWCNTs to deliver ACh into the neurons in brain. A The schematic illustration of the mechanism for AD development and treatment with ACh. ACh is the natural transmitter of cholinergic nervous system. It is systhesized with acetyl (A) from acetyl-COA and choline (Ch) re-absorbed from the synaptic cleft and released from the neurite end of the presynapric neurons when nervous electric pulse comes. The released ACh binds to the acetylcholine receptor (AChR), which open the ion channel of the receptor to give rise to active potentials, completing the message transmission between neurons. The ACh is disintegrated by acetylcholinesterase (AChE) into A and Ch after ACh finish their mission, which keeps the sensitivity of transmission of neural pulses. A and Ch may re-enter the presynaptic knob to be re-used in the synthesization of ACh. In the patients of AD, the neurons lost their capability to synthesize ACh so that the message can't be transmitted between neurons, leading to the dementia. Obviously, supplying foreign ACh will be an effective therapeutic but ACh can't be used as drugs because of its strong polarities which prevent it from entering the neurons in the brain through blood-brain barrier. B: the preparation of the drug delivery system of ACh targeting to neurons in brain. ACh may be adorsorbed onto SWCNTs by hydrophobic interactions or electrostatic interactions, which has been verified by Raman spectrum although the mechanism remains to be elucidated. C: The preparation of AD models. D: treatment of AD with single SWCNTs, free ACh and SWCNT-ACh. The results demonstrated that both single SWCNTs and free ACh cannot improve the dementia of AD mice but SWCNT-ACh can make the learning and memory capabilities of AD mice recover to normal (Yang et al. 2010), demonstrating SWCNTs can be effectively used as drug carriers of ACh in the treatment of AD.

Color image of this figure appears in the color plate section at the end of the book.

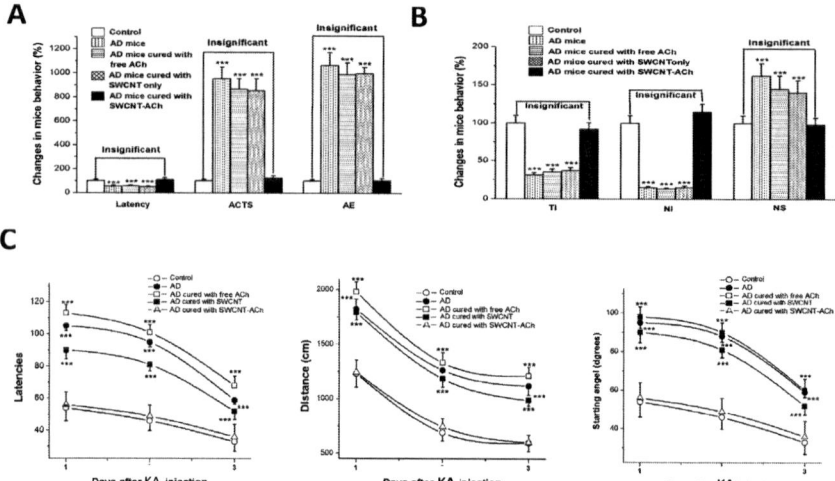

Fig. 4 Effects of SWCNT-delivered ACh on the learning and memory in AD mice A: The effects of SWCNT-ACh on learning and memory of AD mice measured by the step-down test. Latency: the time of the AD mice stay on the plate preventing the mice from electric stimulating; TS: the accumulated time of the mice being stimulated; NE: the number of the errors made by mice; C: the effects of SWCNT-ACh on the learning and memory of AD mice measured by the shuttle-box test. TI: The latency of active escape; NI: The numbers of active escape; NS: The numbers of passive evidence; D: The effects of SWCNT-ACh on the learning and memory of AD mice measured by the Morris water maze test. D3, D4 and D5 are the 3th, 4th and 5th day after the injection of kainic acid. ***$p < 0.001$ compared with control.

THE ELECTROPHYSIOLOGICAL PHARMACOLOGY OF CNTs ON NEURONS

As carriers targeting neurons, their effects on the functional activities of neurons must be made clear. Many excellent studies have provided rich knowledge on this issue. Ion channels are important functional nanostructures in neurons. Park et al. (2003) have provided evidence that CNTs can affect the function of ion channels although their experiments were done with non-nerve cells. They exposed Chinese hamster ovary cells expressing a variety of potassium ion channels to unmodified SWCNTs of varying diameter and the effects of CNTs were determined by whole-cell current recordings. SWCNTs affected all the types of potassium ion channel that were expressed, with human ether-a-go-go-related gene channels showing the greatest inhibition on current. Small-diameter CNTs (0.9 nm), such as SWCNTs, had the greatest blocking ability and the effects were concentration-dependent and reversible, while large-diameter MWCNT have no blocking effects on the channel activity. It is speculated that the CNTs may sit in or on top of the channel pore and interrupt ion flux and transition of the channel between the open and inactive states.

Chhowalla et al. (2005) functionalized SWCNTs to change their ability of channel-blocking. 2-Aminoethylmethane thiosulfonate (2-AEMTTS), of which a sulfhydryl probe binds to cysteine residues, was attached to carboxy-terminated SWCNTs. In Chinese hamster ovary cells expressing hyperpolarization activated cyclic nucleotide-gated ion channels were shown reversible blockage by whole-cell current recordings when exposed to carboxy-terminated SWCNTs. However, the block became irreversible when 2-AEMTTS-conjugated SWCNTs were applied, presumably because of the influences of the 2-AEMTTS-cysteine interactions on the hyperpolarization-activated cyclic nucleotide-gated cation channels, indicating the possibility to target channel-blocking SWCNTs to specific channels by functionalization with particular groups. What must be pointed out is that CNTs may inadvertently affect the activity of cells they come into contact with. Lovat et al. (2005) performed a series of experiments to examine the effects of CNTs on the electrical activity of the neurons cultured on CNT substrates. The activities of neurons were assessed by single-cell patch clamp recording after being in culture for 8 days. A 6-fold increase in the frequency of spontaneous postsynaptic current generation, was found in neurons grown on MWCNTs compared to those on plain glass. However, the group was not able to record significantly larger spontaneous postsynaptic currents from the neurons grown on MWCNTs in comparison with that from those recorded from the neurons grown on plain glass, although there was such a tendency. There was also a 6-fold increase in spontaneous action potential frequency in the neurons grown on MWCNTs (Fig. 5). It was believed that this increase in activity was not due to cell density or survival rate of the neurons on the MWCNT substrate, since there was no significant difference in cell density found on both substrates. The cell density was determined by using immunolabeling of cells with antibodies against neuron-specific microtubule-associated protein 2 and astrocyte-specific glial fibrillary acidic protein. At the same time, the effects did not appear to be caused by gross differences in the morphology of neurons, since the number of neurites per neuron was similar. Besides, Both sets of neurons also displayed a similar resting membrane potential, input resistance and capacitance. Additionally, it also seems unlikely that the CNTs affected channels in the neurons directly, since they were MWCNTs which, as stated previously, should not have much effect on channels. The authors speculated that the high conductivity of the CNT substrate might have affected voltage-dependent membrane processes resulting in the increased activity. The results of works of Liopo et al. (2006) seemed to be in consistent with that of Lovat et al. (2005). They showed that the conductive nature of a CNT substrate could be harnessed as a method to stimulate neurons growing on top of them. SWNTs were deposited onto polyethylene terephthalate films and

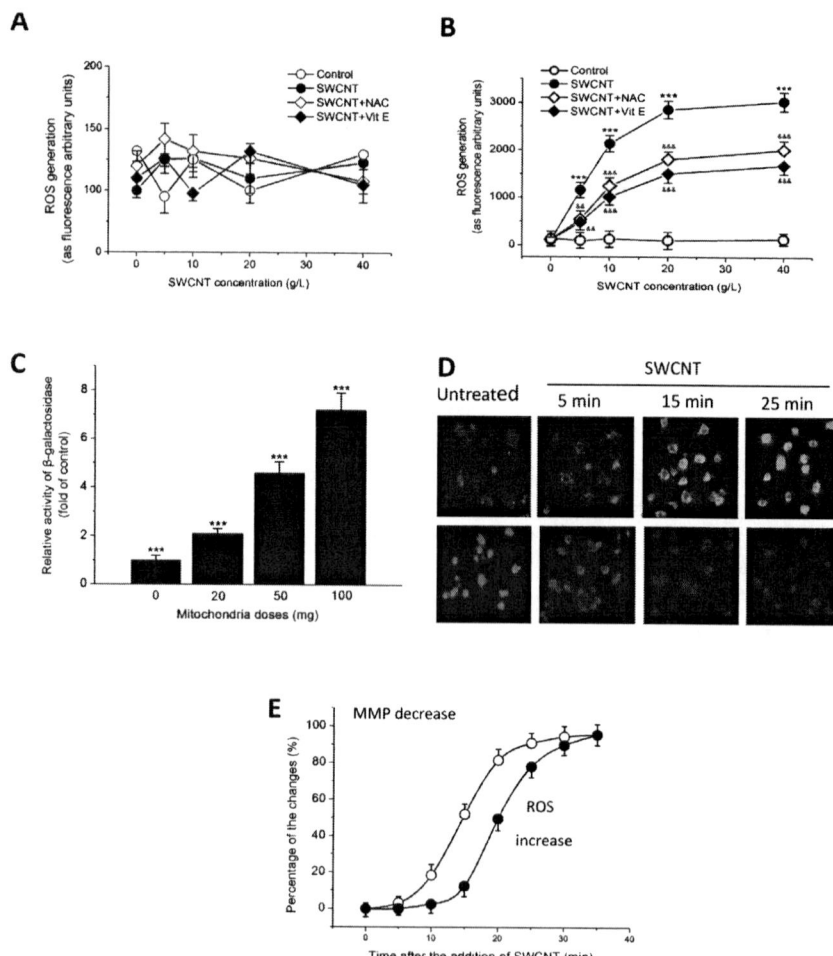

Fig. 5. **Relationships between lysosomal damage and mitochondrial damage studied *in vitro*.** (A) Effects of SWCNTs on the production of reactive oxygen species (ROS) in lysosomes. (B) Effects of SWCNTs on the production of ROS in mitochondria. NAC, N-acetylcysteine (10 µmol L^{-1}); Vit E, vitamin E (10 µmol L^{-1}). (C) The influence of mitochondria on the lysosomal damage by SWCNTs. Mitochondria were added at the indicated concentrations, and the activities of β-galactosidase were measured. Lysosomal damage was expressed as the activity of the β-galactosidase leaked from lysosomes. (D) Fluorescence images showing the influences of SWCNTs on the mitochondrial membrane potential (MMP). Green fluorescence increased and red fluorescence decreased with incubation time. (E) Changes in the red-to-green fluorescence ratios in the mitochondrial membranes over 25 minutes of incubation with SWCNTs. SWCNTs significantly decreased the ratio time-dependently. (F) The time courses for the changes of both MMP and mitochondrial ROS production. ***P < 0.001 compared with control. &&&P < 0.001 compared with SWCNT.

Color image of this figure appears in the color plate section at the end of the book.

a stimulation chamber was created by gluing a ring in the middle of the film to contain dorsal root ganglion neurons with electrodes attached to the CNT substrate outside the ring. An inward current that closely resembled currents induced by depolarizing voltage steps was recorded in the neurons by whole-cell patch clamp when a 1A current was applied through the CNTs.

BIOCOMPATIBILITY OF CNTs WITH NEURONS

For the use of nanomaterials as drug carriers targeting the nervous system, biocompatibility with neurons is a prerequisite. There are many studies demonstrating that CNTs are excellently compatible with neurons. CNTs have sizes and shapes similar to that of neuronal processes and are strong yet flexible and can be made conductive, which are all advantageous qualities for creating scaffolds for neuronal growth, showing a bright promise in promoting the regeneration of neurons. Mattson et al. (2000) provided the first demonstration that CNTs could be used as substrates for neuron growth. They grew hippocampal neurons on glass coverslips coated with the permissive substrate polyethyleneimine (PEI) and overcoated with MWCNTs. It was discovered that modifying the MWCNTs with a biologically active compound, 4-hydroxynonenal could improve the growth of neurons; this modification was done simply by physiosorption. Following this study, Hu et al. (2004) grew neurons on MWCNT substrates with different charges to investigate whether CNT substrates could be modified systematically in order to affect neurite outgrowth and branching. They found that neurons on as-prepared MWCNT substrate did not grow as well as those on the commonly used substrate, PEI, although MWNTs did act as a permissive substrate. Then Hu et al. (2005) modified the charge of the MWCNT substrate by functionalizing with carboxyl groups, poly-m-aminobenzene sulfonic acid or ethylenediamine to create negatively, zwitterionic or positively charged nanotubes, respectively. It was found that as the substrate became more positive the length and branching of neurites increased. In a subsequent study, it was found that neurons cultured on substrates made of a graft copolymer with SWCNTs functionalized with branched PEI to "dilute" PEI's positive charge, resulted in neurite outgrowth and branching intermediate to that of as-prepared MWNTs or PEI alone (Hu et al. 2005). Bio-compatibility of CNTs can be decreased by some modifications. The studies of Liopo et al. (2006) showed that functionalized SWCNTs with 4-tertbutylphenyl or 4-benzoic acid were less supportive of neural cell attachment and growth than unfunctionalized CNTs. Covalent modifications increase the retention of an attached compound/group on CNTs, which can be achieved with proteins that play a role in brain signaling. Neurotrophins are protein growth factors that promote the survival and differentiation of neurons. MWNTs covalently modified with two such proteins, nerve growth factor (NGF) and brain-derived neurotrophic factor (BDNF), could be used to regulate the growth of neurons (Matsumoto et al. 2007). MWNTs coated with NGF or BDNF added to the culture medium of embryonic chick dorsal root ganglion cultures promoted neurite outgrowth similar to soluble NGF or BDNF. These experiment results demonstrate that biologically active molecules like neurotrophin bound to CNTs did not loss their activity and interacted with cells to promote their function. Beyond depositing CNTs onto planar glass coverslips, Gheith et al. have created freestanding films of a poly(Ncetyl-4-vinylpyridinium bromide-co- N-ethyl-4-vinylpyridinium bromide-co-4-vinylpyridine) SWNT copolymer and used them as substrates for neural cell growth (Gheith et al. 2005). Furthermore, it has been reported that directionally oriented MWNTs in the form of sheets or yarns offer an alternative presentation of CNTs to cells (Galvan-Garcia et al. 2007). Cell attachment, differentiation and cell growth were promoted significantly by these CNT forms. Additionally, highly purified CNTs also allowed neurons to extend processes of which the number and length were comparable to those of neurons grown on a planar permissive substrate, represented by polyornithine pre-treated glass. Thus, these experiments demonstrated that purity of CNTs as well as the 3-dimensional organization of the CNT substrate/scaffold can affect the interaction between neurons and CNTs. Interestingly, carbon threads and ribbons up to 30 cm in length can be made from SWNTs using a particle coagulation spinning process (Dubin et al. 2008). These nanomaterials were also shown compatible with neural cells by culturing hippocampal neurons and PC12 cells on them. Cultured for a week, neurite outgrowth on the nanothread surface was found by immunolabeling against a neuron specific label,β-III tubulin. NGF that can induce neurite outgrowth from PC12 cells can exert such an action on PC12 cells grown on CNT threads. These experiments demonstrated the biocompatibility of these nanothreads, indicating that they may be suitable materials for the construction of electrodes or nanowires in implantable devices. Gabay et al. (2005) created a micropatterned array of CNT islands to study self-organization of neural networks, in which they used poly (dimethyl siloxane) stamp to imprint a pattern of iron nanoparticle islands on quartz substrates and used chemical vapor deposition to grow CNTs on the iron catalyst islands. The neurons cultured on these substrates accumulated on the CNT islands within 4 days. Neuronal processes were seen to bridge the nonpermissive quartz gaps to form adjacent island connections. Similarly, a SWCNT patterned substrate was created by Sorkin et al. (2006) applying iron nitrate catalyst to coverslips with a polydimethylsiloxane stencil and then growing CNTs by chemical vapor deposition. The cultured hippocampal neurons on these substrates grew spontaneously into islands with neurites connecting each island after 2 to 3 days. These patterned networks showed promises to study neuronal networking with CNT islands as electrical connections for sensing or stimulation.

Besides the above outlined investigations on the interactions between CNTs and neurons, some authors have reported their attempts to use CNTs in stem cell research. The growth and differentiation of neural stem cells on CNT substrates was also demonstrated recently (Jan and Kotov 2007). Coverslips coated with six layers of SWNTs dispersed in poly(sodium 4-styrene-sulfonate) and PEI were used to culture neural stem cells, in comparison with poly-L-ornithine-coated coverslips. Cells were cultured in epidermal growth factor (EGF)-depleted medium to induce differentiation. Cells showed good behaviors of adhering to wall and differentiating into neurons, astrocytes and oligodendrocytes as demonstrated by immunoreactivity of nestin, microtubule-associated protein 2 (anti-MAP2), glial fibrillary acidic protein (anti-GFAP), and oligodendrocyte marker O4 (anti-O4). These cells were shown just as viable on the CNT substrate as on poly-ornithine and differentiated and extended processes similarly. There was no adverse effect of growing on nanotube substrates, indicating these materials could be used for applications involving nerve stem cells.

SAFETY PROBLEMS OF CNT AS DRUG CARRIERS

However, the application of CNTs in clinical neuroscience is hindered by their toxicity. In particular, ability of fullerenes to induce lipid peroxidation has caused serious concern. Largemouth bass exposure to pure unmodified fullerenes showed that C60 fullerenes can enter into the brain and led to an increase in lipid peroxidation, as well as a decrease of glutathione in the gill (Cui et al. 2010; Oberdorster 2004). The problem may be related partly to their hydrophobicity, aggregation propensity and interaction with the membranes of cells. There are experiments showing indications that these problems may be mitigated by some procedures. For example, the surface modifications that decrease surface hydrophobicity and increase solubility of the CNTs were also shown to decrease CNTs cytotoxicity (Sayes et al. 2006). Furthermore, there are reports suggesting that ability of fullerenes to generate free radicals may be due to metal ion impurities that can be removed from fullerene samples, which can abolish the toxicity of CNTs (Pulskamp et al. 2007). It was also shown that simple filtering of SWCNTs can decrease their toxic effects on smooth muscle cells (SMC), as measured by growth inhibition (Raja et al. 2007). Some studies showed that the free-radicals can be scavenged by purified MWCNTs (Fenoglio et al. 2006). Hence, purification and chemical modification of fullerenes aimed to increase solubility and decrease toxicity will be needed for their successful application in medicine and clinical neuroscience.

It may be seen that there exist some paradoxes about the biocompatibility of CNTs, which gives rise to the issue about the toxic mechanisms. If the mechanisms behind the toxic effects can be made clear, the application of SWCNTs in biomedicine will become possible: methods can be taken to reduce or decrease the toxic effects. For this purpose, we investigated the mechanisms behind the biocompatibility of CNTs in mice (Yang et al. 2010). *In vitro* experiments revealed that SWCNTs induced the increase of reactive oxygen species (ROS) in mitochondria while induced no significant changes in ROS level in lysosomes. SWCNTs did not increase the leaking rate of β-galactosidase without the existence of mitochondria while they did increase it when mitochondria were added into the reaction systems (Fig. 4). The effect of SWCNTs to increase ROS in mitochondria is believed attributable to the blocking of the electron transmission chains (ETC) for bio-oxygenation, the key biochemical reaction to produce energy usable for cells in mitochondria, the power plant in organism bodies, because the increase of ROS closely followed the collapse of ETC (Fig 5). *In vivo* experiments demonstrated that there would be no lysosome damage without mitochondria damage and the SWCNT-containing rate (SCR) in lysosomes was much more before and higher than the damage rate of lysosomes while SCR in mitochondria almost overlaid to damage rate of mitochondria (Fig. 6). The *in vivo* experiment results strongly support that of the *in vitro* experiments.

The results of both *in vivo* and *in vitro* experiments demonstrated that there are different organelles to mediate the pharmacological effects and the toxicological effects; lysosomes are the pharmacological organelles while mitochondria are the toxicological organelles of SWCNTs. This means the toxic effects may be avoided or decreased by preventing SWCNTs from entering into mitochondria. Further investigations found that some drugs and energy-supplying status can promoting the selectivity in the distribution of SWCNTs in the two organelles. In fact, SWCNTs themselves have the propensity to get into lysosomes: they almost did not enter into mitochondria but lysosomes in the doses of below 300 mg/kg. This means the toxic effects of SWCNTs can be avoided or decreased simply by the controlling of doses.

In the animal models, it was found that SWCNTs have the feature of insidiousness, which means the animals still manifested a good health status although there have been ultrastructural damage in the level of cells. At 400–500 mg/kg they caused significant subcellular damage

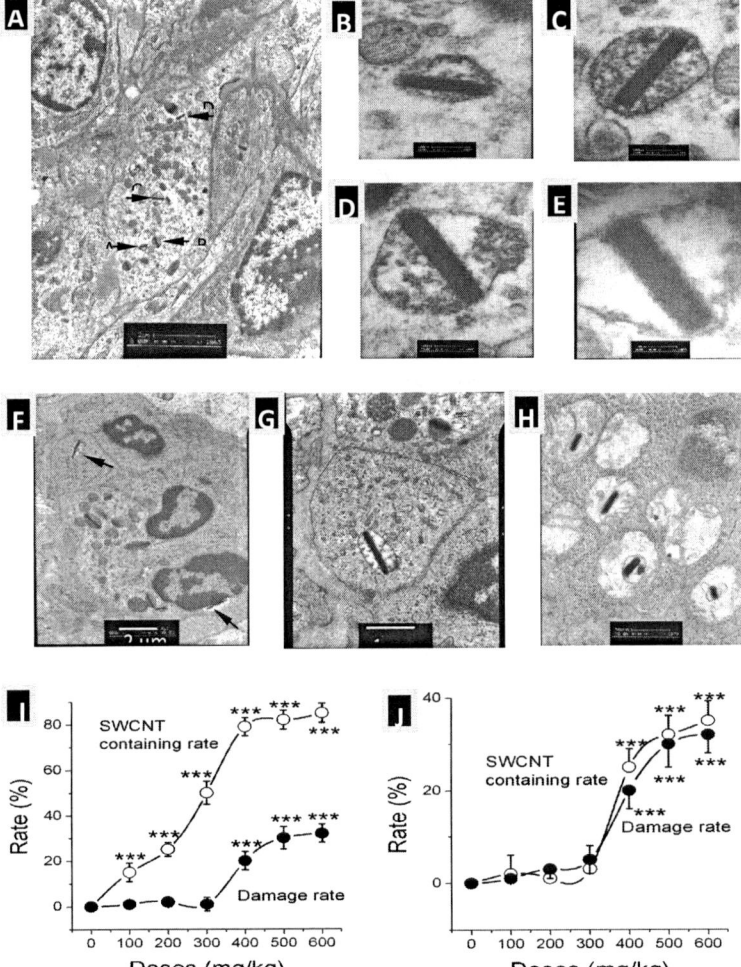

Fig. 6. The damage effects of single-walled carbon nanotubes (SWCNTs) on lysosomes and mitochondria under transmission electron microscopy. (A) A cell with SWCNT-damaged lysosomes (arrows). (B) Magnification of the SWCNT-damaged lysosome indicated by arrow A in A. (C) Magnification of the SWCNTdamaged lysosome indicated by arrow B in A. (D) Magnification of the SWCNT-damaged lysosome indicated by arrow c in A. Some contents are being excreted (arrow). (E) magnification of the lysosome indicated by arrow d in A. The whole lysosome became a large cavity. (F) The SWCNT-damaged mitochondria (arrow) begin to appear in cells. (G) Mild mitochondrial damage: the cristae are smaller. (H) Severe mitochondrial damage. The cristae are nearly completely destroyed or have disappeared. The residual cristae reveal that the structures are mitochondria. SWCNTs were found in some of them. (I) The relationship of SWCNT dose to percentages of lysosomes containing SWCNTs and of lysosomes showing damage. ***$P < 0.001$ compared with control. (J) The relationship of SWCNT dose to percentages of mitochondria containing SWCNTs and of mitochondria showing damage. ***$P < 0.001$ compared with control.

revealed as ultrastructural pathological changes but did not affect overall physiology and well-being of the animals. Animals showed no signs of abnormality by symptom observation and growth assessment (Yang et al. 2010). Traditionally, drug toxicity in drug development is largely evaluated by symptoms, histopathological observations, and blood panels including hematology. In animal experiments toxicity is often evaluated as the dosage for the death of half of the animals (LD50), the symptoms, and the organ injuries. The ultrastructural changes have been seldom used in the safe evaluation of drugs. Obviously, those traditional indexes are not sufficient for the evaluation of the toxicity of SWCNTs because of the insidiousness of their toxicity. Based on the unique pathological features of SWCNTs, the safety range of some nanomaterials should be re-evaluated by the ultrastructural pathological changes at the subcellular level in preference to using traditional histopathological and symptomatic indexes.

THE SAFETY RANGE OF CNTs AS DRUG CARRIERS

As mentioned above, the main problems that hinder the biomedical applications of CNTs are the bio-safety-related issues because of the cytotoxicity of them while the promises of CNTs in the treatment of the disease of nervous system are attractive. Can CNTs be safely used in the therapies for nervous diseases? This is one of the questions for those scholars who are interesting in solving the problems by using CNTs. Recently, we have performed several experiments to tried to investigate it (Yang et al. 2010). The results demonstrated that the damage of SWCNTs can be significantly decreased and controlled under such a degree that SWCNTS did not result in any ultrastructural changes. As shown by the experiments on the treatment of AD mice, the effective doses of SWCNT-ACh were determined to be in the range of 20–50 mg/kg, corresponding to the doses of ACh at 4–10 mg/kg. The ED_{99} value of SWCNTs was determined to be 25 mg/kg, corresponding to that of ACh at 5 mg/kg. Based on ED_{99} and the maximal safe dose of 300 mg/kg for the damage of mitochondria and lysosomes (which is taken as the TD_1 although no organelle damage is observed at this dose because it is very difficult to find TD_1 dose), the safe range for SWCNTs was proposed as 12 in the treatment of AD in mice, which is high enough for their *in vivo* use (Yang et al. 2010).

Conclusion Remarks

The studies up to now have provided a good quantity of evidences for the potentials of CNTs to be used in the neuroscience as well as the possible negative effects. As we know, however, their toxicity does not necessarily militate against their application in nanomedicine, because all substance may be toxic in a strict sense; the key is to control the dosage. Only at high dosages did SWCNTs cause the pathological changes in the ultrastructures of lysosomes and mitochondria; they are highly safe at low doses, providing the basis for their use in neuroscience. It is believed that CNTs will practically play roles in neuroscience in not too long a time along with the understanding of their pharmacological and toxicological profiles.

Summary

- Carbon nanotubes (CNTs) are believed to be one kind of important classical nanomaterials, with structures composed of sheets of graphene that form into a cylinder, either with a single wall, single-walled carbon nanotube (SWCNT), or multiple wall, multi-walled carbon nanotubes (MWCNT).
- CNT can be used as drug carriers in targeting drug delivery systems and SWCNT can deliver acetylcholine into the neurons in the central nervous system. SWCNT-based acetylcholine delivery system (SWCNT-ACh) has good effects on experimental Alzheimer disaese.
- CNTs have good biocompatibility with neurons with physiological influences on the electrical activities of neurons.
- CNTs also have negative effects on cells but the pharmacological effects and toxicological effects are mediated by different organelles and there are differences in the doses for the two kinds of effects. Based on the pharmacological and toxicological mechanisms, the toxic effects of CNTs can be avoided while the curative effects can be sufficiently used.
- CNTs have great promise as drug carriers in the drug delivery systems targeting to the neurons in central nervous system.

Dictionary

Drug carriers: Objects such as particles that can carry drugs from the administration site to the therapeutic effects-related sites in the body.

Pharmacological effects: The effects that are related to the favorable action of drugs such as the suppression of disease symptom or the elimination of pathological factors.

Toxicological effects: The effects that are related to the unfavorable action of drugs such as the damage to normal cells, organs or bodies to cause new symptoms or diseases.

Lysosomes: Organelles in the cells with single lipid membrane, functioning as the digestive organ of the cell to eliminate foreign particulate substances and segments from the cell itself.

Lysosomotropic: have the priority to distribute in lysosomes.

Mitochondria: Organelles in the cells for biooxygenation to provide energy in the mode of ATP for cell activities as the power of the cell. Mitochondira contain electron transmission chains (ETC). The leak of free electrons from ETC is an important source for the production of radicals in normal cases.

Electron transmission chains: A complex enzyme system composed of many biomacromolecules with different capabilities to capture and loss electrons. It hands electrons from H atoms to oxygen atoms, finally leading to the production of H_2O. In the process, the energy from the oxygenation of hydrocarbons is transformed into ATP molecules that can be directly used by functional proteins.

Acetylcholine: The earliest found natural neurotransmitter opening the road toward the understanding of the chemical transmission of signals between neurons. It is synthesized in neurons and is released from neurons when nerve pulses come. The site of the brain related to the intelligent activities such as learning, memory and thinking uses acetylcholine as their neurotransmitter.

Neurotransmitter: Molecules with the functions of signal transmission between neurons. They are synthesized in neurons and released when nerve pulses come. The bind to the receptors in the post-synaptic membrane and open the ion channels to arouse active potentials in post-synaptic neurons and thus to carry out their task of signal transmission.

Alzheimer disease: A degenerative disease of the central nervous system with the decrease in the levels of acetylcholine because of synthesis impairment. The main symptoms are the impairment of learning, memory and thinking.

ED99: The doses for the drugs effective on 99% animals.

TD1: The doses for the drug toxic on 1% animals.

Safe range: The ratio of TD1/ED99. The larger is this ratio, the safer are the drugs.

Acknowledgements

This work was supported by 973 program of China (2006CB933304, 2010CB933904).

Abbreviations

TDDS	:	target drug delivery system
CNT	:	carbon nanotubes
SWCNT	:	single-walled carbon nanotubes
MWCNT	:	multi-walled carbon nanotube
DWCNT	:	double-walled CNTs
PEI	:	polyethyleneimine
NGF	:	nerve growth factor
BDNF	:	brain-derived neurotrophic factor
EGF	:	epidermal growth factor
anti-MAP2	:	microtubule-associated protein 2
anti-GFAP	:	glial fibrillary acidic protein
anti-O4	:	oligodendrocyte marker O4
CgA	:	chromogranin A
PABA	:	poly(aniline boronic acid)
FET	:	field-effect transistor
2-AEMTTS	:	2-Aminoethylmethane thiosulfonate
PFF	:	particulate-free food
Ach	:	Acetylcholine
AD	:	Alzheimer disease
BBB	:	blood-brain barrier
KA	:	kainic acid
SWCNT-ACh	:	SWCNT-based ACh delivery system
SMC	:	smooth muscle cells
ROS	:	oxygen species
ETC	:	electron transmission chains
SCR	:	SWCNT-containing rate
ED_{99}	:	the effective dose on 99% animals
TD1	:	the toxic dose on 1% animals

References

Bao, C., F. Tian and G. Estrada. 2010. Improved visualisation of internalised carbon nanotubes by maximising cell spreading on nanostructured substrates. Nano Biomed. Eng. 2: 201–210.

Bekyarova, E., Y. Ni, E.B. Malarkey, V. Montana, J.L. McWilliams, R.C. Haddon and V. Parpura. 2005. Applications of carbon nanotubes in biotechnology and biomedicine. J. Biomed. Nanotechnol. 1: 3–17.

Cai, D., M.M. Jennifer, Z.-H. Qin, Z. Huang, J. Huang and T.C. Chiles. 2005. Highly efficient molecular delivery into mammalian cells using carbon nanotube spearing. Nat. Methods 2: 449–454.

Chhowalla, M., H.E. Unalan, Y.B. Wang, Z. Iqbal, K. Park and F. Sesti. 2005. Irreversible blocking of ion channels using functionalized single-walled carbon nanotubes. Nanotechnology 16: 2982–2986.

Cherukuri, P., S. Bachilo, S. Litovsky and R. Weisman 2004. Near-infrared fluorescence microscopy of single-walled carbon nanotubes in phagocytic cells. J. Am. Chem. Soc. 126: 15638–15639.

Chen, D., X. Wu, J. Wang, B. Han, P. Zhu and C. Peng 2010. Morphological Observation of Interaction between PAMAM Dendrimer Modified Single Walled Carbon Nanotubes and Pancreatic Cancer Cells. Nano Biomed. Eng. 2: 61–66.

Cui, D., H. Zhang, J. Sheng, Z. Wang, A. Toru, R. He, O. Tetsuya, F. Gao, H. Sung, C. Huth, H. Hu, G.M. Pauletti and D. Shi. 2010. Effects of cdse/zns quantum dots covered multi-walled carbon nanotubes on murine embryonic stem cells. Nano Biomed. Eng. 2: 246–256.

Dubin, R.A., G. Callegari, J. Kohn and A. Neimark. 2008. Carbon nanotube fibers are compatible with mammalian cells and neurons. IEEE Trans Nanobioscience 7: 11–14.

Fenoglio, I., M. Tomatis, D. Lison, J. Muller, A. Fonseca, J.B. Nagy and B. Fubini. 2006. Reactivity of carbon nanotubes: Free radical generation or scavenging activity. Free Radic. Biol. Med. 40: 1227–1233.

Gabay, T., E. Jakobs, E. Ben-Jacob and Y. Hanein. 2005. Engineered self-organization of neural networks using carbon nanotube clusters. Phys. a-Stat Mech. Appl. 350: 611–621.

Galvan-Garcia, P., E.W. Keefer, F. Yang, M. Zhang, S. Fang, A.A. Zakhidov, R.H. Baughman and M.I. Romero. 2007. Robust cell migration and neuronal growth on pristine carbon nanotube sheets and yarns. J. Biomater. Sci. Polym. Ed. 18: 1245–1261.

Gao, L., L. Nie, T. Wan, Y. Qin, Z. Guo and D. Yang. 2006. Carbon nanotube delivery of the GFP gene into mammalian cells. Chem. BioChem. 7: 239–242.

Gheith, M.K., V.A. Sinani, J.P. Wicksted, R.L. Matts and N.A. Kotov. 2005. Single-walled carbon nanotube polyelectrolyte multilayers and freestanding films as a biocompatible platform for neuroprosthetic implants. Adv. Mater 17: 2663–2670.

Hu, H., Y.C. Ni, V. Montana, R.C. Haddon and V. Parpura. 2004. Chemically functionalized carbon nanotubes as substrates for neuronal growth. Nano Lett. 4: 507–511.

Hu, H., Y. Ni, S.K. Mandal, V. Montana, B. Zhao, R.C. Haddon and V. Parpura. 2005. Polyethyleneimine functionalized single- walled carbon nanotubes as a substrate for neuronal growth. J. Phys. Chem. B. 109: 4285–4289.

Iijima, S. 1991. Helical microtubules of graphitic carbon. Nature 354: 56–58.

Jan, E. and N.A. Kotov. 2007. Successful differentiation of mouse neural stem cells on layer-by-layer assembled single-walled carbon nanotube composite. Nano Lett. 7: 1123–1128.

Kam, N.W.S. and H.J. Dai. 2005. Carbon nanotubes as intracellular protein transporters: Generality and biological functionality. J. Am. Chem. Soc. 127: 6021–6026.

Kostarelos, K., L. Lacerda, G. Pastorin, W. Wu, S. Wieckowski and J. Luangsivilay. 2007. Cellular uptake of functionalized carbon nanotubes is independent of functional group and cell type. Nat. Nanotechnol. 2: 108–113.

Liang, Z., F. Liu, I. Grundke-Iqbal, K. Iqbal and C.X. Gong. 2007. Down-regulation of cAMP- dependent protein kinase by over-activated calpain in Alzheimer disease brain. J. Neurochem. 103: 2462–2470.

Liopo, A.V., M.P. Stewart, J. Hudson, J.M. Tour and T.C. Pappas. 2006. Biocompatibility of native and functionalized single-walled carbon nanotubes for neuronal interface. J. Nanosci. Nanotechnol. 6: 1365–1374.

Liu, J., A.G. Rinzler, H.G. Dai, J.H. Hafner, R.K.P. Bradley and J. Boul. 1998. Fullerene Pipes. Science 280: 1253–1256.

Liu, Y., D.-C. Wu, W.-D. Zhang, X. Jiang, C.-B. He and T.-S. Chung. 2005. Polyethylenimine-grafted multiwalled carbon nanotubes for secure noncovalent immobilization and efficient delivery of DNA. Angew. Chem. Int. Edn. 44: 4782–4785.

Lovat, V., D. Pantarotto, L. Lagostena, B. Cacciari, M. Grandolfo, M. Righi, G. Spalluto, M. Prato and L. Ballerini. 2005. Carbon nanotube substrates boost neuronal electrical signaling. Nano Lett. 5: 1107–1110.

Malarkey, E.B. and V. Parpura. 2010. Carbon Nanotubes in Neuroscience. Acta Neurochir. Suppl. 106: 337–341.

Malarkey, E.B. and V. Parpura. 2007. Applications of carbon nanotubes in neurobiology. Neurodegener. Dis. 4: 292–299.

Matsumoto, K., C. Sato, Y. Naka, A. Kitazawa, R.L. Whitby and N. Shimizu. 2007. Neurite outgrowths of neurons with neurotrophin–coated carbon nanotubes. J. Biosci. Bioeng 103: 216–220.

Mattson, M.P., R.C. Haddon and A.M. Rao. 2000. Molecular functionalization of carbon nanotubes and use as substrates for neuronal growth. J. Mol. Neurosci. 14, pp. 175–182.

Oberdorster, E. 2004. Manufactured nanomaterials (fullerenes, C60 induce oxidative stress in the brain of juvenile largemouth bass. Environ Health Perspect 112: 1058–1062.

Park, K.H., M. Chhowalla, Z. Iqbal and F. Sesti. 2003. Single-walled carbon nanotubes are a new class of ion channel blockers. J. Biol. Chem. 278: 50212–50216.

Pulskamp, K., S. Diabate and H.F. Krug. 2007. Carbon nanotubes show no sign of acute toxicity but induce intracellular reactive oxygen species in dependence on contaminants. Toxicol. Lett. 168, pp 58–74.

Raja, P., J. Connolley, G.P. Ganesan, L. Ci, P.M. Ajayan, O. Nalamasu, D.M. Thompson. 2007. Impact of carbon nanotube exposure, dosage and aggregation on smooth muscle cells. Toxicol. Lett. 169, pp 51–63.

Sato, Y., A. Yokoyama, K. Shibata, Y. Akimoto, S. Ogino and Y. Nodasaka. 2005. Influence of length on cytotoxicity of multi-walled carbon nanotubes against human acute monocytic leukemia cell line THP-1 *in vitro* and subcutaneous tissue of rats *in vivo*. Mol. Biosyst. 1, pp 176–182.

Sayes, C., F. Liang, J.L. Hudson, J. Mendez, W. Guo, J.M. Beach, V.C. Moore, C.D. Doyle, J.L. West, W.E. Billups, K.D. Ausman and V.L. Colvin. 2006. Functionalization density dependence of single-walled carbon nanotubes cytotoxicity *in vitro*. Toxicol. Lett. 161, pp 135–142.

Sorkin, R., T. Gabay, P. Blinder, D. Baranes, E. Ben-Jacob and Y. Hanein 2006. Compact self-wiring in cultured neural networks. J. Neural. Eng. 3, pp 95–101.

Yang, Z., Y. Zhang, Y. Yang, L. Sun, D. Han, H. Li and C. Wang. 2010. Pharmacological and toxicological target organelles and safe use of single-walled carbon nanotubes as drug carriers in treating Alzheimer disease. Nanomedicine 6, pp 427–441.

Index

1,2-dimethyl-3-hydroxyl-4-pyridinone 350, 361
1,3-dipolar cycloaddition 186, 188, 189, 196
2,5-dihydroxybenzoic acid (DHB) 109, 111
8-hydroxy-2'-deoxyguanosine (8-oxo-dG) 355

A

α-cyano-4-hydroxycinnamic acid (CHCA) 104, 105
α-Synuclein 139–157
α-synucleinopathy 147
Abnormal clumps 359
Action potential 121, 122, 125, 134
AD 347–352, 354, 355, 358–360
Adenosine 68
Adhesion 184, 185, 195, 196
Adsorption 119, 127, 128, 132, 134
Adult stem cells 165, 173, 178
Affinity 353
Aggregation 76–78, 80–83, 87, 89, 91, 92, 94, 348–350, 354, 360
Alkaline earth metals 129
Alpha-synuclein 82, 93, 94
Alternative therapeutic approach 359
Alzheimer's and Parkinson's diseases 98, 265, 272
Alzheimer's disease 76, 91, 93, 94, 111, 114, 347, 348, 359, 360, 372, 384, 385
Amide groups 353
Amyloid fibrils 139, 141, 144–158
Amyloid plaques 359
Amyloid-β (Aβ) 348, 349
Antibody 76, 81, 83, 85–88, 90–92, 94
Antigen 76, 81, 83, 85, 87, 90, 92
Apolipoprotein E (ApoE) 351
Aptamers 140, 156
Argine-glycine-aspartate (RGD) 319, 327
Aricept 359
Arylsulfatase A (ASA) 112

As-grown VACNF 42, 43, 48, 49, 50
As-grown VACNF MBA 39, 50, 51–53, 57
Assay 355–357
Astrocytes 68
Atomic force microscopy 75, 77–79, 91, 93, 186, 202
Atrophy 312, 325
Attenuated 350
Autograft 311–313, 325
Axon 122, 134, 191, 201
Axon regeneration 311–314, 317, 321–324

B

β- and γ-secretases 350
Ba^{2+} 130
Backbones 353
Back-propagating action potentials 193, 194
BBB model 358
Behavior 349
Benabid 62
Beta amyloid 76, 81, 93, 94
Bi-dentate 350
Bi-directional crossing 351
Binding 126–128, 130–132, 134
Binding interactions 76, 81, 83, 89
Bioavailability 124, 349–351, 359
Biocompatibility 185, 187, 191, 195
Biomaterial 120, 123, 331, 333–335, 338, 340, 343
Biomaterial microstructure 313
Biomimicry 130
Biomolecular Engineering 139
Blending 314, 318, 319
Blocking 350, 354
Blood-brain barrier 286, 288, 289, 304, 305
Blood brain barrier penetration 348
Bluetooth 68
Brain infections 279
Brain tumours 267, 275, 278, 286–288, 294, 295, 304, 305

C

Ca^{2+} 130
Cadmium-selenide 164, 180
Cadmium-telluride 180
Capacitance 64
Carbon nanofiber (CNF) 38–40, 57, 58, 64, 68, 70, 73
Carbon nanomaterials 119, 120, 123, 129, 132, 134
Carbon nanotube 73, 367, 368, 381, 383, 385
Catalytic particles 123
Cd^{2+} 130
Cell migration 313–315, 319, 321
Cell tracking 209, 211, 213, 215
Cell transplantation 209, 211, 213, 214, 218, 222, 223
Cellular excitability 120
Central nervous system 367–369, 371, 372, 383, 384
Chalcogenides 165
Channel blockers 124
Chelating agents 347–352, 359
Chelation therapy 347, 350
Chelator-iron complexes 349
Chitosan 317
Cholinesterase 348, 349
Chromatolysis 312, 325
Chronic pain 62
Clioquinol 350
Closed-loop feedback 64
Co^{2+} 130
Cochlear 64
Cognitive 349
Collagen 313, 314, 316, 317, 319, 327
Combination 349, 351
Conformations 76, 81
Conjugates 347, 348, 352–358, 360, 361
Constant phase element (CPE) 49
Convection-enhanced delivery (CED) 295, 304
Coordination 353
Core-shell 168
Covalent chemical bonds 360
Covalently 351, 352
Cu^{2+} 130
Cyclic voltammetry (CV) 48, 55, 68, 70
Cytotoxicity 147, 148, 153, 157, 158, 354–356

D

Deep Brain Stimulation (DBS) 40, 60, 61, 72
Delayed rectifier potassium current 124
Deliver 348, 349, 351

Delivery technology 351, 359
Dementia 348, 359
Dendrimer 168
Dendrites 185, 191, 194, 201, 202
Dentate gyrus 112
Depolarization 121–124, 134
Depression 62
Desferrioxamine (DFO) 350
Development 348, 349, 352, 358–360
Dextran 164, 167, 168, 175
Diagnosis 358
Disease-modifying Drug (DMD) 347–350, 352, 358, 360, 361
Dissolved oxygen 128
DNA breakdown 355
Donepezil 359
Dopamine 66, 68, 70–73
Dorsal root ganglion 189, 314, 327
Drug carriers 367, 369, 373, 377, 379, 382, 383
Drug delivery system 373, 385
Drug interactions 349
Dysregulation 350
Dystonia 62

E

Efflux system 351
Electric Double Layer (EDL) 48, 56
EDL capacitance 48
Electrical conductive polymers (ECPs) 40
Electrical stimulation 190
Electrochemical impedance spectroscopy (EIS) 48, 49, 57
Electrode 194, 195, 202
Electrode impedance 39, 48, 50, 56, 57
Electrode Voltage 52, 53
Electrode-electrolyte interface 48
Electrolysis of water 38, 39, 48, 55–57
Electrospinning 310, 311, 314–317, 321, 323–326
Electro-spray ionization (ESI) 100, 115
Eletrospun nanofiber 337
Embryonic stem cells 165, 167, 180
Endocytosis 351
Endorem 167, 173
Endothelial 351
Enthalpy of hydration 129, 130, 135
Epilepsy 62, 64–66, 71
Ethylenediaminetetraacetic acid (EDTA) 350, 361
Exelon 359
Extracellular matrix (ECM) 43, 44, 58, 190, 203, 311, 325, 327

ECM-coated VACNFs 43
Extracellular Stimulation 38, 50

F

Faradaic charge transfer 49, 56
Ferric iron stain 353
Ferrocyanide granules 353
Ferromagnetism 164
Fibrillar forms 354
Fibroblasts 313
Field Potential 53, 56
Fluid phase 75
Fluorescence microscopy 355
Fluorinated Nanoparticles 167, 171
Foreign body response 321, 325
Free radical 349, 355
Friedrich ataxia 358
Fullerence 354
Functional Electrical Stimulation (FES) 40, 50, 58
Functional nanoparticles (fNPs) 107, 109, 110
Functionalization 185, 187, 189, 197, 201, 202

G

GABA 66, 72
Galantamine 359
Gating 122, 131, 132, 134
Gelatin 317
Gene therapy 287, 302, 304
Glial 61, 68, 71
Glioblastomamultiforme 178
Gliosarcoma 175
Globus pallidus 62
Glutamate 66, 68, 72
Glycine-argine-glycine-aspartate-serine (GRGDS) 319, 327
Granular cell layer (GCL) 112–115
Grounded collectors 316
Growth cones 188, 189, 196

H

Hallervorden Spatz syndrome 358
Hard ions 129
Headache 62
Hexadentate 352, 353
Highly sensitive protein marker detection 18
Hippocampal neurons 187, 188, 193

Hippocampus 112, 113
Human cortical neuronal cells (HCN-1A) 354
Huntington's diseases 358
Hydrogen peroxide 355, 357
Hydrophilicity 348
Hyperthermia 177, 179
Hypointensive 173, 174, 176

I

Images 354, 355, 357
Immunoassay 17, 18, 20, 23, 24, 26, 27, 29, 31–35
Immunocytochemistry 355, 357
Impedance 64
IMS; also referred to as mass spectrometry imaging [MSI] 97
Imaging mass spectrometry (IMS) 97–104, 107–116
in silico-designed peptides 140, 152
in utero electroporation 170
Inactivation 124, 125, 131
Induced pluripotent stem cells 165, 180
Inhibitors 348, 349
Initiation 348, 349, 358
Inner molecular layer (IML) 112–115
Intranasal drug administration 277
Intravenous administration 269, 277, 279
Iodochlorhydroxyquin 350
Ion Channels 119–123, 125, 127–130,
Ion conductance 122
Ion homeostasis 120, 134
Iron 347–361

K

K^+ 130
Kinetic stability 353

L

L1 350, 353, 361
Label-free 75, 83, 89, 92
Laminin 317–320
Lanthanoids 129
Layer-by-layer 190, 200, 203
Lewy Body 139, 158
Lipid-membrane peroxidation 355
Lipophilicity 352
Live-imaging 175
Low-density lipoprotein (LDL) 351, 357, 361
Low-threshold region 64, 67
Lysosomes 368, 371, 376, 380–384

M

Magnetic cell delivery 209
Magnetic nanoparticles 164, 166, 167, 173, 177, 179
Magnetodendrimers 167–169, 180
MALDI-TOF/TOF 110
Mass spectrometry (MS) 99
Mass-to-charge (m/z) 100
Matrix 98–101, 104, 105, 107–111, 114–116
Matrix-assisted laser desorption/ionization (MALDI) 98, 99, 105, 108
MEA 39, 50–53, 56–58
Mechanical probe 46
Mechanotransduction 39, 46, 47
Mediated processes 350
Medtronic 63
Metachromatic leukodystrophy (MLD) 112
Metal binding ability 348
Metal impurities 123
Metal ions 119, 127–132, 134, 135
Metal-chelator complexes 360
Microbrush Arrays (MBAs) 39, 43, 50
Microelectrode array (MEA) 39, 50–53, 56–58
Microelectrode arrays 70, 195
Microenvironmental 173
Microtubule-associated protein 2 191, 203
Middle molecular layer (MML) 112
Mimicking 351
Mitochondria 368, 376, 380–384
Mitochondria electron transmission chains 380, 384
Mitochondria membrane potential 368, 376
Mode 349
Molecular engineering 140, 141
Mood disorders 62, 66, 71, 72
Multielectrode array 124
Multifunctional 347–352, 358, 360
Multi-nucleated giant cells 321
Multiple sclerosis 175, 180
Multi-Walled Carbon Nanotube 135
MWCNTs 185, 187, 190, 191, 193, 196
Myelin 167, 175, 176
Myelin associated glycoprotein 314, 327
Myelin basic protein 314, 327
Myelination 167

N

Na^+ 130
Nanocarriers 286–294, 296, 298–305
Nanocrystals 125
Nanoelectrode arrays 60, 61, 64, 66, 72
Nanofiber scaffolds 335
Nanofibres 310, 311, 314–327
Nanomaterial interaction 119, 127
Nanomaterials 119, 120, 123, 127–134
Nanometals 119, 120, 124, 128, 129
Nano-PALDI 97–99, 105, 107, 108, 110–112, 114–116
Nanoparticle (NP)-assisted laser desorption/ionization (nano-PALDI)-based IMS 98, 114, 115
Nanoparticle (NP) 17–35, 98, 110, 111, 114, 115, 348, 351–361, 378
Nanoparticle-iron Chelator 347, 348, 351, 352, 354–356, 358, 360, 361
Nanoparticles (AuNPs) 105
Nanoscale 75, 93
Nanosilver 125, 128
Nanostructure 139–141, 148, 151, 152, 156, 157
Nanostructured Lipid Carriers (NLC) 264, 265, 280
Nanotechnology 288, 294
Nanotubes 123, 124, 129, 132
Natively unfolded proteins 139
Nerve autograft 311
Nervous System 310, 313, 327
Neural Electrical Interface (NEI) 38–41, 46, 53–55, 57, 58, 60, 73
Neural network 120
Neurite 185, 187, 188, 190, 191, 195–197, 202
Neurodegeneration 348, 349, 354
Neurodegenerative disease 75, 78, 91, 93
Neuroendocrine disorders 71
Neurofibrillary tangles 359
Neuro-imaging agents 360
Neurological diseases' 358
Neuroma 312, 313, 326
Neuromodulation 60–62, 66, 70–72
Neuronal growth 123
Neuronal network 184, 187, 192, 195
Neurons 61, 66, 68, 71
NeuroPace 65
Neurotransmission 122, 127
Neurotransmitter 61, 69, 72, 73
Ni^{2+} 130, 132
Nickel 123, 124, 127
Nickel ions 127
NMDA (N-methyl-D-Aspartate) 348
Nucleus accumbens 66

O

Obesity 62, 71
Occlusion 123, 132, 133

Olfactory ensheathing cells 313
Oligodendrocyte progenitor cells (OPCs) 167
Oligomer 139, 149, 153, 156
Overexpressing transgenic mice 350
Oxidation 128, 129
Oxidative damage 348, 352, 355, 357, 360
Oxidative stress 119, 126–128, 133, 134, 348–350

P

Paramagnetism 164
Parkinson's disease 76, 62, 64, 66, 71, 72, 91, 139, 140, 157, 158, 358
Pathologies 349
PC12 190, 191, 195–197, 200, 203
PC 12 cells 43
Peak amplitude 125
Peripheral 310–314, 325–327
Perls method 353
P-glycoprotein 351
pH 128, 129
Pharmacokinetic 358
Pharmacology 367, 374
Phase separation 335, 337, 338, 343
Phosphotidylcholine (PC) 111
Photobleaching 165, 178
Poly(lactic-co-glycolic acid) 318, 327
Poly(L-lactic acid) 318, 320, 327
Poly(ε-caprolactone) (PCL) 314, 318, 319, 321, 322, 327
Poly(ε-caprolactone-co-ethyl ethylene phosphate) 318
Polypyrrole (PPy) 39, 64
Polysorbate-80 351
Polystyrenesulfonate 43
Polysulphone conduits 323
Pore block 130
Post synaptic currents 191
PPy-coated VACNF 39, 40, 42–44, 46, 48, 50, 55, 57
PPy-coated VACNF MBA 39, 41, 50–53
Prevention 347, 348, 352, 358–360
Progression 349, 350, 358, 359
Proliferation 355, 356
Protease-resistant oligomeric 354
Protein absorption patterns 357
Protein folding 76
Protein morphologies 76, 77, 83, 84, 86, 89–92
Protein oxidation 355
Protein–antibody complexes 76, 88

Prussian Blue 169, 173, 175, 176
Pseudocapacitance 42, 48, 56
Psychiatric disorders 280

Q

Quantum confinement 163, 165
Quantum dots 135, 164, 167, 170, 171, 177–179

R

Rat hippocampal brain slices 41, 50
Reactive oxidative species 347, 349
Reactive Oxygen Species 135, 376, 380
Receptor 348, 349, 351
Receptor-mediated 351
Redox silence 349
Redox-active centers 349
Regeneration 189, 330–336, 338–340, 342, 343
Reminyl 359
Repolarizing 121
Resting potential 122, 123
Retina 64
Rivastigmine 351, 359

S

Safety range 368, 382
Scaffolds 184, 187, 189, 192, 193
Scanning Electron Microscopy 186, 203
Scavenge 349
Schwann cell 311, 312, 314, 315, 319, 321, 326
Secondary ion mass spectrometry (SIMS) 108
Self-assembling peptide nanofiber scaffolds 335, 342–344
Self-assembly 349
Serotonin 66, 68
Side-effects 350
Silver 125, 127
Simultaneously 349
Single-cell electrophysiology 193
Single-Walled Carbon Nanotube 135
SiO_2 70
Sodium-channel band 67
Soft bases 130
Soft ions 129
Solid Lipid Nanoparticles (SLN) 264, 265, 280
Space charge layer 56
Spectroscopy 186

Spinal cord 183, 197
Spinal cord injury 162, 175, 179, 180, 209, 210, 214, 217, 218, 222, 224, 225, 330–332, 339, 342, 343
Spinneret 316
Sr^{2+} 130
Star-shaped NCO-poly(ethylene glycol)-stat-poly(propylene glycol) 327
Stealth 287, 289, 290, 304
Stereotaxy 286, 294, 295, 303, 304
Stroke 209–211, 213–216, 222, 224
Structures 75, 76, 80, 84, 85, 88, 89, 91, 92
Subthalamic nucleus 62
Sulfatide 106, 111, 112, 113
Superparamagnetic iron oxide nanoparticles 164, 176, 178, 180, 209, 210, 212, 222, 225
Superparamagnetism 163, 164
SWCNTs 184, 185, 187, 190, 191, 193, 195–197
Synapse 122, 127
Synaptic activity 185, 191

T

Tangled bundles 359
Target 347, 349, 352, 358
Targeted drug delivery systems (TDDS) 265
TAT-peptides 164
Temperature 128, 132
Thalamus 62
Therapeutics 348, 349, 358–360
Thermogravimetric analysis 186, 203
Thiol 127
Three pronged 349
Three-dimensional frameworks 342
Tissue engineering 209–211, 219, 222, 331, 333, 335, 338, 343
Titanium dioxide (TiO_2) 106
Topography 313, 314, 321
Toxicants 129
Toxicity 348–355, 359, 360
Toxicology 133, 367, 374
Transcranial magnetic stimulation 72
Transcytosis 351
Transferrin (Tfr) receptors 167
Transient outward potassium current 124
Transition metals 129
Transmission 186, 203
Transmission electron microscopy (TEM) 353
Transport mechanisms 357
Transport vehicles 348
Traumatic brain injury 331, 338, 342, 343
Treatment 347–352, 358–360
Tremor 62, 64
Truncated 140–143, 147–150, 153
Truncated α-synuclein 140, 141, 147–150, 153
Tungsten wire electrodes 39, 50–53
Two-dimensional polyacrylamide gel electrophoresis (2-D PAGE) 356, 361

U

Ultrasound biomicroscopy 170, 180
Upregulation 350

V

Vertically Aligned Carbon Nanofibers (VACNFs) 39
Voltage sensor 121, 126, 128, 131
Voltage-Gated Calcium Ion Channel 135
Voltage-Gated Ion Channel 135
Voltage-Gated Potassium Ion Channels 135
Voltage-Gated Sodium Ion Channels 135

W

Wallerian degeneration 312, 326
Warburg diffusion element 49
Whole-cell patch clamp 123
WINCS 69
Wireless 68–70

Y

Y^{3+} 130, 132
Yttrium 123, 124, 127, 130, 131

Z

Zn^{2+} 130, 132
ZnO, CuO and Ag nanoparticles 124

About the Editors

Professor Colin R. Martin is a qualified Nurse and Chair in Mental Health at the University of the West of Scotland and Adjunct Professor at the Royal Melbourne Institute of Technology (RMIT), Melbourne, Australia. Professor Martin is a Chartered Health Psychologist and a Chartered Scientist. Having originally trained in psychiatric nursing, he specialised in the addictions and following further training, worked as a community psychiatric nurse and then as an addictions counsellor in the NHS. On completion of BSc and PhD degree's in psychology, Professor Martin worked in senior management posts in the NHS followed by academic posts in the UK and the Far East, during which he conducted original research in both the addictions and the mental health aspects of chronic disease. He has published many scientific papers in psychology, biology, medical and nursing journals. Professor Martin is honorary Consultant Psychologist to The Salvation Army, UK and Eire Territories and was instrumental in formulating the addictions policy of the Salvation Army (UK and Eire) over recent years to develop high quality and evidenced-based clinical care and services.

Victor R. Preedy BSc, PhD, DSc, FIBiol, FRCPath, FRSPH is Professor of Nutritional Biochemistry, King's College London, Professor of Clinical Biochemistry, Kings College Hospital and Director of the Genomics Centre, King's College London. Presently he is a member of the Kings College London School of Medicine. Professor Preedy graduated in 1974 with an Honours Degree in Biology and Physiology with Pharmacology. He gained his University of London PhD in 1981 when he was based at the Hospital for Tropical Disease and The London School of Hygiene and Tropical Medicine. In 1992, he received his Membership of the Royal College of Pathologists and in 1993 he gained his second doctoral degree, i.e., DSc, for his outstanding contribution to protein metabolism in health and disease. Professor Preedy was elected as a Fellow to the Institute of Biology in 1995 and to the Royal College of Pathologists in 2000. Since then he has been elected as a Fellow to the Royal Society for the Promotion of Health (2004) and The Royal Institute of Public Health (2004). In 2009, Professor Preedy became a Fellow of the Royal Society for Public Health. In his career Professor Preedy has carried out research at the National Heart Hospital (part of Imperial College London) and the MRC Centre at Northwick Park Hospital. He has collaborated with research groups in Finland, Japan, Australia, USA and Germany. He is a leading expert on the mechanisms of disease and has lectured nationally and internationally. He has published over 570 articles, which includes over 165 peer-reviewed manuscripts based on original research, 90 reviews and 20 books.

Ross J Hunter AKC BSc MBBS MRCP PhD trained in medical sciences at King's College London (Times University ranking 11th in UK). He spent a further year at Imperial College London (Times University ranking 3rd in UK) and was awarded his BSc in Cardiovascular medicine in 1998. Since returning to his medical training at King's College School of Medicine, he has remained an honorary research fellow at The Department of Nutritional Sciences, researching the effect of different nutritional states and alcoholism on the cardiovascular system. He was awarded his bachelor of medicine & surgery (MBBS) with distinction in 2001. He trained in general medicine in London and Brighton and was made a member of the Royal College of Physicians (UK) in 2005. He trained as a Registrar in cardiology and general internal medicine from 2005–2008 in the London Deanery. Since 2008 he has been a research fellow at the Department of Cardiology & Electrophysiology at St Bartholomew's Hospital London, conducting clinical research and clinical trials in cardiology and electrophysiology, and was awarded his PhD in 2011. He was a young investigator of the year finalist at the Heart Rhythm Society (USA) in 2011. He has published over 60 scientific articles of various kinds and is Editor of 4 books.

Color Plate Section

Chapter 2

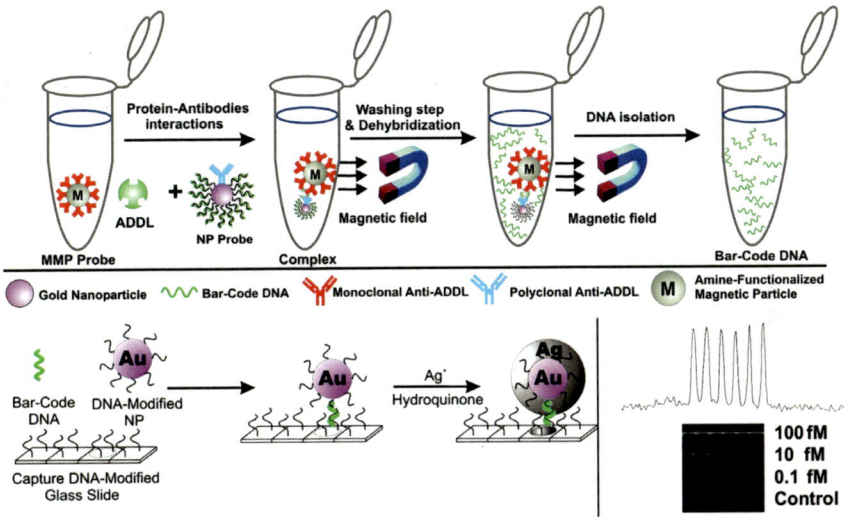

Fig.1. Schematic of the bio-bar-code assay (BCA) procedure for detecting ADDL in cerebral spinal fluids. Magnetic Micro-particles (MMPs) were modified with Amyloid beta derived diffusible ligands (ADDL) capture antibody and gold nanoparticles (AuNP) probes were modified with ADDL detector antibody and Bar-code DNA strands. The sandwich structures (magnetic beads-target protein-AuNP probes) were formed when the target proteins were present. After several wash steps, the bar-code DNA was eluted into the solution and was then detected with silver enhancement and microarray or PCR. Down to 0.1 fM ADDL protein can be detected with this assay. Reprinted with permission from reference (Georganopoulou et al. 2005) © 2005, The National Academy of Sciences.

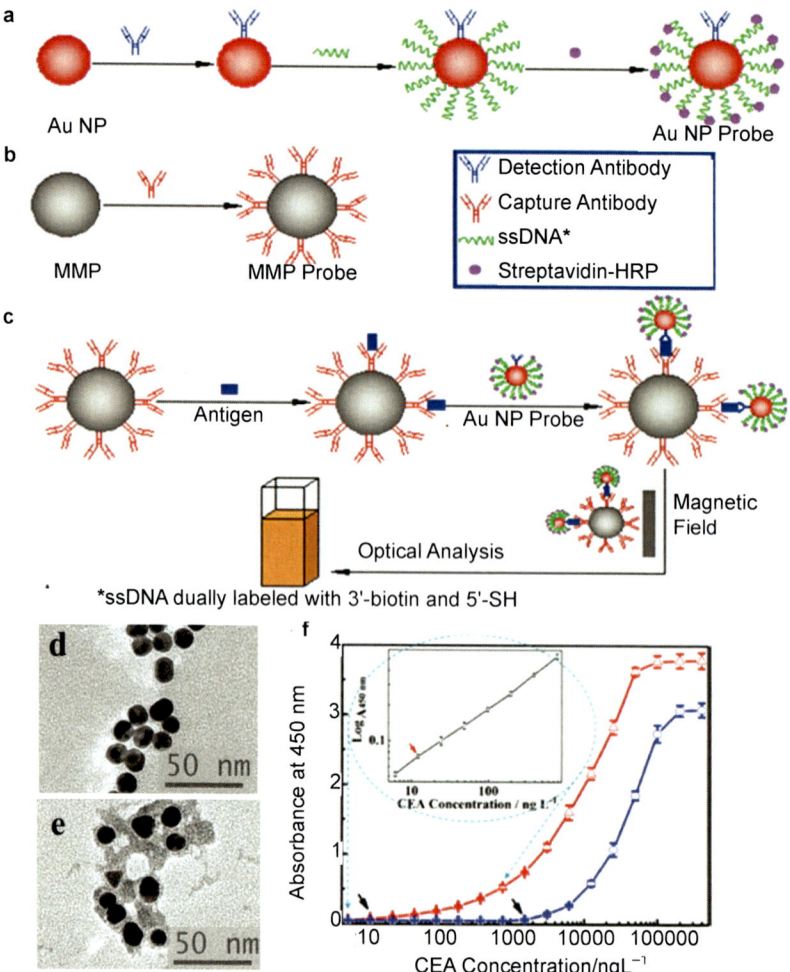

Fig. 2. Schematic of preparation of multi-component Au NP probes (a) and MMP probes (b), Nano-ELISA procedure (c), Transmission electron micrograph (TEM) of gold nanoparticles(d) and AuNP probes(e), and calibration curves for spectrophotometric detection of CEA protein using the multi-component Au NP probe based immunoassay (red light) and the conventional ELISA (blue light) (f). MMPs were modified with capture antibody and AuNP probes were modified with detector antibody and horseradish peroxidase (HRP), which was immobilized onto AuNPs through DNA strands and the streptavidin-biotin reaction. Sandwich structures (magnetic beads-target protein-AuNP probes) were formed and incubated with 3,3′,5,5′-tetramethylbenzidine (TMB) solution. Inset: log-log calibration curve of the multi-component AuNP probe-based immunoassay at a low CEA concentration; CEA concentration was varied from 6 to 781 ng/L. Reprinted with permission from reference (Liu et al. 2010) © 2010, The Royal Society of Chemistry.

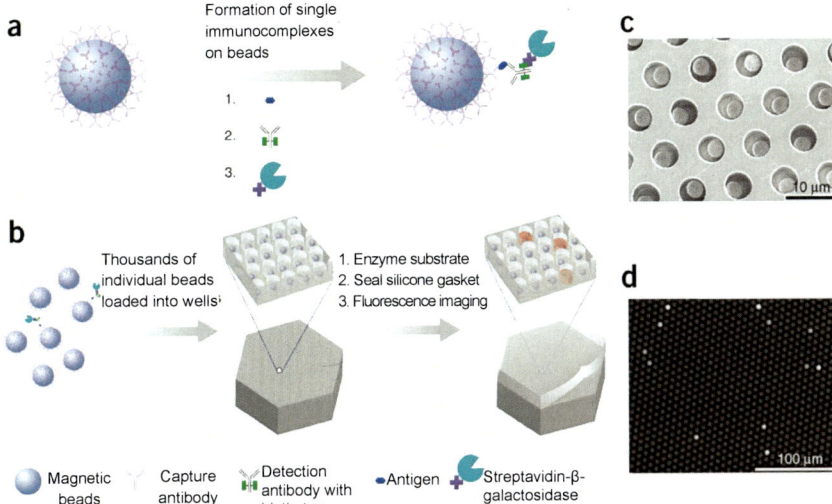

Fig. 3. **Schematic of single-molecule ELISA, based on well arrays.** (a) The captured antibody was immobilized onto magnetic beads and single proteins were captured through the formation of sandwich structures (magnetic beads-single protein molecules-second antibody with beta-galactosidase). (b) The beads, with or without sandwich structures, were loaded onto micro-wells and the fluorescence of beta-galactosidase catalyzed substrates was scanned with fluorescence imaging. (c) Result of scanning electron micrograph (SEM). There is one bead in the majority of wells. (d) Result of fluorescence imaging. Fluorescent signals in one well means that the bead has captured the single protein molecules. Reprinted with permission from reference (Rissin et al. 2010) ©2010, Nature Publishing Group.

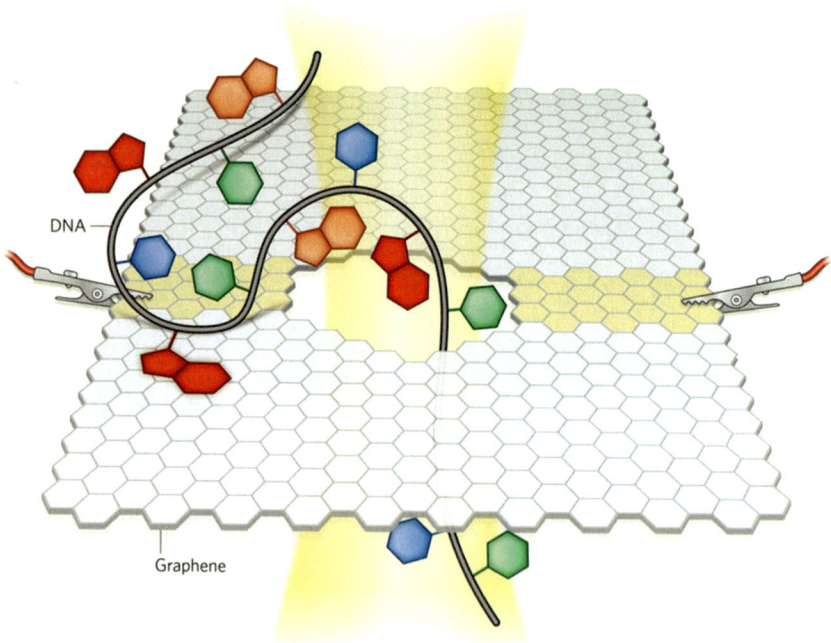

Fig. 4. **Illustration of a graphene-based nanopore for single DNA sequencing.** DNA fragments pass through the nanopore on graphene. Different colors represent different nucleotides. Reprinted with permission from references (Garaj et al. 2010 and Bayley 2010) © 2010, Nature Publishing Group.

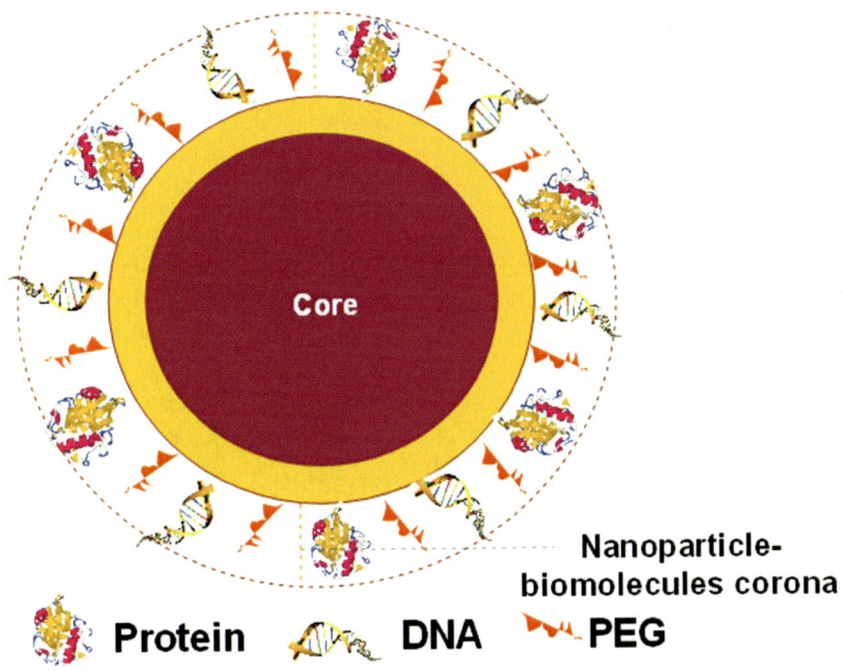

Fig. 5. Schematic of nanoprobes. Protein, oligonucleotide, PEG and other molecules, immobilized specifically or adsorbed non-specifically to the surface of nanomaterials, compete for the nanoparticle surface and result in a corona around the nanoparticles (unpublished material of author).

Chapter 3

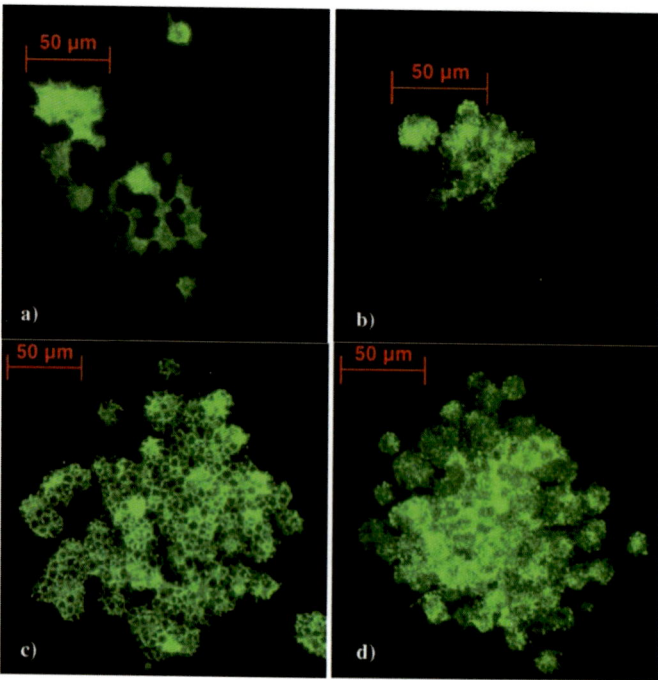

Fig. 3. **Actin Stained PC12 Cell Cultures on VACNF at Day 7.** Fluorescence images of VACNF arrays with (+) and without (-) PPy and collagen ECM coatings: a) PPy- and ECM-, b) PPy+ and ECM-, c) PPy- and ECM+, and d) PPy+ and ECM+.

Color Plate Section 403

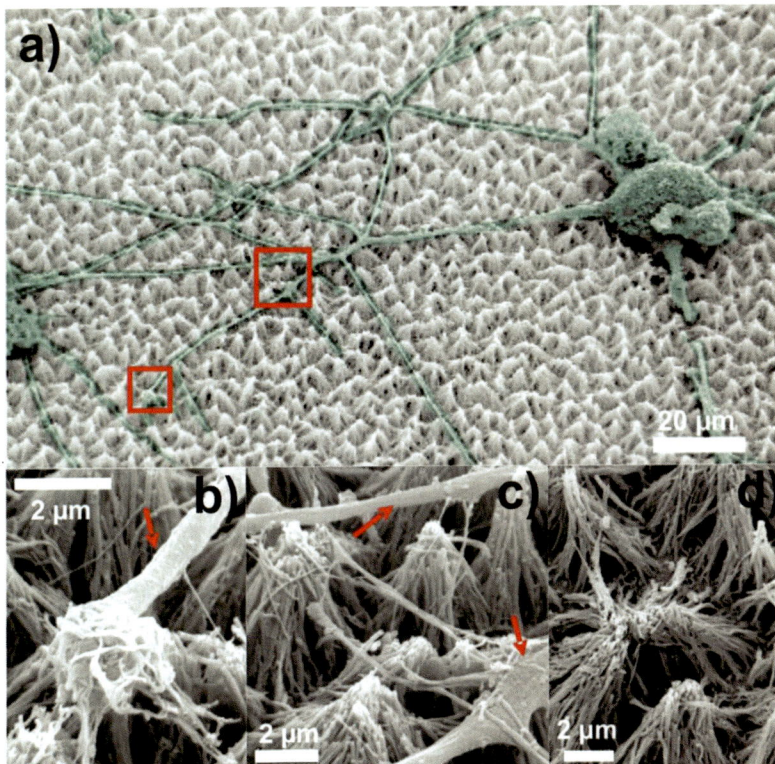

Fig. 4. **PC12 Cells Cultured on As-grown VACNF Microbundles.** SEM images of a) differentiated PC12 cells forming an extensive neural network, b) the fibril filapodia at the extended neurite terminus anchoring on CNF microbundles, c) neural nanofibrils bridging the submicron neurite branches, d) the substrate without subjecting to cell culture.

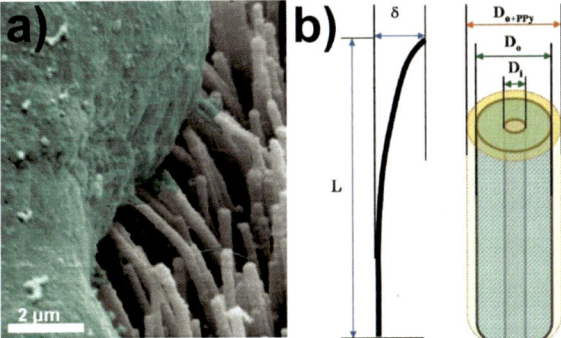

Fig. 6. **Cell Adhesion Force Derived from the Bending of VACNFs.** a) SEM images of PPy-coated VACNFs bent under the force exerted by the cell body and b) the definition of the parameters of the CNF: L (the length), δ (the lateral deflection of the tip from the original vertical position), D_o (the outer diameter), and D_i (the inner diameter).

Chapter 4

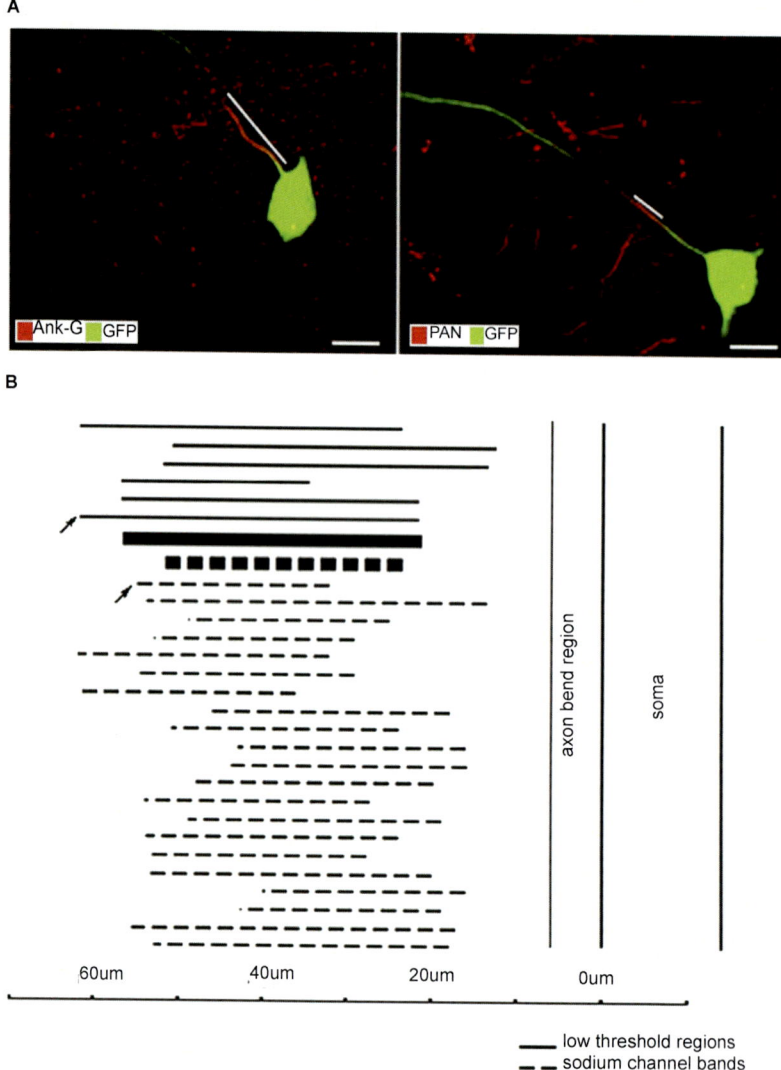

Fig. 4. A. Examples of sodium-channel bands from two unidentified ganglion cell types. Ank-G: ankyrin G (protein associated with high-density sodium channels). GFP: green fluorescent protein. PAN: pan sodium antibody. Scale bar: 25 μm. B. The regions of high-density sodium-channels and low-threshold (to electrical stimulation) are coextensive. Each thin line corresponds to the length and position of a low-threshold region (solid lines) or high-density sodium-channel region (dashed lines). Each thick line represents the mean of the low-threshold regions (thick solid line) or the high-density sodium channel regions (thick dashed line). (from Fried et al. 2009, with permission).

Color Plate Section 405

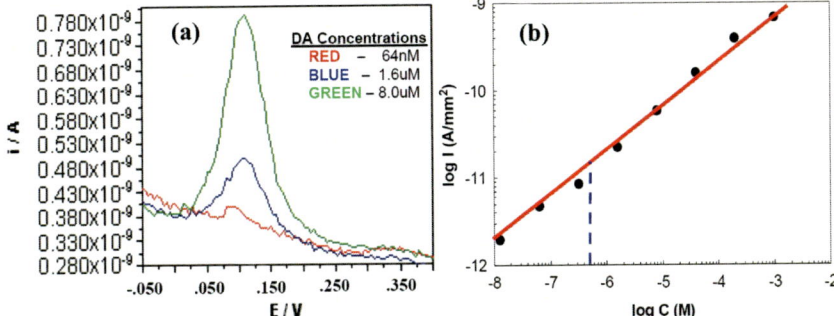

Fig. 8. (a) The signal of differential pulse voltammetry for detection of dopamine with a CNF NEA in 64 nM, 1.6 µM, and 8.0 µM dopamine in PBS solutions. (b) The calibration curve of the peak current in differential pulse voltammetry vs. the concentration of dopamine from 13 nm to 1.0 mM, with the blue vertical line indicating the detection limit of a typical carbon microelectrode at ~500 nM. (unpublished data, NASA Ames Research Center–J.L.).

Chapter 6

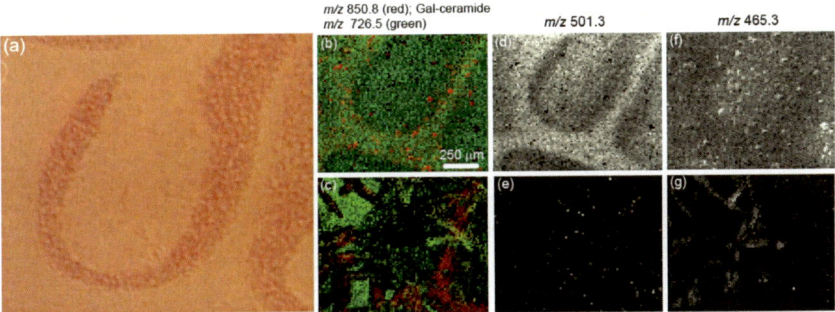

Fig. 9. Nanoparticle-assisted laser desorption/ionization based imaging mass spectrometry (nano-PALDI-IMS) of low molecular weight compounds showing improved ion distribution image quality. Optical images of rat cerebellum tissue before spraying with nanoparticles (NPs) (a) and 2,5-dihydroxybenzoic acid (DHB) solution, and ion images obtained with NPs (b, d, f) and DHB (c, e, g) are shown. Visualized ions were identified as galactosylceramide (C24h:0) and phosphotidylcholine (PC) (diacyl-34:2) by tandem mass spectrometry on both DHB and NP coated sections. Reprinted from Taira et al. 2008 with the permission of ACS Publications.

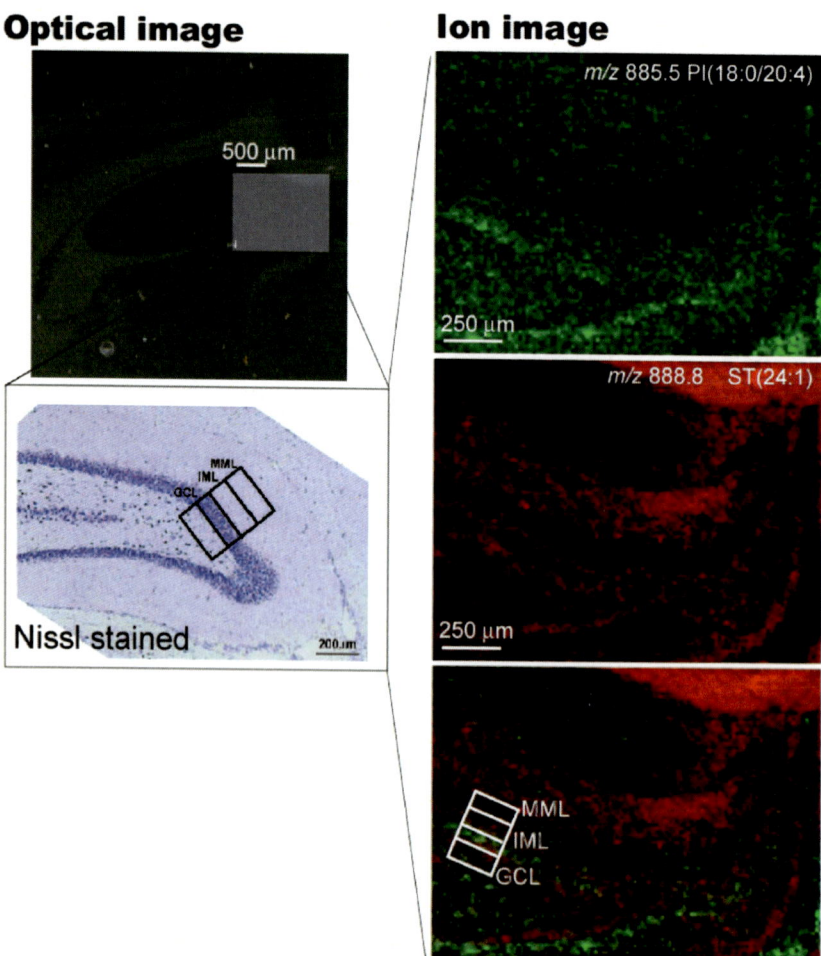

Fig. 11. **Nanoparticle-assisted laser desorption/ionization improves spatial resolution in imaging mass spectrometry (IMS).** Optical image of rat hippocampus indicating measurement area for nanoparticle-based IMS is shown along with Nissl-stained section indicating fine layer structure of rat hippocampus (left panels). Ion images, which reveal hippocampal layer specific distribution of phosphatidylinositol (18:0/20:4) at m/z 885.5 and sulfatide (24:1) at m/z 888.8, are presented (right panels). GCL = granular cell layer, IML = inner molecular layers, and MML = middle molecular layers. Reprinted from Ageta et al. 2009 (left column) and Sugiura and Setou 2010b (right column) with the permission of Springer.

Chapter 11

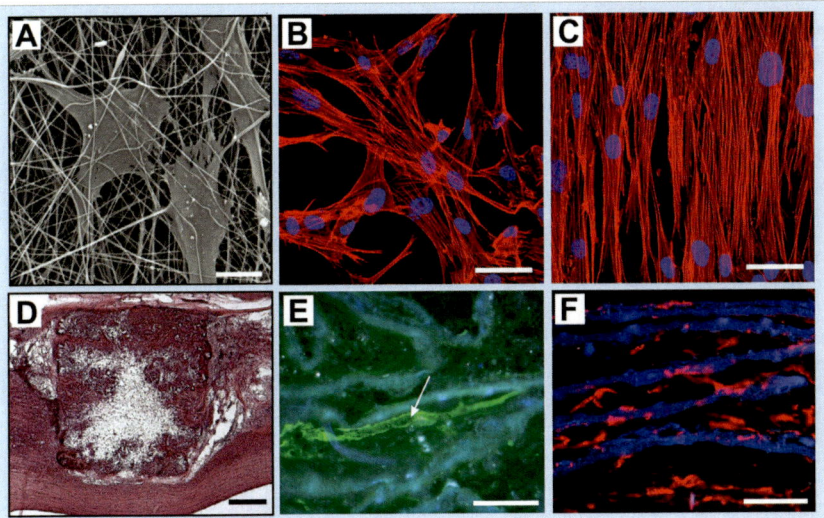

Fig. 5. Electrospun nanofibers as a scaffold for cell culture and implantation into the spinal cord. (A) Representative scanning electron micrographs of human mesenchymal stem cells (hMSCs) cultured on gelatin nanofibers. (B) Confocal micrographs of hMSCs grown on random and (C) aligned gelatin nanofibers. Cells were stained for F-actin cytoskeletal filaments and cell nuclei. (D) A rolled nanofiber scaffold was implanted into a hemisection of a rat spinal cord and evaluated 4 weeks after implantation. The ingrowth of tissue into an implanted rolled nanofiber scaffold stained with hematoxylin-eosin. (E) Blood vessel (RECA staining) ingrowth (arrow) into the nanofibrous scaffold. (F) The ingrowth of Schwann cells (p75 staining) into a nanofibrous scaffold. Scale bar: 20 μm (A), 50 μm (B, C, E, F) and 500 μm (D). Unpublished results.

Chapter 12

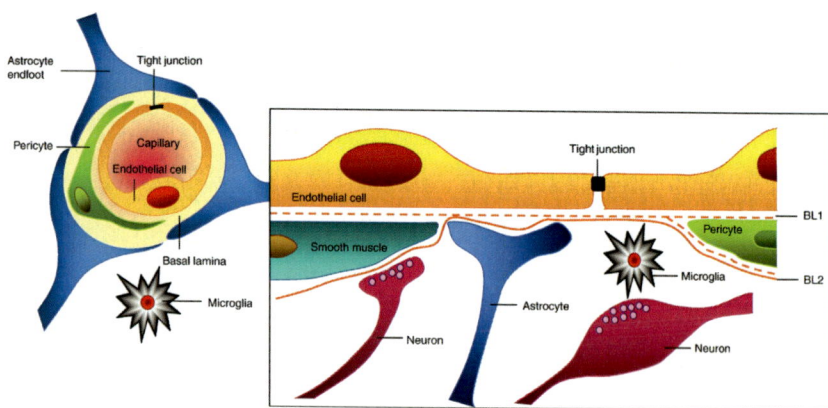

Fig. 2. **The cellular structures forming the BBB.** The cerebral endothelial cells form tight junctions at their margins, which seal the aqueous paracellular diffusional pathway between the cells. Pericytes are distributed discontinuously along the length of the cerebral capillaries and partially surround the endothelium. Both the cerebral endothelial cells and the pericytes are enclosed by and contribute to the local basement membrane, which forms a distinct perivascular extracellular matrix (basal lamina 1, BL1) different in composition from the extracellular matrix of the glial endfeet bounding the brain parenchyma (BL2). Foot processes from astrocytes form a complex network surrounding the capillaries and this close cell association is important in induction and maintenance of the barrier properties. Axonal projections from neurons onto arteriolar smooth muscle contain vasoactive neurotransmitters and peptides and regulate local cerebral blood. BBB permeability may be regulated by release of vasoactive peptides and other agents from cells associated with the endothelium. Microglia are the resident immunocompetent cells of the brain. The movement of solutes across the BBB is either passive, driven by a concentration gradient from the plasma to the brain, with more lipid-soluble substances entering most easily or may be facilitated by passive or active transporters in the endothelial cell membranes. Efflux transporters in the endothelium limit the CNS penetration of a wide variety of solutes (based on Abbott et al. 2006) (Abbott et al. 2010).

Color Plate Section 409

Chapter 13

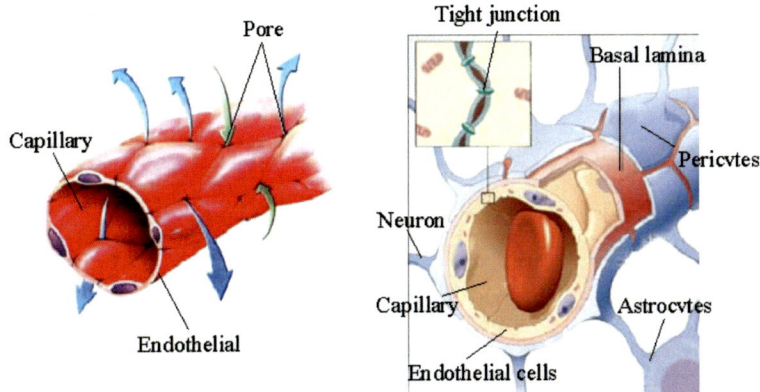

Fig. 1. General capillary (left) and brain capillary (right). Brain capillaries are made of endothelial cells sealed with tight junction surrounded by astrocytes and pericytes.

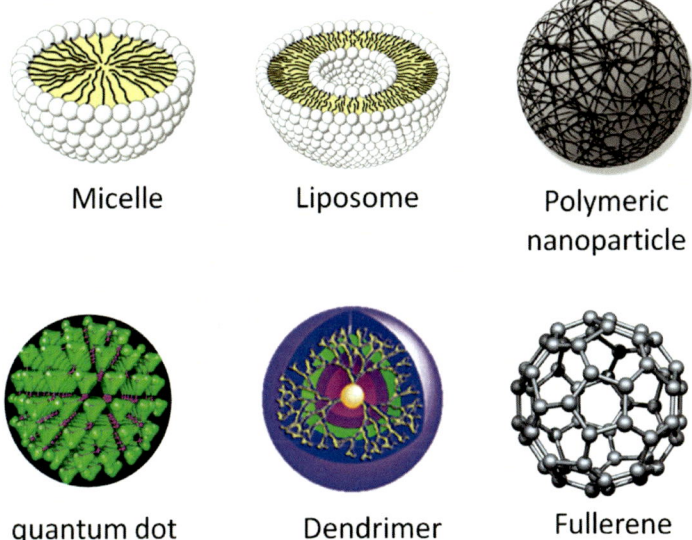

Fig. 3. Different types of nanoparticles. There are different type of articial nanoparticles for drug delivery system.

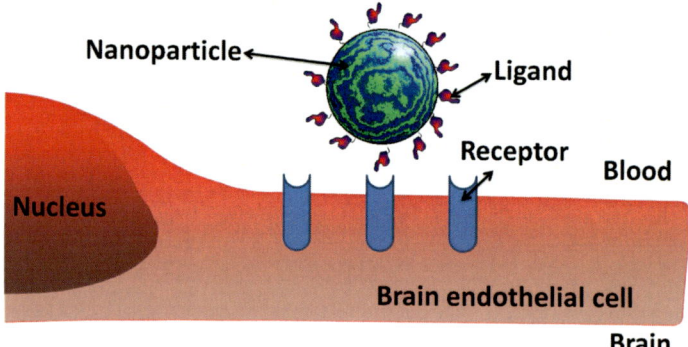

Fig. 4. **Receptor mediated targeting of nanoparticles.** The targeted nanoparticles are functionalized with an outer layer of receptor-specific ligands. After binding the receptor-ligand complex is endocytosed and potentially transcytosed to the brain.

Chapter 14

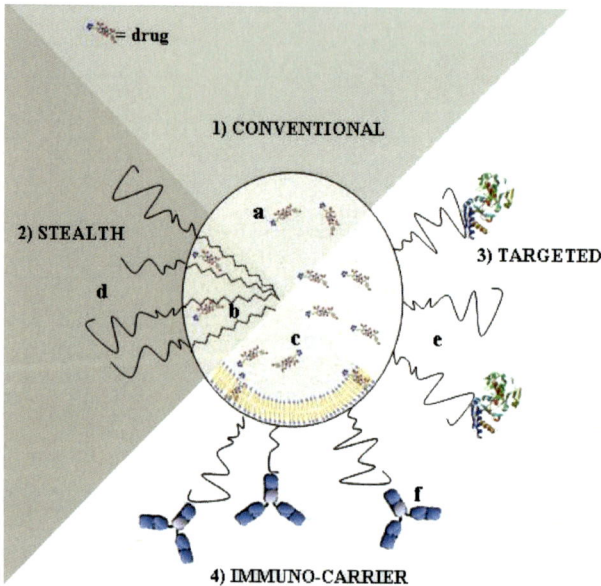

Fig. 1. **Pharmaceutical nanocarriers.** The schematic structure of pharmaceutical **nanoparticulate carriers** (a–polymeric or lipid nanoparticles; b–polymeric micelles, and c–liposomes). 1–Conventional nanocarrier; 2–long-circulating nanocarrier (d–surface-attached protecting polymer, usually PEGs, allowing for prolonged circulation of the nanocarrier in the bloodstream); 3–targeted nanocarrier (e–specific targeting ligand attached to the carrier surface); 4–immunocarrier (f–specific monoclonal antibody linked to the carrier surface). Adapted with permission from (Craparo et al. in press). Copyright (2010) Blackwell.

Chapter 15

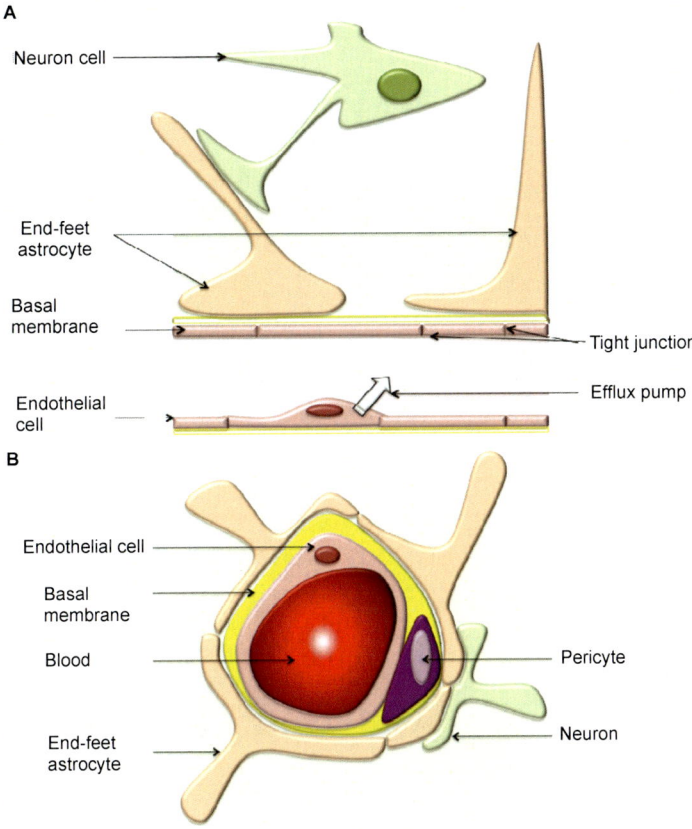

Fig. 1. Schematic representation of the blood-brain barrier (A: longitudinal and B: transversal). In the brain, tight junctions bind endothelial cells; they are surrounded by pericytes and basal membrane. Moreover, end-feet of astrocytes create a second layer that surrounds basal membrane, increases the non-permeability of the BBB and allows neuron protection.

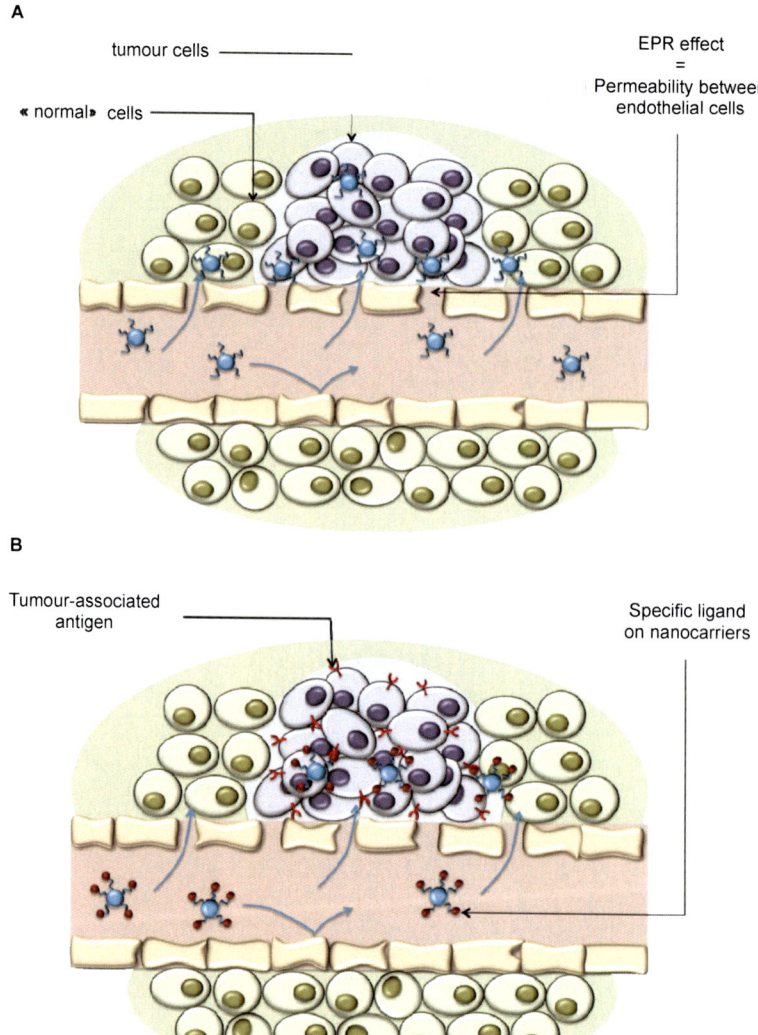

Fig. 6. **Passive and active targeting of nanocarriers in brain tumour.** (A) Passive targeting is based on the longevity of the nanocarrier in the blood and its accumulation in the tumour site *via* the EPR effect. (B) Active targeting is based on the functionalisation of the nanocarrier with ligands that bind to tumour-associated antigens over-expressed on tumour cells.

Chapter 17

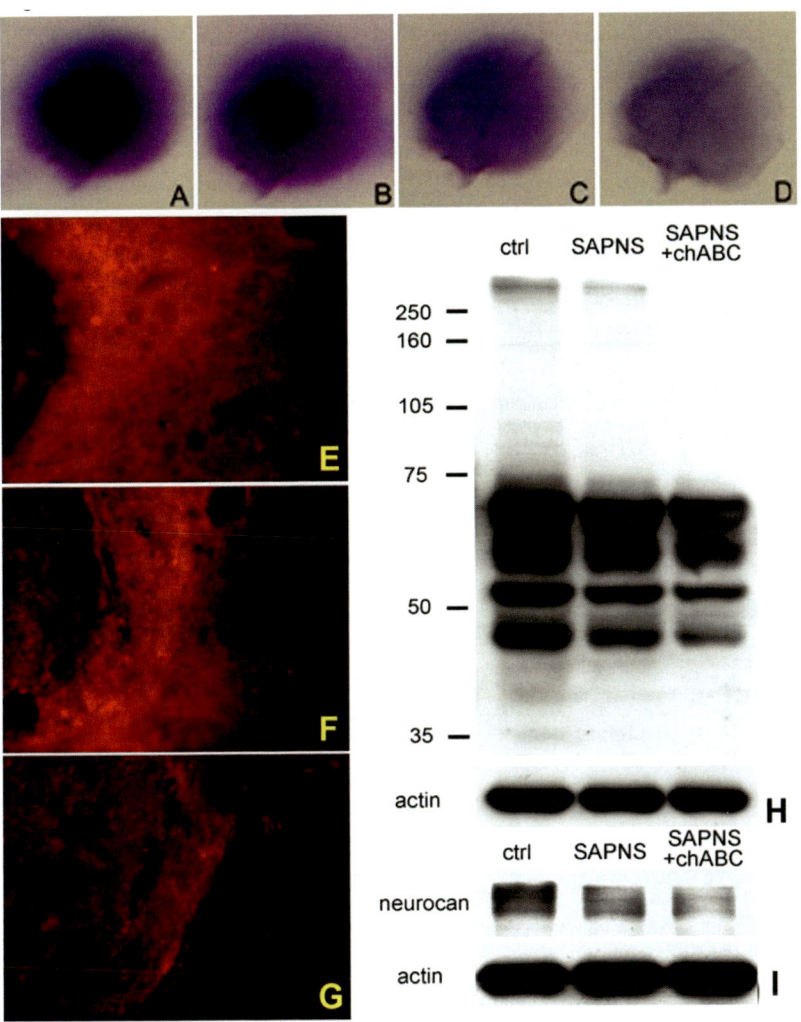

Fig. 1. SAPNS is a good material for chemical control release (unpublished data). (A-D) the mixture of SAPNS and typan blue was merged in saline for 1h, 1d, 1w, and 4w respectively, showed the dye was released slowly. When the injured spinal cord was treated with the mixture of SAPNS and Chondroitinase ABC (chABC), which is an enzyme for digesting chondroitin sulfate proteoglycans (CSPGs), the CSPGs concentration in the host tissue was decreased significantly. (E-G) Immunohistochemistry showed the CSPG expression in control (E), SAPNS alone treated (F) and SAPNS with chABC treated (G). (H, I) Western blotting showed the total CSPGs (H) and neurocan (I), which is a component of CSPGs, in different treated injured spinal cord.

414 Nanomedicine and the Nervous System

Chapter 18

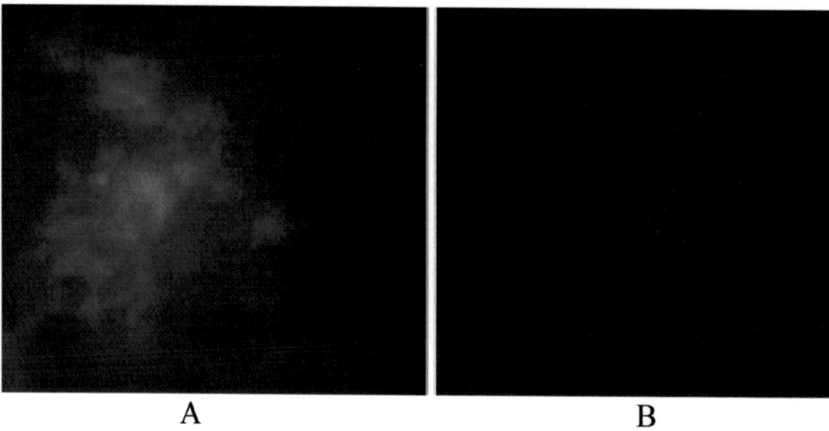

Fig. 3. Images of fluorescence microscopy. (A) Precipitates of Aβ aggregates readily form in PBS without nanoparticle iron chelator conjugates. (B) No Aβ precipitates form in PBS-containing nanoparticle iron chelator conjugates (Modification with permission from Liu et al., 2009b. Copyright Elsevier Limited).

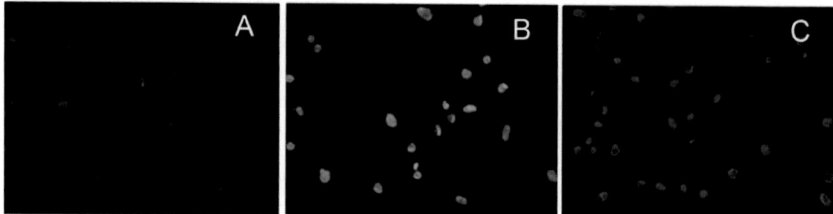

Fig. 5. Immunocytochemistry assay of 8-oxo-dG. A, B and C show the untreated neuronal cells, the cells treated with hydrogen peroxide and the cells treated with both hydrogen peroxide/Nano-N2PY, respectively. The cells of Fig. B are much brighter than that of Fig. A and C, which indicates that there is more oxidative damage of DNAs in the cells treated with hydrogen peroxide than the untreated cells and the cells treated with hydrogen peroxide/Nano-N2PY (Unpublished material).

Chapter 19

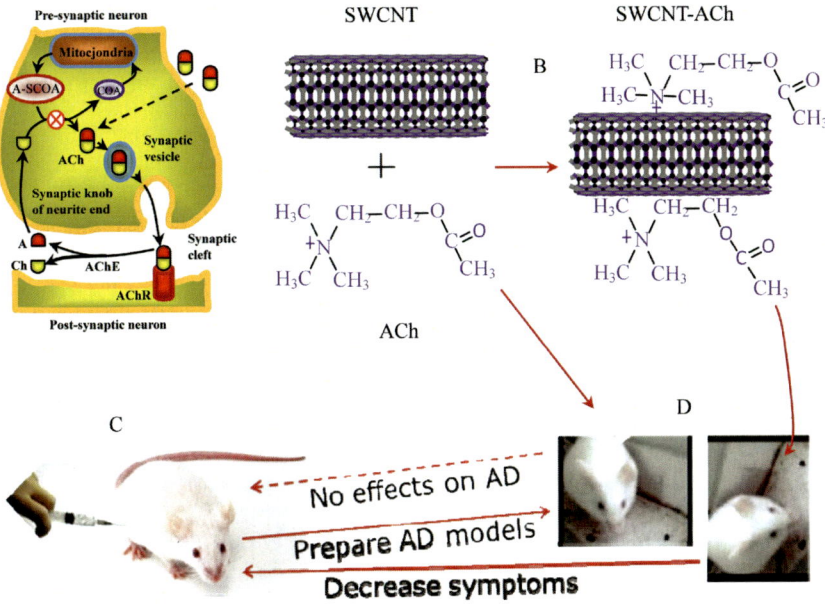

Fig. 3. The Mechanism of Alzheimer's Disease to develop and the protocol of the experiments to verify the capability of SWCNTs to deliver ACh into the neurons in brain. A The schematic illustration of the mechanism for AD development and treatment with ACh. ACh is the natural transmitter of cholinergic nervous system. It is systhesized with acetyl (A) from acetyl-COA and choline (Ch) re-absorbed from the synaptic cleft and released from the neurite end of the presynapric neurons when nervous electric pulse comes. The released ACh binds to the acetylcholine receptor (AChR), which open the ion channel of the receptor to give rise to active potentials, completing the message transmission between neurons. The ACh is disintegrated by acetylcholinesterase (AChE) into A and Ch after ACh finish their mission, which keeps the sensitivity of transmission of neural pulses. A and Ch may re-enter the presynaptic knob to be re-used in the synthesization of ACh. In the patients of AD, the neurons lost their capability to synthesize ACh so that the message can't be transmitted between neurons, leading to the dementia. Obviously, supplying foreign ACh will be an effective therapeutic but ACh can't be used as drugs because of its strong polarities which prevent it from entering the neurons in the brain through blood-brain barrier. B: the preparation of the drug delivery system of ACh targeting to neurons in brain. ACh may be adorsorbed onto SWCNTs by hydrophobic interactions or electrostatic interactions, which has been verified by Raman spectrum although the mechanism remains to be elucidated. C: The preparation of AD models. D: treatment of AD with single SWCNTs, free ACh and SWCNT-ACh. The results demonstrated that both single SWCNTs and free ACh cannot improve the dementia of AD mice but SWCNT-ACh can make the learning and memory capabilities of AD mice recover to normal (Yang et al. 2010), demonstrating SWCNTs can be effectively used as drug carriers of ACh in the treatment of AD.

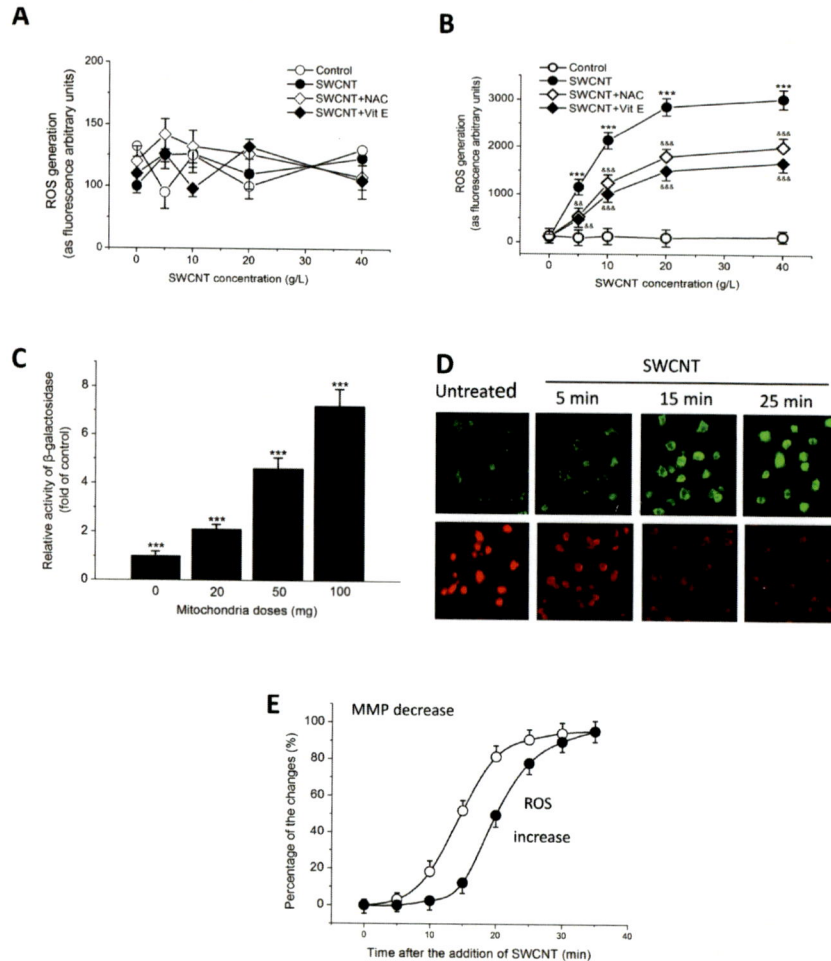

Fig. 5. **Relationships between lysosomal damage and mitochondrial damage studied in vitro.** (A) Effects of SWCNTs on the production of reactive oxygen species (ROS) in lysosomes. (B) Effects of SWCNTs on the production of ROS in mitochondria. NAC, N-acetylcysteine (10 µmol L^{-1}); Vit E, vitamin E (10 µmol L^{-1}). (C) The influence of mitochondria on the lysosomal damage by SWCNTs. Mitochondria were added at the indicated concentrations, and the activities of β-galactosidase were measured. Lysosomal damage was expressed as the activity of the β-galactosidase leaked from lysosomes. (D) Fluorescence images showing the influences of SWCNTs on the mitochondrial membrane potential (MMP). Green fluorescence increased and red fluorescence decreased with incubation time. (E) Changes in the red-to-green fluorescence ratios in the mitochondrial membranes over 25 minutes of incubation with SWCNTs. SWCNTs significantly decreased the ratio time-dependently. (F) The time courses for the changes of both MMP and mitochondrial ROS production. ***$P < 0.001$ compared with control. &&&$P < 0.001$ compared with SWCNT.